# OIL BURNING

# OIL BURNING

BY

## IR. H. A. ROMP
MECHANICAL ENGINEER

WITH 262 FIGURES AND 7 TABLES

SPRINGER-SCIENCE+BUSINESS MEDIA, B.V.

1937

First edition April 1937

© Springer Science+Business Media Dordrecht 1937
Originally published by Martinus Nijhoff, The Hague, Netherlands in 1937
Softcover reprint of the hardcover 1st edition 1937

Additional material to this book can be downloaded from http://extras.springer.com

ISBN 978-94-017-5721-8          ISBN 978-94-017-6069-0 (eBook)
DOI 10.1007/978-94-017-6069-0

# CONTENTS

## FIRST SECTION

## THE HISTORICAL DEVELOPMENT OF OIL BURNING

### CHAPTER I

### CHAPTER II

### CHAPTER III

### CHAPTER IV

### CHAPTER V

## SECOND SECTION

### BASIC PRINCIPLES OF OIL BURNING

#### CHAPTER I

#### CHAPTER II

# CHAPTER VII

# CHAPTER VIII

# CHAPTER IX

# CHAPTER X

# CHAPTER XI

# CHAPTER XII

# CHAPTER XIII

## THIRD SECTION

### MODERN FORMS OF CONSTRUCTION OF OIL BURNING DEVICES

# CHAPTER I

# CHAPTER II

# FOURTH SECTION

## FUTURE DEVELOPMENT OF OIL BURNING

### CHAPTER I

Ten points — Increased use of lower-grade fuel oils — Mechanical atomization, pressure atomizing and centrifugal principle — Low pressure air — Automatic regulation of flame capacity — Use of water vapour as soot and carbon preventor — Domestic oil burning — Increased use of kerosene in range burners — Gas oils — Mechanically assisted evaporation — Gas oils for high speed Diesel traction — Fully automatic mechanical burners.

### CHAPTER II

81 different conditions for oil burning application — A field for every sound type of oil burner — At least six types — Fully automatic electric oil burner — Non or semi-automatic electric oil burner of fire-pot type — Non or semi-automatic oil burner using natural draught — Small capacity electric oil hearth burner — Small capacity oil hearth burner using no electricity — Small capacity oil burner of range burner type for cooking and hot water supply.

### CHAPTER III

Climatic conditions — Frequency of out-door temperatures — Required heat with various out-door temperatures — Full load less thans 3 % of total heating season required — Domestic oil burners should be very economical at about half load — Range of regulation required at least 1 to 6, preferably 1 to 10 — "On-of" regulation Safety Devices "Vis-a-flame" — Flame guards — Protectostats — Stack stats — Time switches — Gas clearing period — Re-ignition — Influences of intermittent heat supply on room temperatures — Time lags — Sensitiveness of room thermostats + or — 1° F. not sufficient — Air temperatures and draught in heated rooms — The "cold 70" phenomenon — Means to avoid it — Graduated or semi-graduated regulations — N. F. — CUÉNOD — HEIL COMBUSTION — PROGRESSIVE "High-low-off" double nozzle unit — Temperatures taken as a basis for automatic regulation— FULTON-SYLPHON — DUO-STAT — MINNEAPOLIS-HONEYWELL.

### CHAPTER IV

Development of high speed Diesel traction will probably necessitate the use of heavier oils for domestic purpose — "200 sec. fuels" — English development — CLYDE — AUTOMESTIC — English QUIET MAY — BRITISH OIL BURNER — KALOR-

## LIST OF ILLUSTRATIONS

---

---

## PREFACE

Four lectures called: *History, Present Theories, Present Forms* and *Future of Oil burning*, were delivered at Delft in April 1934 before members of the „Koninklijk Instituut van Ingenieurs" (Royal Institute of Engineers — The Hague, Holland) on four afternoons consecutively. Although the subjects discussed are substantially the same, a certain amount of rearrangement was considered desirable for publication while moreover several problems, especially those raised in questions asked after the lectures, have been entered into more fully in this book, which therefore covers a good deal more ground than the actual lectures.

There is such rapid progress in the development of oil burning that it became an exceedingly difficult task for the author to keep pace with the inventions and numerous changes in design made in the interval between the reading and the publication of the lectures. He was obliged to cry „halt" towards the end of 1935, but he will be very grateful to his readers for any information they may be able and willing to furnish on later inventions, new features and errors of commission or omission. Such communications will be duly acknowledged and the information utilized at the earliest opportunity.

The author long hesitated as to whether to quote names of manufacturers of the burners described in this book or not. On the one hand the omission of names, which has been accepted as a principle in most handbooks on oil burning, may look more impartial, but, on the other, knowledge of the trade-names of the burners is such a welcome help to those concerned with the practice or study of oil burning that the author finally decided in favour of quotation. As he is fully aware that in certain cases this might perhaps give rise to difficulties, he wishes to point out that any comments from burner manufacturers on possible mistakes in figures or descriptions concerning their past and present products will be gratefully received in order that such discrepancies may be corrected in the event of a reprint.

Finally the author wishes to express his thanks to those who assisted him by constructive criticism and for other help in translating and preparing drawings.

Amsterdam, January 1937.                                        HENDRIK A. ROMP

# SECTION I

## THE HISTORICAL DEVELOPMENT OF OIL BURNING

———

# CHAPTER I

EARLY COMBUSTION METHODS FOR FUEL OIL

Although the use of mineral oil and its residues for burning and other purposes has been recorded since ancient times, we shall only deal here with the application of fuel oil for technical purposes, which dates back to about 1860.

It is not possible to give any one person the credit of having "invented" oil burning, for as a matter of fact about 1860 there were several simultaneous attempts, so that it is clear that already at that time the problem was a "burning question". Neither can we look for priority to the dates of descriptive publications or patents, as in those times much of the research work was not published. Nor is it possible to give any one country the honour of the invention, although it must be said that Russia is the only country in which oil burning has been a marked success from the very beginning, leading to a rapid and successful development of its application for several industrial purposes.

In the early development of oil burning from about 1860 to 1880 there were roughly three spheres of activity, the Russian, American and West-European spheres, the last-mentioned being divided into the English and French spheres. Up to about 1880, owing to the difficulty of communication, there could be no intensive exchange of experience between these parts of the world, but from that time onward there came to be more and more coordination.

The history of the later development may be divided into two periods, one from 1880 to about 1900 and the other from 1900 up to the present day. Until about 1900 it is fairly easy to follow the development of oil burning, but after that year such rapid advances were made and so many devices invented that it is almost impossible and at any rate confusing for the reader to try to give a chronological account of the development. Moreover, a great many of the devices of that time (beginning of 20th century), are based on what might be called technical superstition rather than on sound principles, which at that time were known incompletely and only to a few.

Consequently for an account of the history after 1900 only a rough sketch can be given of some milestones in the development of the industrial and domestic oil burners, while a detailed description of the various forms of oil burning devices must be left for later discussion in Section III.

---

## CHAPTER II

### EARLY DEVELOPMENT IN RUSSIA

This stage, which is probably the oldest, was characterized by an abundance of heavy unsaleable residue (waste oil, astatki, pacura, mazouth) for which some practicable use had to be sought (the main product of mineral crude oil being lamp oil) and the means to this end were found in *pulverization with saturated steam* by means of devices mainly of the so-called *slot type* (outside mixing). This method was from the very first successful and ultimately led in Russia to a rapid and wide-spread application of oil for steam-ships and locomotives.

It is curious to note that in Russia, although very heavy fuels had to be dealt with, there were but very few failures, contrary to the experience in the early attempts in other countries, where initial difficulties even with light fuels were so serious that development was retarded for several decades. The success attained in Russia was due to the general *adoption of steam* as medium for pulverization, since steam has not only a *physical* action (pre-heating and pulverization) but also a *chemical* action as a *soot-preventer*, rendering it possible to burn very heavy fuels without smoke, even when rather primitively designed burners are used. This soot-preventing action of

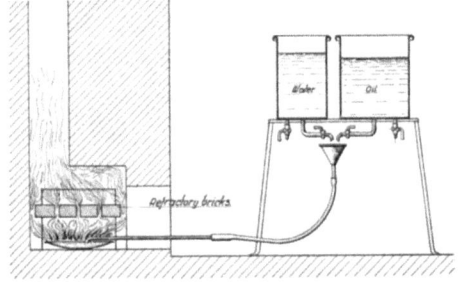

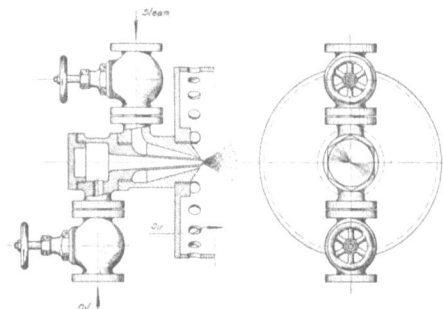

Fig. 1. Astrakan oil stove                    Fig. 2. Spakowski's oil burner

steam (due to the so-called "water gas reaction", which will be discussed later on) seems to have been common knowledge in Russia long before 1860, it being recorded, for instance, that for house-heating a so-called Astrakan *burner* (see fig. 1) was used, consisting of one or more flat dishes placed in the bottom of the Russian stove, into which oil flowed from a reservoir by gravitation. This oil was heated and evaporated by radiation from refractory bricks placed in the oil flame above. Apparently because combustion was very smoky when oil alone was used, water was allowed to flow from a second reservoir to the dishes, where it evaporated spontaneously and threw up the oil in drops, thus giving a much better combustion with a less smoky flame.

It seems that it was this primitive forerunner of steam pulverization that led to practically all subsequent experiments for industrial purposes being made with steam, thus contributing largely towards the successful development in Russia. One of the pioneers to be mentioned is SPAKOWSKI, who patented a *steam oil burner* in 1866, which was followed in 1870 by an improved design (see fig. 2). This burner, in which the oil was admitted to the centre and surrounded by steam as a concentric ring, was a remarkable success. The steam, being the velocity agent, was in intimate contact both with air from the outside and with oil on the inside. The Spakowski burner, probably by mere chance, gave a flame which was very favourable for marine boilers and avoided damage by impingement of the flame on the boiler surface. This was the first representative of the *nozzle type* of oil burner.

Another inventor, named LENZ, designed his first burner in 1870 (fig. 3), which he called the *"Forsunka"* — this name afterwards became a class name for Russian steam-pulverizing oil burners — comprising two slots, an upper one for the oil and a lower one for the steam, arranged in such a manner that four flat streams of oil dropped on to a flat stream of steam. This might be called the first representative of the *slot type* of oil burner, more particularly kown as *"drooling burner"*. The capacity of the oil burner could be regulated by changing the free area of the slots, which was done

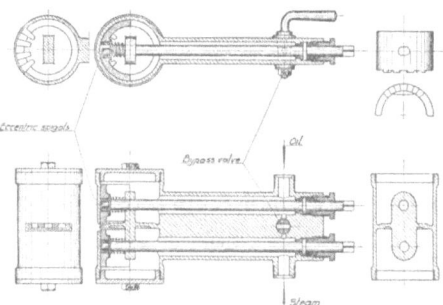

Fig. 3. LENZ's first "Forsunka"

by means of adjustable slides moved up and down by turning spindles with eccentric spigots. This method of regulating by *changing the outflow area of the steam* is theoretically better than the pinching of the steam with a valve in the steam line before the burner, because in the former case the steam pressure at the mouth of the burner and, therefore, the speed of the steam will be practically the same at all loads, which is of the utmost importance, since the steam is the velocity agent of both oil and air; in the second method of regulating, the steam pressure and its velocity at the burner mouth decrease with decreasing loads, which may result in poor combustion at low loads.

Although theoretically better, the Lenz burner gave much trouble in practice, because the movable slides were exposed to the radiant heat of the flame and were thus apt to seize. This led to a

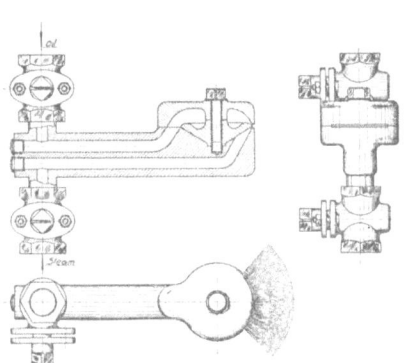

Fig. 4. ARTEMEV's oil burner

similar construction being designed by ARTEMEV (see fig. 4) based upon the same principle but without adjustable slots, and this met with more success because of its simplicity, cheapness and easy cleaning.

Another special feature of the Lenz burner, subsequently copied in almost all other "Forsunka's", was the *bye-pass valve* for cleaning the oil slots by blowing steam through them.

The first Lenz burners were not at all satisfactory, the shape of the flame being unsuitable, but Lenz improved on this later by using larger semi-circular slots. Thus it is seen that from the very beginning of oil burning the difficulty has been to obtain the *right shape of flame*, a problem which even nowadays may give us much trouble. There was a time when oil flames were considered dangerous, being apt to give short blow-flames that damage the boiler walls, and consequently a long, fluttering, "soft" flame was considered to be the ideal form.

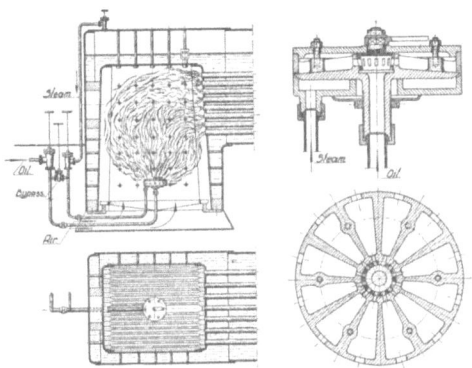

Fig. 5. BRANDT's oil burner

A burner giving this type of flame was, for instance, the BRANDT *burner* (fig. 5), consisting of a flat round box with circular slots around for steam and oil placed in the centre of the fire-box of a locomotive boiler. This burner demonstrates clearly a mistake that was very often made in the early stages of oil burning: the burner is placed *inside* the fire chamber and, therefore, is too much exposed to radiant heat, resulting in the formation of oil coke, whilst it is very difficult to clean. Later on he improved on his design by adopting the *nozzle type* of fig. 6.

The same mistake was at first made by URQUHART, a Russian railway director, who

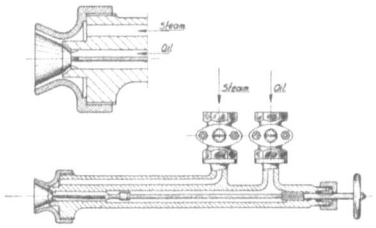

Fig. 6. BRANDT's nozzle
oil burner

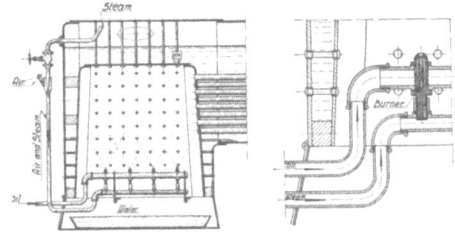

Fig. 7. URQUHART's first oil burner
for locomotives

must be regarded as one of the most successful pioneers of the application of oil burning for locomotives. His first construction (1874) (see fig. 7), which was not at all successful, consisted of a number of nozzles placed at the bottom of the fire-box. This he subsequently improved on in 1882 (see fig. 8) by placing the burner *outside* the fire-box and adopting the *nozzle type*, but with steam in the centre and oil around it, as contrasted with Spakowski, who applied oil in the centre. This type of burner was so successful that in 1885 143 locomotives of the Russian railways were equipped with Urquhart burners

We have now already mentioned two fundamental types of construction, viz.:
1) *The nozzle type* (Spakowski and Urquhart), and
2) *The slot type* (Lenz, Brandt, Artemev).
A third type was developed by KÖRTING (see fig. 9), who started in 1876 with
3) *the tube type* of oil burner, which was a forerunner of the VENTURI burners,

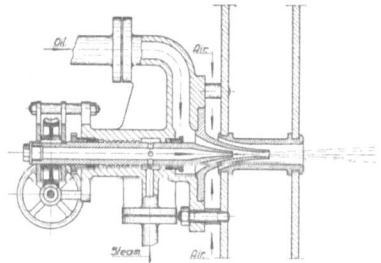

Fig. 8. URQUHART's later nozzle type

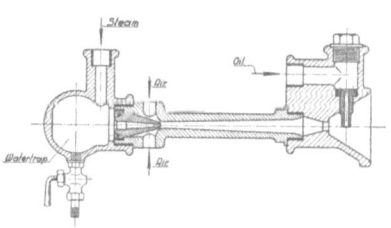

Fig. 9. KÓRTING's tube oil burner

frequently combined with steam injectors for the air necessary for the combustion. This combined system of oil pulverization and air suction promoted a previous thorough mixing of air and steam and, therefore, also facilitated better contact between air and oil. In some instances air suction by steam jets was successfully combined with the slot type too, as for instance in the KARAPETOW *burner* (see fig. 10).

It is not the author's intention to describe every Russian oil burner of that period; a lengthy description may be found in "Die Feuerungen mit flüssigen Brennstoffen", LEW (Kröner — Leipzig 1925). Only typical specimens will be discussed as far as such may be of interest to show the progress of the funda-mental principles of oil burning.

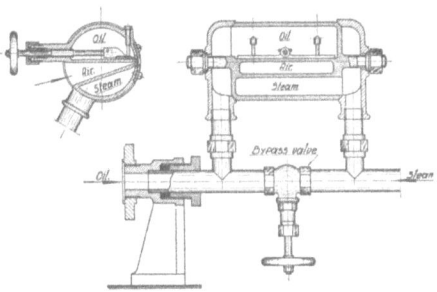

Fig. 10. KARAPETOW's oil burner for locomotives

Before leaving the Russian sphere of development it is of importance to note that as early as 1874 almost every ship of the commercial and naval Caspian Sea and Wolga fleets was running on oil. The oil burners used were exclusively of the steam pulverizing type. The consumption of fresh water for this purpose apparently caused no serious difficulties, for the low pressure boilers (5 atm.) were fed with salt water, the salt content simply being limited by temporarily blowing off.

Further, by about 1890 nearly all Russian locomotives, except those in coal districts and Siberia, were running on oil fuel. Comparing data from other countries — fuel oil was first used on warships in Italy about 1893, in England about 1895 and in America about 1910 — there is no doubt that Russia was far ahead in this matter.

## CHAPTER III

### EARLY DEVELOPMENT OF OIL BURNING IN AMERICA

The outstanding feature of this stage in America, curiously enough, is that there seems to have been too much theoretical knowledge, this being the cause of several failures. In fact it was generally considered as absolutely necessary to evaporate the oil first and then to burn the oil gas. Although this may be quite right from a theoretical point of view (our modern spray burners produce liquid drops which evaporate by radiation of heat, and the gas thus produced is burnt), the idea of heating and evaporating the fuel oil in a separate retort and burning the oil gas under the boilers led to nothing but complete failures. This unsatisfactory start gave oil burning a bad reputation in America and this is perhaps the reason why America was rather late in the industrial application of oil burning on a large scale, so that it was not until 1890–1900 that the first merchantmen of the Pacific Coast tried oil burning.

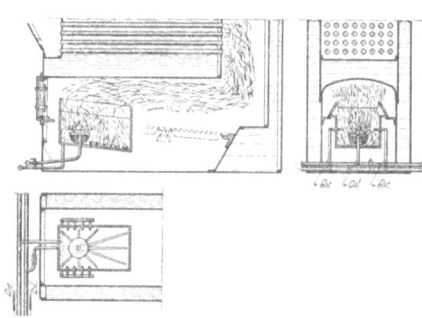

Fig. 11. BIDLE's oil fire for marine use.

In 1860 a burner designed by BIDLE (see fig. 11) was reported, into which the oil was fed through a pipe passing through a cast-iron bucket or basket of glowing anthracite to evaporate the oil; needless to say, this burner was not a success.

In 1862 a gas retort burner was constructed by SHAW & LINTON (see fig. 12), in which the oil was evaporated by the heat of the flame itself and partially cracked by precombustion. The safety valve on the combustion space speaks for itself!

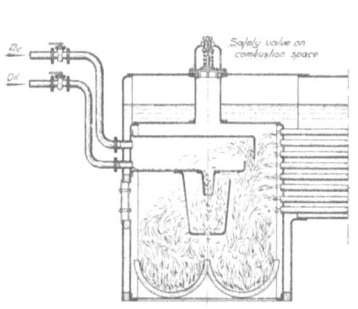

Fig. 12. SHAW and LINTON's oil gas fire.

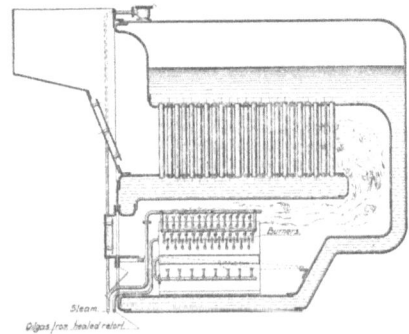

FIG. 13. FOOTE's oil gas furnace for marine use.

In 1863 a gas furnace was invented by Colonel FOOTE; the oil was heated in a separate retort by anthracite fire and the oil gas burnt under the boiler (fig. 13). A curious illustration of a failure due to too much theoretical knowledge is the fact that Foote, seeing that the combustion was very smoky, argued that this was due to *lack of hydrogen in the fuel* and he therefore passed steam over iron filings, thus producing hydrogen with which to "dilute the oil gas". This, of course, involved two "combustive" materials, fuel oil and also iron filings. But this practical objection never became felt as Foote's retort was very soon choked with carbon and burnt through. Later designs by SIMMS and BARFF, EAMES and DÜRR, other representatives of the gas retort burner type, all proved to be failures.

In 1863 BRIDGE-ADAMS invented a spray burner for locomotives and although it was not itself any great success it was the first step in the right direction. About that time an American patent was filed in the name of COOK, who seems to have had the idea of pulverizing the oil by means of centrifugal force, but nothing can be found of any application in this direction.

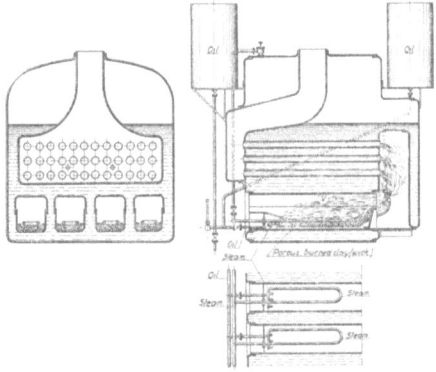

In 1864 another type, differing from the foregoing, was constructed by RICHARDSON, who was inspired thereto by the ordinary wick-lamp. The bottom of the fire-box of the boiler was covered with a layer of a porous kind of burned clay with ducts for the passage of the oil, which layer acted as a large wick

Fig. 14. RICHARDSON's "oozing furnace".

(fig. 14). This so-called "oozing furnace" drew the attention of the British Admiralty and some experiments were made with it, but of course the pores were very soon choked with carbon and the mixing of oil vapour and air was very bad, resulting in a very smoky fire.

In 1875 Comm. ISHERWOOD of the American Navy was instructed to study the possibilities of the new fuel for naval purposes. His conclusion was: "*Atomization is the only way*". It is not because his report was of much consequence that it is mentioned

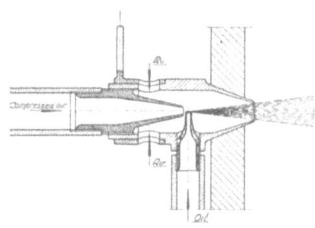

Fig. 15. WALKER's scent-spray burner.

here, but he was probably the man who introduced the word "*Atomization*", which, as a matter of fact, is a very exaggerated term for spraying or pulverization, but nevertheless it is now used in all English-speaking countries.

Between 1870 and 1880 some other constructions appeared for a time (SALISBURY, ROGER, SADLER, ORVIS) and in addition there was WALKER's construction (fig. 15) of the *scent-spray type*, which had some temporary success and the principle of which reappears even in some modern applications (Caloroil).

In 1887 the Pennsylvanian Railway Co. equipped some locomotives with oil burning devices on the Russian principles, but on the whole oil burning was not a great succes in America at that time.

---

## CHAPTER IV

### EARLY DEVELOPMENT IN WESTERN EUROPE

### A. *In France*

In this country there was a remarkably clear insight into the problems concerned with oil burning, resulting in reasonably good performance. The idea was to find a use for the tar oil from the gas works and very good work in this direction was done especially by the gas technologists AUDOUIN, DUPUY DE LÔME and SAINTE-CLAIRE DEVILLE, who devised the gas-lighting of Paris.

In 1867 AUDOUIN constructed his furnace consisting of a number of *vertical fire bars* (see fig. 16) covered with a film of oil evaporated by the radiant heat of the flame itself.

In 1868 AUDOUIN and SAINTE-CLAIRE DEVILLE constructed a similar furnace (see

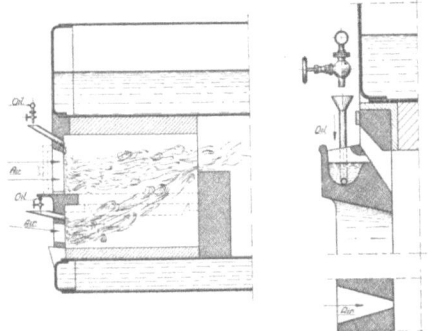

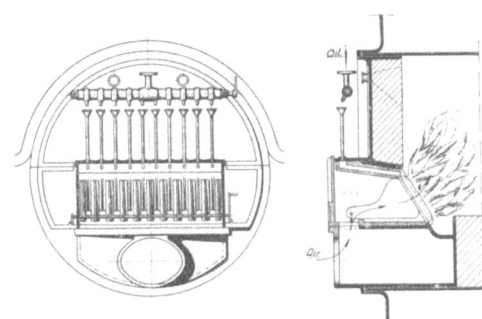

Fig. 16. AUDOUIN's oil grates.          Fig. 17. AUDOUIN and SAINTE-CLAIRE DEVILLE's oil grate.

fig. 17) of *inclined fire bars* with adjustable air registers to give the highest possible velocity to the air at the point of evaporation of the oil film, resulting in thorough mixing and clean and efficient combustion. It is interesting to read the description of these burners (Annales de Chimie et de Physique, 1867, Tôme XV page 30), where Audouin discusses the problem "Why can't we burn oil in a more simple way without the complications and cost of the American burners?".

He was the first to introduce *natural draught* as a motive force for oil burning in combination with evaporation of the oil spread out into a film, and he was fully aware of the necessity of reducing all resistance to the air to the utmost and of focussing the air velocity on the evaporating oil film. As a tech-nologist Audouin pointed out the importance of *mini-mum excess air* and its influence on the flame temper-ature, perfection of combustion and efficiency, which shows that he had a very clear theoretical knowledge combined with good practice.

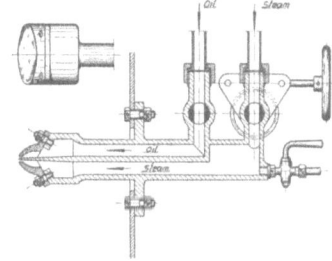

Fig. 18. JENSEN's oil burner.

The yacht "Le Puebla" of Napoleon III and several locomotives were equipped with so-called "DEVILLE *grates*" and everything looked very promis-ing, but the war of 1870 put a stop to this progress and the subject of oil burning was not taken up again till 1883. At that time all the activity of the French was concentrated upon revenge. A friendly advance towards Russia led to the exchange of experience in oil burning, especially for marine work. Thus the JENSEN *burner* (see fig. 18), which was adopted as a standard burner for the Russian Black Sea fleet, was also introduced in the French Navy.

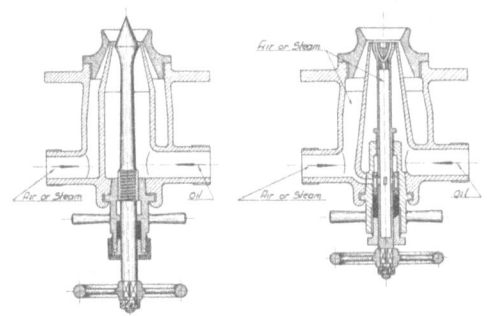

Fig. 19. D'ALLEST oil burners for steam or compressed air.

However, the extra consumption of fresh water for oil burning on board was prohibitive, so that other means had to be sought and were in fact found by D'ALLEST (engineer of FRAISSINET Comp., Marseilles), who made a series of exhaustive studies of different steam oil burners and also designed several *compressed air pulverizing* burners for torpedo boats (fig. 19). He showed that even with the necessary addition of steam-driven air compressors the radius of action of the torpedo boat destroyers could be considerably increased by conversion from coal to oil, without increasing the total dead-weight. Moreover oil produced less smoke, which was of course soon recognized as an important factor for warships. The GUYOT burner (see fig. 20), which was afterwards adopted as a standard by the French Navy, is in the main a further development of the principles of the D'ALLEST burners.

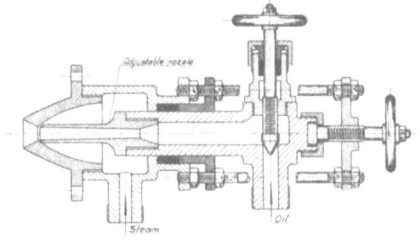

Fig. 20. GUYOT's oil burner (French Navy).

An illustration of how French technologists' clear grasp of combustion problems in about 1865 completely deserted them after the war in 1870 is furnished by the gas

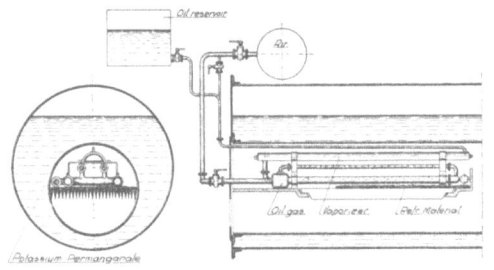

retort burner designed by DE BAY and DE ROSETTE (fig. 21). By this system compressed air and oil are heated together in a retort heated by the oil flame itself and the gas thus produced is burnt. But now in their patent the following remarkable feature is claimed: To prevent incomplete combustion, which was apparently due to the *poor oxygen content of the air*, a

Fig. 21. DE BAY and DE ROSETTE's burner.

number of tubes filled with potassium permanganate or barium peroxide were also heated in the flame, which gave off oxygen "to enrich the air". It is for this "invention" that the patent D.R.P. 31,962 was granted (1885).

Another characteristic of this period is the *combined fires*, i.e. oil sprayed on coal fires, generally resulting in poorer combustion of both fuels. To avoid the cost of coal in some cases a layer of "coal" made of refractory material was laid on the grates and oil was sprayed on this, evaporated and burnt (fig. 22).

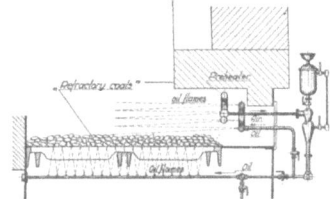

Fig. 22. MORTH's oil fire.

The fundamental idea of this can be traced in the modern centrifugal domestic oil burners of the wall flame type.

## B. *Early development in Great Britain*

In Great Britain a very good start was made, with clear insight into combustion problems, whilst the advantages of the application of *superheated steam* were soon recognized. As an exception must be mentioned the erroneous idea with regard to the function of the steam: it was thought that steam dissociated by the heat of the oil flame was able to produce extra heat by the burning of the dissociation products.

The British Admiralty showed a keen interest in the new fuel and stimulated experiments considerably. However, notwithstanding the good results obtained with various types of oil burners, no general application of them was made until about 1880, owing to the high cost of fuel oil and the opposition raised by coal-mining concerns.

To quote some details of the period it may be mentioned that in 1865 AYDON, WISE and FIELD carried out their first experiments at South Lambeth with a burner (fig. 23) in which the oil dripped at right angles onto a steam jet.

The dates of AYDON's and SPAKOWSKI's British patents on oil burning differ only by a few months, but to AYDON must be given the honour of the first application of superheated steam, which largely contributed to the success of his and subsequent experiments of the English school. The advantages of the use of superheated steam lie not only in the greater heat of the steam causing a better preheating of the oil, but also in the reduction of steam consumption with equal velocity effect because of the larger specific volume of superheated steam, and, last but not least, in a more constant flow and steady flame due to the avoidance of the water drops of the jet of saturated steam, which at times may even extinguish the flame.

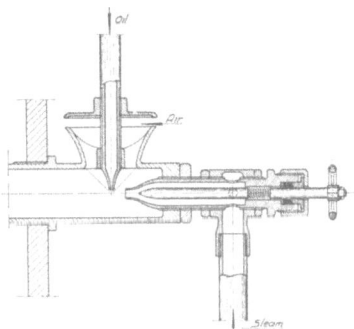

Fig. 23. AYDON, WISE and FIELD's first burner.

The British Admiralty was very interested in these experiments, and Admiral SELWYN, who had already experimented with a less successful construction of his own (concentric oil and steam flow, fig. 24) got into touch with AYDON, with the result that a combined construction was

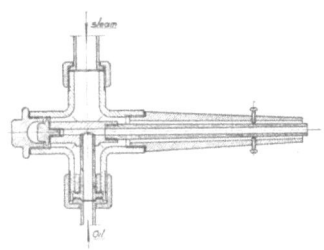

Fig. 24. SELWYN's first oil burner.

designed, the AYDON-SELWYN *burner*, in which the oil and steam met under an acute angle (see fig. 25). With this burner a 10$1/2$-fold evaporation was obtained with tar residue, against a 7$1/2$-fold evaporation with coal, on an installation on board H.M.S. "Oberon". Very good results were also obtained in heating heavy armour plates.

In 1868 an oil-gas retort furnace of DORSETT & BLYTHE (fig. 26) was installed on board the "Retriever", consisting of a retort filled with tar oil first heated by a coal fire until a pressure of 20 lbs/sq. inch was obtained, when the oil gas was used to heat the retort further until the pressure reached 50 lbs/sq. inch. After this the gas was admitted to the furnaces under the boilers, the burners consisting of a number of nozzles placed in a pipe as shown in fig. 26.

This system had the special attention of the Admiralty because it needed no fresh water, it soon being realized that oil burners using steam could not be very promising for ocean-going ships. Although as regards combustion the results of the experiments with this oil-gas furnace seem to have been very good, difficulties were encountered with the retort, which soon got choked with carbon and burnt badly, obviously owing to the very high preheating of the fuel (500° C.). Another point was the difficulty in

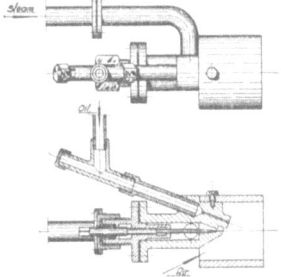

Fig. 25. AYDON and SELWYN's oil burner.

keeping the pipe connections tight at 50 lbs and 500° C., which even nowadays with our modern heat-resisting packing materials would not be a pleasant job.

Also the oozing furnaces of RICHARDSON already mentioned, and of PATERSON were tried out under the direction of the Admiralty, but were found to be less efficient, and

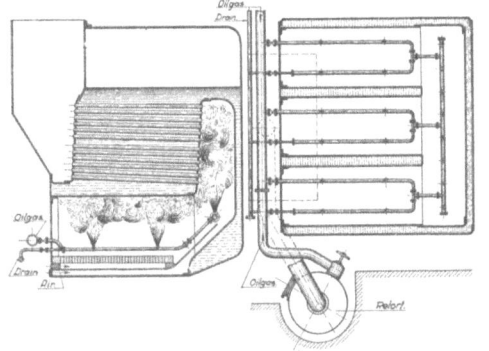

Fig. 26. DORSETT and BLYTHE's oil gas burner.

it was clear that only the spray and jet method, either by pulverizing liquid drops or by means of a jet of gasified oil, constituted an efficient way of burning oil.

As already stated, the British Admiralty did very good work in developing oil burning, but the price of tar and oil and the difficulties in obtaining mineral fuel oil stood in the way of any practical applications for naval purposes.

It was about 1883 that in oil burning matters a new pioneer came to the fore, in the person of Mr. HOLDEN, locomotive superintendent of the Great Eastern Railway. His object was to construct an oil burning device which would make it possible to use oil on locomotives without any brickwork or internal alterations of the fire box, so that oil burning could be almost instantaneously changed over to coal firing and vice versa or even used in combination one with the other. After numerous experiments he arrived at the design

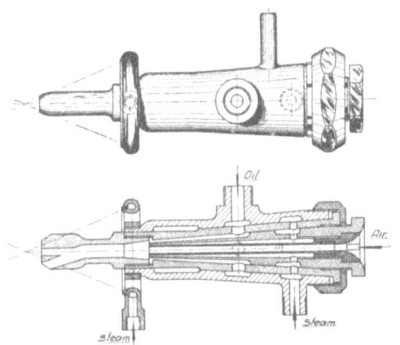

Fig. 27. HOLDEN's steam-jet oil burner.

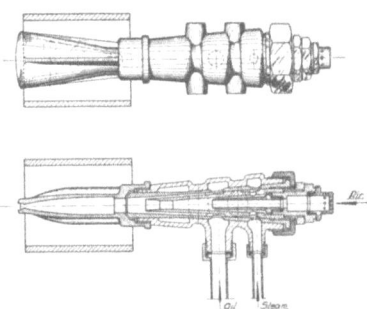

Fig. 28. HOLDEN's oil burner (Austrian Railways).

shown in fig. 27, a special feature of which is the hollow ring jet, which blows extra steam around the nozzle and induces a strong current of atmospheric air, ensuring a complete combustion with a short flame and making it possible to dispense with any refractory lining inside the fire box. The success of the Holden burner was so complete that later on numerous railway companies all over the world took out licences for its use, and even nowadays several *modified* HOLDEN *burners* are in use (e.g. on the Austrian Railways) (fig. 28).

The main objection to the use of oil at that time was the *uncertainty* of its always being available, which drawback, however, did not count so seriously for the railway companies, which very often had their own coal mines connected up with gas works producing tar oil as a by-product, so that it can readily be understood that the first application of oil burning on a large scale was on locomotives and not on ships.

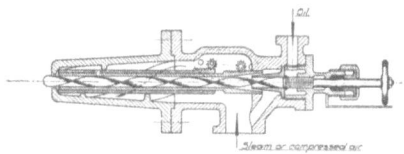

Fig. 29. KERMODE's steam-jet oil burner.

There were still two factors standing in the way of the general application of oil burning on steamships: *firstly* the *scarcity of fuel oil* and *lack of a world-wide distribution of fuel oil bunker stations*, and *secondly* the *lack of an efficient way of burning oil without steam or compressed air*, which problem was not solved until 1902, by KÖRTING's so-called *pressure atomizing burner*.

Among others who have done much to develop oil burning may be mentioned: KERMODE, who introduced a rotation element (fig. 29) to give the steam a whirling motion and obtain a hollow conical flame and good mixing, and further: THORNYCROFT, WHITE, RUSDEN-EELES (fig. 30), ORDE (fig. 31), ARMSTRONG and WHITWORTH.

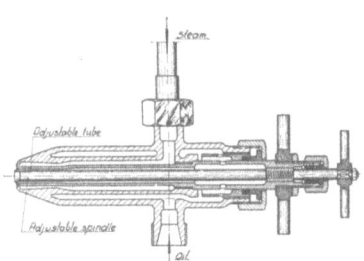

Fig. 30. RUSDEN-EELES' oil burner.

Especially the ORDE burner is a good example of the application of the *Venturi principle*, which gradually came to be applied by oil burner designers. This principle affords efficient means of converting pressure into velocity by using a suitable combination of conical tubes and it is quite natural that it found a successful use by designers of a velocity apparatus, which, as a matter of fact, an oil burner is. This Venturi principle also affords the best means of creating underpressure from velocity, and very often Venturi cones are used at the same time to suck in air, to produce a thorough mixture and complete combustion (see figs. 17, 28 and fig. 31).

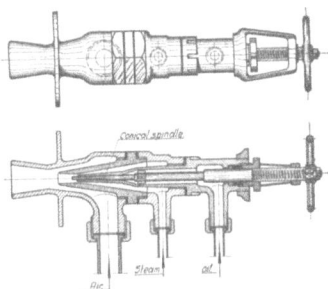

Fig. 31. ORDE's oil burner.

## CHAPTER V

### PERIOD OF COORDINATION IN DEVELOPMENT

#### A. *Development of steam pulverizing oil burners*

It has now been seen that the work done in Russia, America, France and Great Britain, each with its own individual characteristics or "school" so to speak, was more or less independent. Between 1890 and 1900, however, owing to a better exchange of thought and experimental results, research work gradually came to be coordinated. The first part of this period, up to 1905, witnessed an enormous development of the steam pulverizing principle, and an idea of the number of possibilities is given by the table in fig. 32, called *Classification of steam pulverizing oil burners about 1904* (taken from the U.S. Naval Liquid Fuel Board Report of 1904).

Oil burners using steam or compressed air may be divided into *outside mixing* and *inside mixing*, according to whether the contact of oil and steam takes place *outside* or *inside* the burner. The outside mixing burners, which are of the older class, generally have a somewhat higher steam consumption, but are less sensitive to variation of the steam pressure and to moisture of the steam than the inside mixing burner, since in the latter the steam exerts a certain back pressure on the oil flow.

The class of *outside mixing burners* may be divided into:

1) *Drooling burners*, in which the oil dribbles onto the steam jet, and

2) *Shearing burners*, in which the oil is forced onto the steam at right angles, assisted more or less by a syphon-action, this form constituting an improvement on the first, the oil being torn up or sheared as it were, whereas in the first type of this class the oil drops are more or less projected as such into the furnace without intensive pulverization.

The *inside mixing type* may be divided into:

3) *Chamber burners*, comprising a chamber in which a preliminary mixing of oil and steam takes place, resulting in a better pre-heating, lower viscosity of the oil and therefore better pulverization (with less steam consumption) than would be possible with an outside mixing type.

4) *Venturi burners*, which as a matter of fact are also chamber burners but in which the chambers are formed according to the *Venturi principle*, i.e. built up from cones and tapering nozzles. This fourth class may be considered as a development of the third, because of its better conversion of pressure into velocity, so that it gives often better results at lower steam pressure. Generally speaking, modern efficient steam pulverizing oil burners belong to the fourth class, although this does not mean that equal efficiencies may not be obtained with burners of the other classes.

The thorough study made by the U.S. Naval Liquid Fuel Board, under the direction of Adm. MELVILLE, of the numerous steam pulverizing oil burners used about

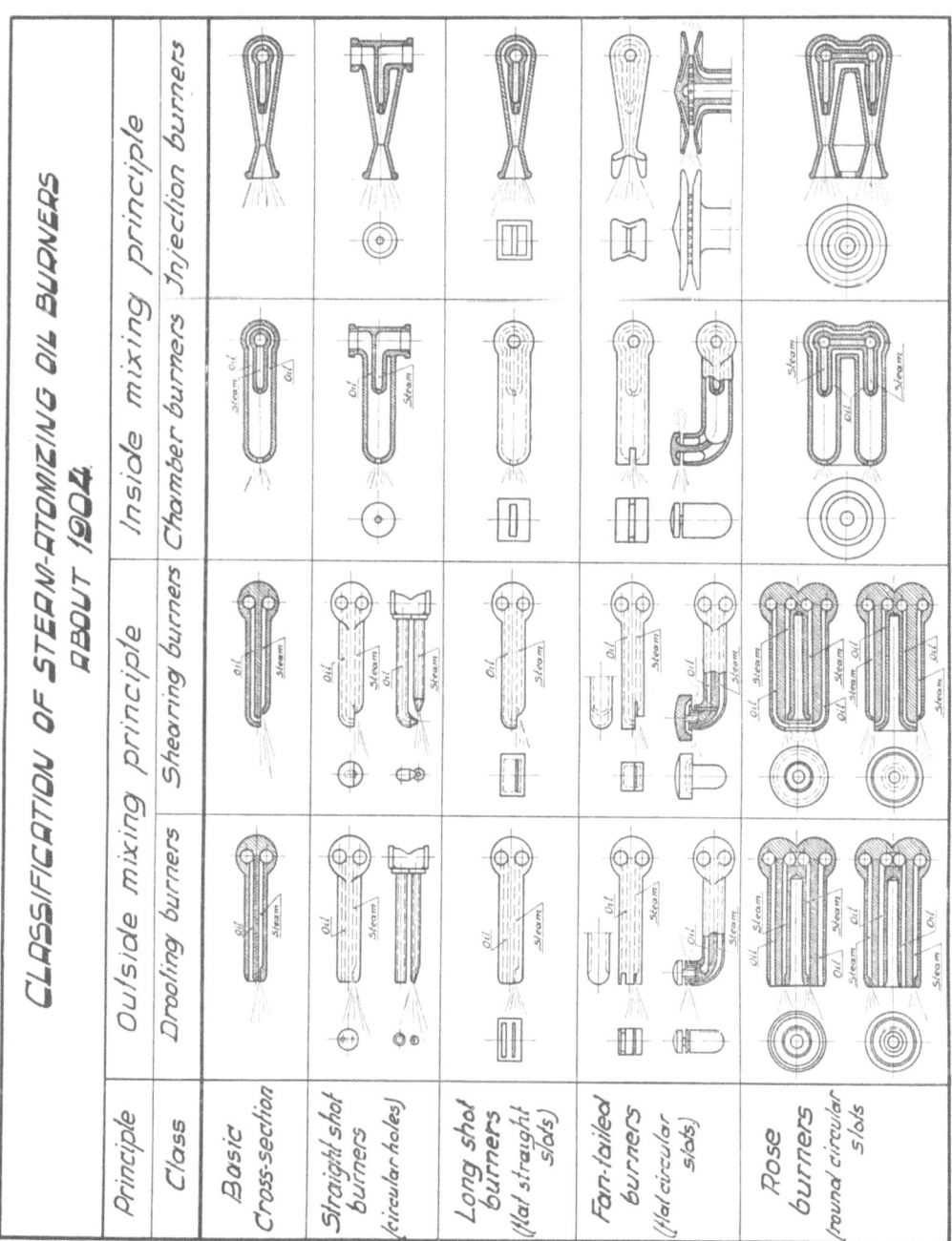

Fig. 32.
Classification of steam-atomizing oil burners about 1904.

1900–1905 showed that each of these classes may be worked out according to four main forms, viz.:

a) *Round separate holes* (straight shot burners),
b) *Flat straight slots* (long shot burners),
c) *Flat circular slots* (fan-tailed burners),
d) *Concentric circular slots* (rose burners),

thus giving 16 sub-classes, of each of which numerous representatives may be found.

It is impossible to say which form is the best, since more or less good constructions could be made of nearly all of them, depending largely upon the form of flame obtainable and the use for which the burner is destined; in this respect very small alterations to the burners may be of great consequence.

## B. *Compressed air oil burners*

The same classification may be applied to the *compressed air oil burners* of about 1900. Although this kind of burners, in which a part of the combustion air itself is utilized for pulverization without the aid of other media, must be, theoretically, the most efficient and was thought for a time to be the best solution of the oil burning problem on board of ships — the steam used for driving the air compressors could be condensed and the loss of fresh water thus avoided — it did not prove to be a great success, mainly for the following reasons:

1) For compressed air a *compressor installation* was needed, which involves not only yet another undesired complication but also a considerable consumption of power, extra weight and space.

2) Only a relatively *small part of the combustion air* served as pulverizing agent, the remaining part having to be fed to the furnace by other means, e.g. by means of a fan, blower or natural draught, because the air-oil jet is less capable of sucking in sufficient air than a steam-oil jet would be.

3) At that time *compressed air oil flames* usually were very *short and hot*, thus involving danger of damage to the walls by impingement; moreover these flames were very noisy, due to a very violent roaring combustion, which even proved to be a serious objection to the application of this burner type on passenger ships.

4) Contrary to the case with steam, the *expanding air cools the oil* instead of preheating it; therefore it is necessary to provide means of preheating the compressed air, which involves yet another complication.

As will be seen later on in the discussion of modern domestic oil heating, oil burning based upon pulverization by air alone could be made a success by using very low air pressures and larger portions of the total air for pulverization.

### C. *Pulverization by oil pressure*

As a mile-stone in the development of oil burning for marine purposes should be mentioned the invention of KÖRTING, who constructed the first so-called *pressure atomizing oil burner* in 1902 (fig. 33). By means of a *rotation element* (now generally known as *"plug"*), he gave the oil a rapid rotary motion, in such a way that when leaving the orifice (*"tip"*) the oil was broken up by the centrifugal force into fine drops. Thus the pulverization could be obtained merely by applying pressure to the oil (50 lbs/sq.in.). As Körting soon found out that the magnitude of the oil drops was strongly governed not only by the

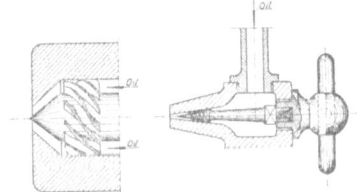

Fig. 33. KÖRTING's first pressure atomizing burner.

pressure but also by the viscosity and capillary constant of the oil, he *preheated* it to about 100° C., which proved to be very successful.

Almost at the same time SWENSSON developed another principle of pulverizing the oil with pressure alone, which method, however, was never applied on a large scale. The oil passed through a fine jet onto the point of a sharp V-shaped metal cutter, which had the effect of pulverizing it into a very fine spray (fig. 34).

Körting's invention led to a large number of similar constructions being made,

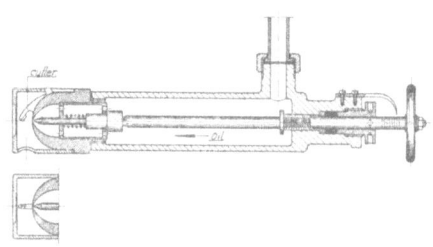

Fig. 34. SWENSSON's pressure atomizing burner.

all based upon the same principle, the oil being given a rotary motion before leaving the orifice, and only differing in the means of producing this rotation. Some well-known names may be mentioned: KERMODE, who used tangential slots at right angles to the axis of the oil burner: SAMUEL WHITE, who used spiral grooves on a conical surface; THORNYCROFT, SMITH, MEYER (Dutch East Indian Lines) and several others, who will be discussed later on.

This invention of the pressure atomizing oil burner in 1902 having removed one of the main objections to the use of fuel oil on ships, its application soon became widely spread, so that the need arose for a chain of *fuel oil bunkering stations* all over the world, such as already existed for coal, and in meeting this need Great Britain played the most important rôle.

### D. *Development of Oil burning in America*

It has often been said that the application of oil burning originated in the U.S.A., but as has been shown this is by no means borne out by the facts. On the contrary,

America played a decidedly secondary rôle in the early development of *industrial* oil burning, viz. that of *producer* rather than of a *consumer* of fuel oil. On the other hand it is certain that the later development of *domestic* oil burning, about 1920, began mainly in the U.S.A.

As a matter of fact, it was not until about 1900, when almost all the European Navies had already partially adopted fuel oil and several European merchantmen too were already running on oil, that oil burning was started on an experimental scale on some U.S.A. merchant-vessels of the Pacific Coast, first for coastal, and later on also for overseas trade.

It will be easily understood why oil burning in the American mercantile marine started on the Pacific Coast and not on the Atlantic Coast. The Pacific Coast was notorious for its high coal prices, there being no coal mines there, whereas on the Atlantic Coast there was plenty of coal. Thus it is recorded, that at that time three ships, the "Sierra", "Ventura" and "Socoma", of the Sydney—San Francisco Line, to mention only a few vessels, in fact never bunkered coal in America but always at Sydney, where they had to fill the forehold entirely with coal for the outward voyage, using the bunker coal and carrying cargo in the forehold during the return voyage. Considering the difficulties of transporting the coal from the forehold to the stokehold and the fact that coal trimmers are the most troublesome of a ship's crew, it is not to be wondered at that fuel oil was tried out on this kind of ships as soon as it was found that California could produce it in plenty. The World's Fair at *Chicago* in 1892 and the tremendous production of some oil wells at that time — e.g. the Spindle Top gusher in Texas (1901) produced about 100,000 barrels per day — roused public interest in oil burning and allayed the fear that supplies of oil would soon be exhausted, but again it was more the *mercantile* marine than the navy which took the initiative in America.

About 1902 a U.S. Naval Liquid Fuel Research Board was formed under the direction of Adm. MELVILLE to study the possibilities of the new fuel. Although the Board must be given the credit for having carried out very useful experimental work, their knowledge of the historical development of oil burning in Europe was rather limited. For instance, in his report of 1902 Adm. Melville stated: "It has only been within the last three years that the exceeding importance of atomizing the oil has been recognized", which is not at all in accordance with the facts, as the reader of this historical survey of the development of oil burning will certainly agree. Further he stated: "There is no record that previous to two years ago, any boiler ever evaporated the amount of water with oil as a combustible that was secured under forced draught conditions with coal as a fuel", whereas the experiments carried out by the British Admiralty on the "Retriever" as early as 1867 already showed a 12½-fold evaporation for oil against a 7½-fold evaporation with coal!

Although Adm. Melville's subsequent report of 1904 was on the whole in favour of the application of fuel oil for naval purposes, he devoted almost all his attention to the steam and air atomizing burners and only very cursorily mentioned mechanical atomization or, as he calls it, "artificial spraying", the importance of which, especially for marine use, he does not seem to have realized; this is proved by the lengthy description he gave of a centrifugal burner designed by WILLIAMS (fig. 35), consisting of a revolving horizontal plate, directly driven by a steam turbine, from the surface of which the oil was projected.

This view is supported by Mr PEABODY, the inventor of the well-known oil burner of that name, who wrote in 1923 ("Fuel Oil", Jan. 1923, *The Mechanical Atomizer. Its History in Use*.) as follows: "We owe to our English cousins and the enterprise of our own Navy Department the inception of the important development which has led to the use of the mechanical atomizer in this country (America). The British Navy being in closer touch with the progress being made in oil burning in Europe, anticipated us in appreciation of the importance of atomizing oil by means of centrifugal force instead of by means of an atomizing medium, such as steam or compressed air." and further:

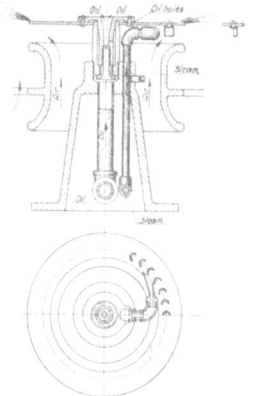

Fig. 35. WILLIAMS' centrifugal burner.

"In the year 1907, however, Captain NORTON of the U.S. Navy made a trip abroad where he found the mechanical atomizer in successful use in the British Navy and elsewhere, and quick to perceive its merit, he was instrumental in having the specifications for the U.S. Battleships of the *Wyoming* class include a paragraph requiring the installation of mechanical atomizing oil burners, to be used in conjunction with coal fires. The attention of American engineers was thus directed to a type of oil burner which was but little known in this country at that time".....
"There was at that time, so far as the writer knows, not a single mechanical atomizer in use on this side of the Atlantic".

Indeed it was in 1907 that Capt. NORTON, on a visit to England, realized at once the advantage of the Körting system, but it was not until 1909 that the first U.S. battleship, the "North Dakota", was equipped with a SCHÜTTE-KÖRTING oil-burning installation (fig. 36), and even then oil was only admitted as an *auxiliary* fuel, which was also the case with the battleship "New York" launched in 1914.

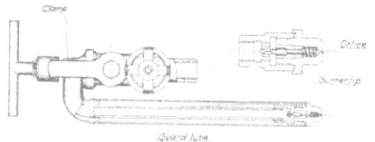

Fig. 36. SCHUTTE-KORTING burner. (U. S. Navy).

In the meantime, however, the mercantile marine had eagerly accepted oil as the main fuel. In 1905 the Bethlehem Shipbuilding Corp. brought on the market the DAHL burner, which was also of the pressure atomizing type and was readily applied on many merchant-vessels. In 1912 a *fuel oil testing plant* was built at the *Philadelphia Naval Yard*, principally due to the initiative of Adm. CONE, according to whose publications the following pressure atomizing burners were in use with the Navy at that time: ENGINEERING DAHL, SCHÜTTE–KÖRTING, PEABODY, NORMAND (French) and THORNYCROFT (Eng.). The merchant fleet was using: DAHL, FORE-RIVER, WHITE, BEST and LASSOV LOVEKING.

In 1917 an impetus was given to naval oil burning by America's entering the Great War, when there was a great need of it for the transport convoys across the Atlantic, with *quicker bunkering* and, last but not least, *smokeless combustion*, which often meant life or death to the crews.

Whereas the capacity limit for pressure atomizing burners in 1912 was about 500 lbs/hour/burner, ten years later, in 1922, it was between 1200 and 1500 lbs/hour/burner and a temporary development of boiler capacities of *three times the rated H.P.* of the boilers could be obtained, which was a very remarkable feature for warships and could never have been obtained with coal. This shows how rapid the progress was in this respect during the war.

However, it was still felt as a serious drawback of pressure atomizing oil burners that they offered *too narrow a range of regulation*, the only way to decrease the capacity being to reduce the oil pressure, which, of course, gave a slower rotary speed of the oil at the tip of the burner with a less satisfactory pulverization, so that this pressure reduction could only be applied to a certain limit. Theoretically it would, of course, have been possible to regulate the capacity of a pressure atomizing oil burner by changing the outflow area of the orifice ("tip"), but as the dimensions of the tip are rather small and, for instance, the insertion of a tapered spindle into the orifice would involve so many friction losses in the rotating oil as to spoil seriously the pulverization, no practical solution was found in that direction.

Generally the capacity was regulated by the *number of burners put into operation*, which was not by any means an ideal way, because the burners that were cut out were generally exposed to the radiant heat of the

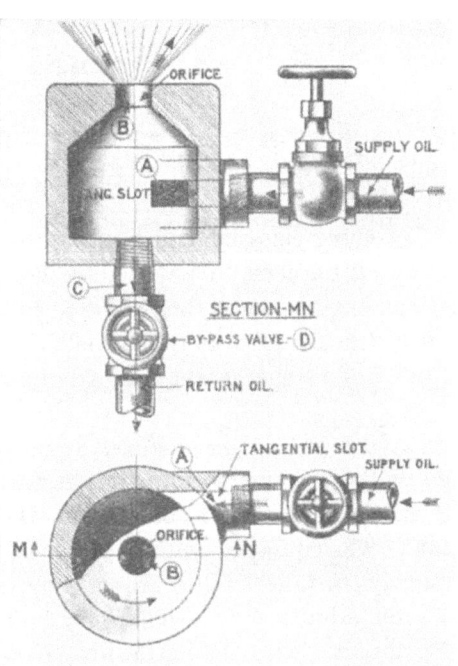

Fig. 37. PEABODY-FISHER "wide-range oil burner".

flames of the other burners and often got choked with carbon, so that before they could be put into service again they had to be cleaned.

It may be regarded as a second mile-stone in the history of the development of pressure atomizing oil burning that PEABODY–FISHER (PEABODY was the founder of the well-known American oil burner firm Peabody Eng. Corp. N.Y., and FISHER was an officer in the U.S. Navy) invented in 1921 their *wide-range burner* (fig. 37).

According to this system the oil from the pump enters tangentially into a swirling chamber in the burner just as it does with several other types, but the difference is that *only a part of the oil leaves the orifice* in front of the burner and is burnt, *the rest of the oil being returned* to the oil tank through a central bore. The quantity of oil returned is regulated by a valve, which at the same time regulates the quantity of oil burnt. By means of this device it is possible not only to increase considerably the rotary speed of the oil at the tip at full load by circulating quantities of oil several times larger than the quantities of oil burnt, but also to *maintain that speed at very low capacities* and thus guarantee an extremely *good pulverization at any load*. The return lines of several burners may be connected together to one line and governed by one single valve, by which means all burners may be regulated simultaneously.

It must be understood that this description has by no means the pretension of being a complete story of the development of the marine oil burner and is only meant as an indication of some mile-stones. It is almost impossible, and anyhow would be beyond the scope of this book, to give a complete historical survey of every burner application, such as for locomotives, the ceramic and metallurgical industries and so on. However, we will make one exception, with regard to the application of oil burning for *domestic heating*, and conclude this historical section with a survey of the development of the domestic oil burner.

# CHAPTER VI

## HISTORICAL DEVELOPMENT OF DOMESTIC OIL BURNING

For a long time the use of liquid fuel for domestic purposes received comparatively little attention, because suitable oils were either very high in price or not obtainable in small quantities, whilst moreover the general idea existed that fuel oil could only be burnt by pulverization with steam or compressed air in specially designed furnaces and, therefore, was restricted to industrial uses.

It has already been mentioned that even earlier than 1860 the so-called ASTRAKAN *oil stove* (fig. 1) was used in Russia to a certain extent for house heating, and it is curious to note that this stove already comprised a means of burning heavy residues (astatki, pacura, mazouth) without mechanical means, a problem which, as a matter of fact, is not at all solved even now. It is true that the present demands with regard to safety, cleanliness and efficiency are not the same as those in Russia at that time.

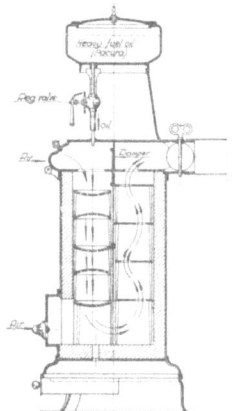

Another simple device for burning oil for domestic heating is shown in fig. 38, representing an oil stove which is used to some extent in Roumania, and which burns heavy residue ('pacura') trickling on a number of perforated plates.

A study of the patents on oil burning shows that there has always been a tendency to reduce the dimensions and capacities of devices for industrial oil burning in such a way that they could be used for domestic purposes, but as may clearly be seen from the former history of industrial oil burning, the simple methods designed to work without pulverization of the oil failed to operate properly on a large scale, and for smaller units the difficulties became worse (e.g. NOBEL's *grates* and *troughs* of fig. 39 and 40).

Fig. 38. ROUMANIAN domestic oil burner for residue.

The method of preliminary evaporation of the oil could not be applied to residue, because these oils could not be entirely evaporated without being cracked, and consequently the burners based on the *principle of previous evaporation* could not be made a success until *distillates* were to be had comparatively cheaply, as these oils, being produced by distillation, could of course also be evaporated completely for a second time in a burner.

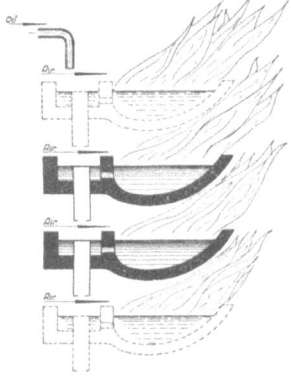

The general method of burning oils without pulverization has always been *to spread the oil out on a surface exposed to the radiant heat of the oil flame itself*, a method used by AUDOUIN (fig. 16 and 17) in 1867, whose sloping grates, grids or bars covered with an evaporating oil film may be found as new inventions — with slight modifications — in patents of all countries at regular intervals

Fig. 39. NOBEL's industrial oil grate.

of say 10 years from that date. However, none of these inventions came to be applied, mainly for the following reasons:

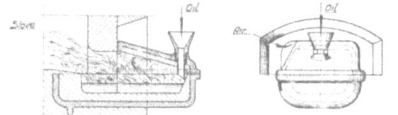

1) Too heavy oils were used, leaving a carbon deposit on the evaporation surface.

2) The oils were not readily available to the public in suitable quantities at moderate prices.

Fig. 40. NOBEL's domestic oil burner.

3) The units had still too large heating capacities and could certainly not be regarded as being safe for direct room heating.

These drawbacks were gradually overcome in America after about 1920, especially when it was found possible to *centralize the heating of various rooms*, so that a larger heating unit could be used.

Whilst domestic oil burning became very popular in America from about 1920 onwards, the development in Europe took place some 5 years later, starting especially in Switzerland, that being a relatively wealthy country with severely cold winters, whereas Switzerland itself produces no coal, which made competition easier for oil. Moreover the cost of transport up to the mountains is less per B.T.U. for oil than for coal, which factor also largely contributed towards the popularity of oil in such mountainous countries.

The oil industry, which in the beginning was based entirely upon the market for kerosene (lamp oil), found a serious competitor for this product in gas-lighting, which, except in the Far East, rapidly gained ground. This urged the oil companies to look for another outlet for their kerosene, and thus the petroleum lamp was developed for house heating and cooking. The use of kerosene vapours in *incandescent mantle lamps* (KITSON, fig. 41) could not be applied on any large scale, principally because the oil gas was too rich and apt to form a carbon deposit on the mantles, so that it had to be diluted with air, which, however, gave all sorts of other troubles. *Carburetted air-gas installations* (aerogene installations) using a light gasoline are still in use to some extent, e.g. in the East Indies.

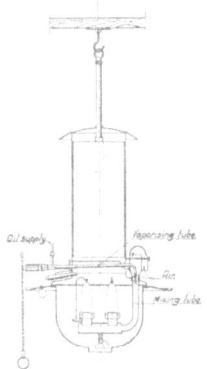

Fig. 41. KITSON's vapour lamp.

During the Great War gas-lighting in turn was replaced by electricity, and the gas works had therefore to look for another outlet for their product, so that it soon replaced kerosene for cooking purposes and to a large extent also for emergency heating of rooms, mainly because of the smell of kerosene and the fire hazard when handling it. Thus the competition with gas caused new demands with respect to the domestic use of oil for heating: *comfort* and *foolproofness*, which are very important factors indeed, whereas with industrial oil burning only efficiency and reliability are asked for.

Even the invention to change the yellow flame of a kerosene wick lamp into a blue one (so-called "*blue burners*") could not prevent the kerosene from being superseded by gas for cooking and coal for room heating: the delivery of exhaust gases of more or less completely burnt oils into living rooms was intolerable and if means were provided to conduct them away to the chimney the heating efficiency of the system proved to be lowered to such an extent and the heating cost raised so high that it was impossible to compete with coal stoves.

Whilst oil burning for house heating with cheap, brownish, unrefined domestic fuel oil is really considered as a higher stage of comfort compared with coal, yet it is curious to note that for a long time lamp oil, which is a *higher-priced, water-white refined product,* has been the heating fuel used by the *poorer* classes, with the characteristic smell of oil stoves in badly ventilated rooms.

Another attempt to stimulate the use of kerosene comprised the class of *vapour*

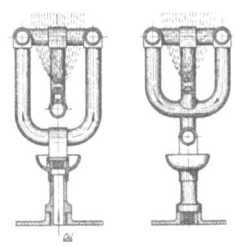

Fig. 42. Original Primus vapour-jet burner.

*jet burners.* But then again there was an objection to the ordinary form of vapour jet burners of the blow lamp type operating with air pressure (Primus fig. 42, Sievert, Wells fig. 43, Butler, etc.), owing to the frequent choking of the nipple, which has but an extremely small aperture and calls for much attention. The vaporizer, too, is apt to get choked with solid deposits from the oil. Consequently this type of burner, although very useful with petrol for such purposes as soldering, melting and blistering paint, has only found application for emergency heating and never formed a serious competitor for gas or coal heating.

Nor could systems in which gas oil is converted into *permanent oil gas* under pressure and distributed in steel containers (Young, Pintsch, Dayton–Faber, Blau-gas, Lowe) find any further application than for the lighting of railway carriages.

As already stated, numerous attempts were made to develop a simple method of burning oil without mechanical means, which were all based upon the *principle of* Audouin, viz. *to expose the surface*

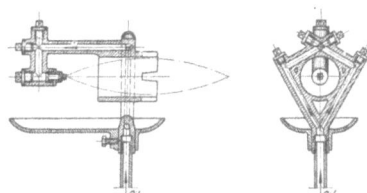

Fig. 43. Wells' vapour-jet burner.

*of an oil film to the radiant heat of the oil flame and burn the oil vapours thus produced.*

As a matter of fact the ordinary kerosene wick lamp is based on the same principle, but it has the advantage that the oil is fed by capillary action of the wick in proportion to its evaporation, thus automatically effecting an equilibrium.

In general the experience has been that an oil burning apparatus based upon the principle of Audouin answered best when the *smallest possible quantity of oil was exposed at a time to the evaporation temperature,* as then the time during which the oil is apt to carbonize was reduced to the minimum; the oil after being admitted into the vaporizer had not to wait to be evaporated. Thus the burners of Audouin and Sainte-Claire Deville with an oil *film* gave less carbon troubles with the same fuel than were experienced with oil *grooves,* for instance, and a burner with an oil *reservoir* fully exposed to the evaporation temperature was still worse in this respect.

A few of the numerous patents may be quoted here:

Class I. *Larger quantity of oil simultaneously exposed to the evaporation temperature.*

1) The Irinyi *burner* (Deutsche Oelfeuerungs Gesellschaft m.b.H. Hamburg), being an example of an oil container in which a *larger* quantity of oil is exposed to the evaporation temperature; described in patents: DRP 254,518; 255,355; 255,356; 267,558; 275,056 and 263,782 (all dating from about 1911) (fig. 44).

2) The BECKER *oil burner* (Becker Feuerungs- u. Maschinen Patent Verwertungs Ges. m.b.H., Berlin-Schöneberg), also having an open horizontal gutter in which a large quantity of oil is exposed to the evaporation temperature; described in patents: DRP 351,967; 359,736; 369,784; 353,574; 355,050; 369,785; 380,934; 385,764; 357,155 and 359,615 (about 1920–1922) (fig. 45).

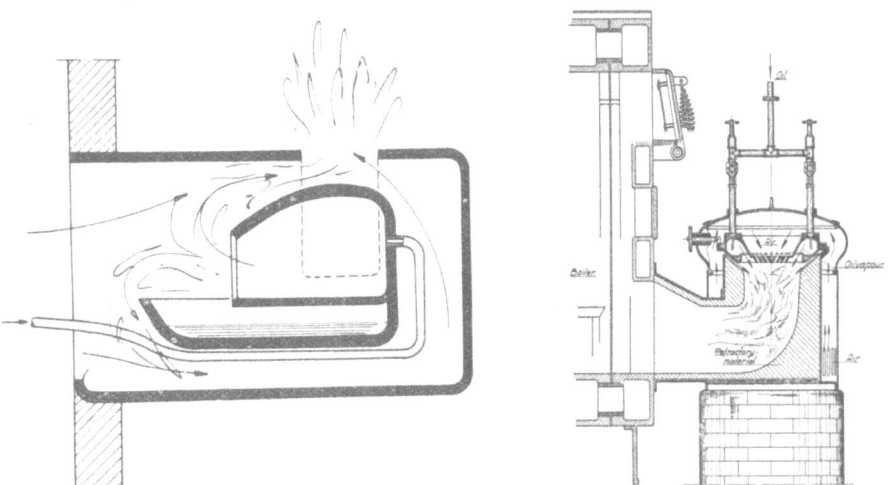

Fig. 44. IRINYI's metallurgical oil burner.          Fig. 45. BECKER's domestic oil burner.

Other inventions of this class are:
3) MARION (Am. Pat. 1,774,937 and 1,755,771 of 1929).
4) BUCHHOLTZ (Br. Pat. 1891 of 1909; Dutch Pat. 1612 of 1913).
5) CAUVET-LAMBERT (Fr. Pat. 666,123 of 1928).
6) POWERS (Am. Pat. 1,782,050 of 1930).
7) SCHIEDECK & OSTER (DRP 518,425 and 518,426 of 1929).
8) COULTAS (Am. Pat. 1,832,280 of 1931).
9) RUPPMANN (DRP 405,192 of 1922).

CLASS II. *Smaller quantity of oil simultaneously exposed to the evaporation temperature.*
1) LOOSER (Sarganz — Switzerland, DRP 533,502 of 1930) (fig. 46).
2) MAYER (Br. Pat. 354,359 of 1930 and Dutch Pat. 28,118 of 1930) (fig. 47).
3) JUNKERS (Dessau, DRP 436,623 of 1924).
4) KREUTZBERGER-GERMAIN (Br. Pat. 214,342 of 1923).
5) FALKENTORP & ERICHSEN (Dutch Pat. 18,293 of 1928).
6) KNUPFFER (Am. Pat. 1,674,282) (fig. 48).
7) GENGELBACH (Am. Pat. 1,698,129 of 1929).
8) HENNEBÖHLE (Am. Pat. 1,555,855 of 1925).
9) SOC. AN. ITALIANA PER LA COMBUSTIONE DEGLI IDRO CARBURI (Fr. Pat 434,420 of 1911).

CLASS III. *Negligible quantity of oil simultaneously exposed to the evaporation temperature.*

1) KRAUSE (Br. Pat. 180,176 — 1922) (oil film on sloping plate with grooves).

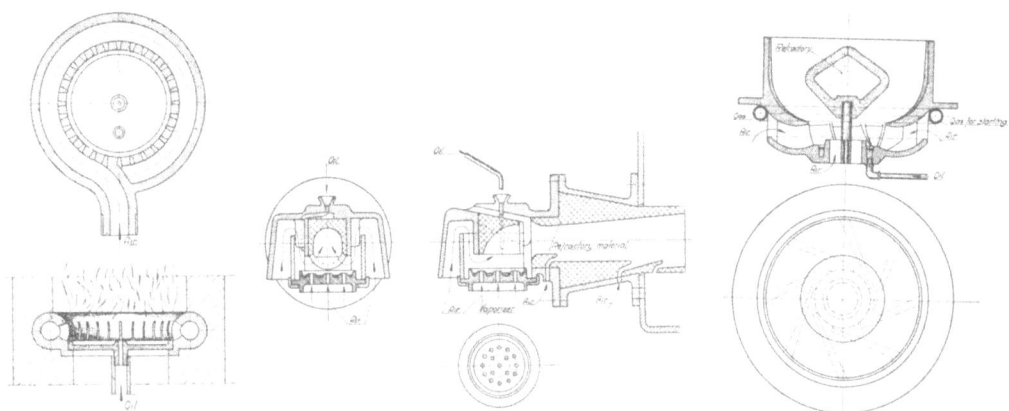

Fig. 46. LOOSER's oil burner for ranges.

Fig. 47. MAYER's domestic oil burner.

Fig. 48. KNUPFFER's oil burner.

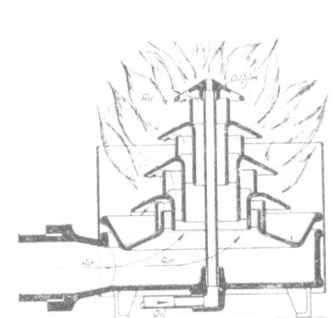

Fig. 49. KERMODE's mushroom oil burner.

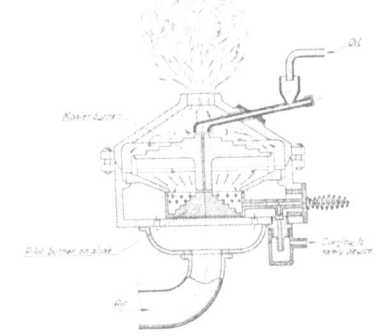

Fig. 50. PIZZI's domestic oil burner.

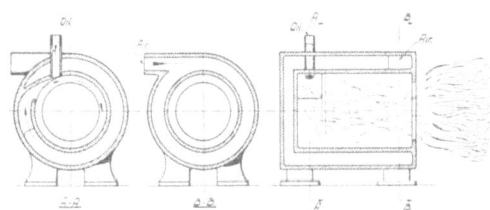

Fig. 51. STRACK's oil burner.

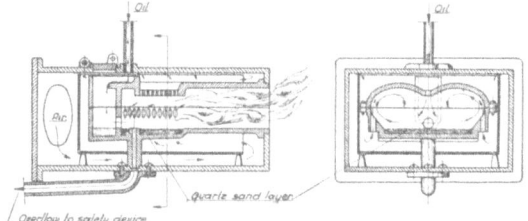

Fig. 52. GIBBS' oil burner with double vortex.

2) KERMODE (Br. Pat. 299,486 — 1927) (oil film on cones like mushroom type) (fig. 49).

3) PIERBURG (DRP 353,300 — 1922) (oil film on vertical plates).

4) BIANCHI (Fr. Pat. 689,829 — 1930) (oil film on sloping grates).

5) MASON (Br. Pat. 320,948 — 1928) (oil film on flat cone).

6) SCHAUMANN(Am. Pat. 1,531,819 — 1924) (oil film on cone, as mushroom type).

7) WHEELER (Am. Pat. 1,691,690 — 1928) (oil film on cone).

8) PIZZI (Am. Pat. 1,901,610 — 1929) (oil film on cone) (fig. 50).

9) STRACK (Fr. Pat. 690,455 — 1930) (oil drops in air suspension) (fig. 51).

10) DUPONT (Am. Pat. 1,427,419 — 1922) (oil film on sloping grate).

11) GIBBS (Am. Pat. 1,624,943 — 1925) (oil film on porous material) (fig. 52).

12) JOHNSON (Br. Pat. 241,317 — 1923) (oil film on porous material).

None of these attempts to develop a simple method for burning oil without mechanical means came to application on any large scale.

---

# CHAPTER VII

### RECENT DEVELOPMENT OF DOMESTIC OIL BURNING

A. *The Development of Domestic Oil burning in America*

About 1919 a more favourable turn took place in America with the development of the so-called *mushroom type* of oil burner.

In this burner, belonging to class III, the oil is also spread out as a film on the outside of a flat cone with or without oil grooves, but a new feature that made it a temporary success was a device for providing *strongly preheated combustion air*, consisting of a number of canals leading through the flame (fig. 53). With suitable light gas oil available at moderate prices and the demand for a convenient, cheap, and simple burner for central hot water heating systems, which gradually came to be wide-spread throughout America, this mushroom type seemed to offer wonderful prospects.

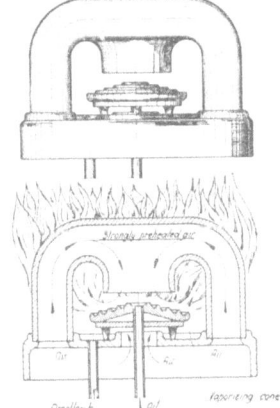

Fig. 53. MUSHROOM-type of oil burner.

In a short time some hundreds of manufacturers of these "natural-draft gravity-feed" burners — so-called because they were operated by "natural draft" only and the oil was fed to the burner "by gravity" — sprang to life; as a witty contemporary expressed it: "Every screw-driver owner builds an oil burner".

Some names of larger manufacturers of this type of burner are: OLIVER, WORTHINGTON, HEATIATOR, OILHEAT, INTERSTATE, VULCAN, EXCELSIOR, WALTHAM, CLIMAX,

FOSTER and LIBERTY, the main difference in their products lying in the number of canals for preheated air, whilst the WILLIAM's HEATING Corp., which afterwards built up an excellent reputation with their OIL-O-MATIC products, at that time came out with the DIST-O-STOVE burner of the mushroom type.

How high expectations rose with regard to the prospects of such burners is illustrated by the following typically American announcement that appeared in the papers as an introduction to one of these burners:

> **WANTED :** A simple, practical and inexpensive fuel oil burner suitable for use in the average home. *A reward equal to the fortune of* JOHN D. ROCKEFELLER *or* HENRY FORD *will be paid* to the inventor of this device. ROCKEFELLER developed *light*, FORD developed *power*, and now AMERICA awaits the rise of the third great genius to complete the development of the last remaining but greatest proporty of the liquid treasure, the *element of heat*. Who shall it be? What fabled wealth shall be his reward?
>
> *The call of the nation is answered!*
>
> **The ..... Oil Burner!** The oil burner anyone can operate. Operated by one valve. Burner for every type of home. No noisy motors. No electricity costs. No moving parts. No friction wear.

After two years the burner in question had disappeared from the market, and the

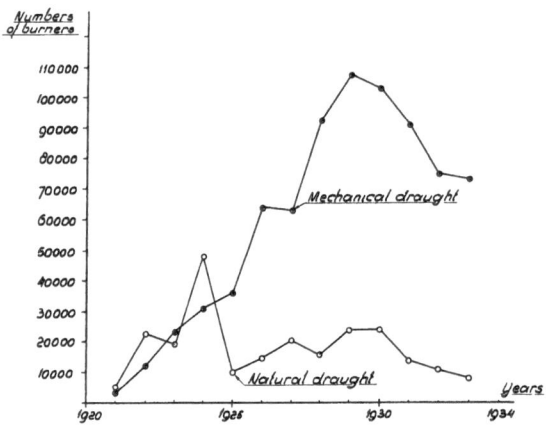

Fig 54. Yearly sales of domestic oil burners in the U.S.A. (From data of Fuel oil Journal 1934).

same happened to almost all of the burners of this *"mushroom type"*, some of which consisted only of some pieces of cast iron worth perhaps not more than $ 20 and were sold for $ 150 to $ 200. The reason of this *"slump in natural-draught gravity-feed"* burners (fig. 54), which took place in 1924/1925, was that the makers, as it was put at the time, „sold a lot but had to take them back too, very often before the instalments had been paid and sometimes with indemnification", mainly for the following reasons:

1) *Natural draught proved to be unreliable as a motive force for oil burners.*

Weather with bad draught conditions and cold chimneys caused smoky combustion and rapid carbonization of the vaporizing cones.

2) *The burners needed too large an excess of air* to obtain a clean combustion, resulting in *too high a consumption*, which will be explained later on in the theoretical part of this book.

3) *The range of regulation was too small.* In cold weather it could not be made to give sufficient warmth and in mild weather and at night it could not be turned low enough. As a result there was also a slow pick-up in the morning when the house was cold.

4) *The burners of this type could not be equipped with automatic controls*, which was essential not only from a point of view of comfort and economy but also of safety.

5) *The burner could not be made sufficiently foolproof*, requiring too much skilful attendance, and consequently was often not approved by the Board of Fire Underwriters.

# At Last! An Oil Burner
## *Every Family Can Afford —*

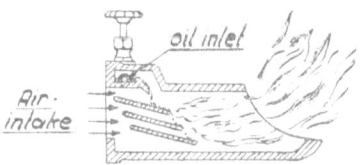

THE ........ Oil Burner is a one piece casting, and its operation is simplicity itself. There are no wicks, no needle valves, no generating tubes, no electric motor or other mechanical parts to wear out or require adjustment. Burns any fuel that is suitable for domestic oil burners. No noise, no smoke, no dirt, no odor. Fully guaranteed in every respect.

$100.$$^{00}$$

Retail price, installed ready for tank

**WANTED:**

*State and County Distributors and Local Dealers*

The ........ opens up an an entirely new field for oil burner sales. Every family is a prospect for this low priced, efficient, fully guaranteed oil burner. Sales are easy—and profitable.

The ........ franchise is valuable. Our exclusive state and county distributor proposition offers tremendous profit possibilities. Sales in territories already allotted are exceeding quotas by far. Write or wire today for full particulars of this remarkable oil burner and of our liberal, profit-producing proposition.

Fig. 55. Revival of AUDOUIN's oil-grate in 1925 (U.S.A.)

The first designers of the so-called *mechanical* domestic burners (CHALMERS (fig. 56) and WILLIAM's *Oil-O-Matic* building pressure atomizing burners, and RAY and AETNA (fig. 56) manufacturing centrifugal atomizing burners), who had always argued in this sense, were thus found to be in the right. Some of the natural-draught gravity-feed burner manufacturers saved themselves just in time by changing to the manufacture of the so-called "mechanically assisted" evaporation burners of the *vaporizing pot type*, consisting of a fire pot in which the oil was admitted, mostly without previous pulverization and evaporated, cracked and burnt by blowing into it a strong current of turbulent air delivered by a fan.

Some names of this type of burner are: LACO (fig. 57), SUPREME, HOFFMANN, BOCK (fig. 58), NORTHERN (fig. 59), BERRYMAN, GRANT, HAAG (fig. 60) and PRIOR (fig. 61) (with the exception of the last two, which are of Swiss origin, all were American products). They form a class by themselves lying between the natural-draught burners and the mechanically atomizing burners, and have been fairly well able to maintain themselves up to the present time for lighter fuels by avoiding the disadvantages of the natural draught mentioned above.

The years 1927 to 1930 saw great activity in the development of the mechanical domestic burners in America, and especially in the *fully automatic form*, because it was then generally felt necessary to *eliminate entirely the influence of human attendance* on the efficiency and safety of operation of the equipment.

Fig. 56. Pioneers of Mechanical Atomization
(AETNA and CHALMERS).

Many manufacturers who were making carburettors, gasoline pumps, shock absorbers, electrical equipment etc., for the automobile industries very soon discovered the new possibilities and started to make fans, oil pumps, filters, pressure regulators, non-return valves, floats, trip buckets (a safety device against leakage, see fig. 62), mercury switches, protectostats, thermostats, stackstats, viscostats, aquastats, pressurestats, vaporstats, pyrostats, duostats and all sorts of other "stats" (rightly or wrongly so named), small motorless bellow or membrane pumps to convey the oil from underground tanks to the burners, special noiseless motors without radio interference, high-tension systems, etc.

Within a short time it was possible to make a fully automatic domestic burner at home merely by buying various parts and assembling them.

A list of 179 American makes of domestic oil burners marketed in 1932 is given in table I, which has been composed from

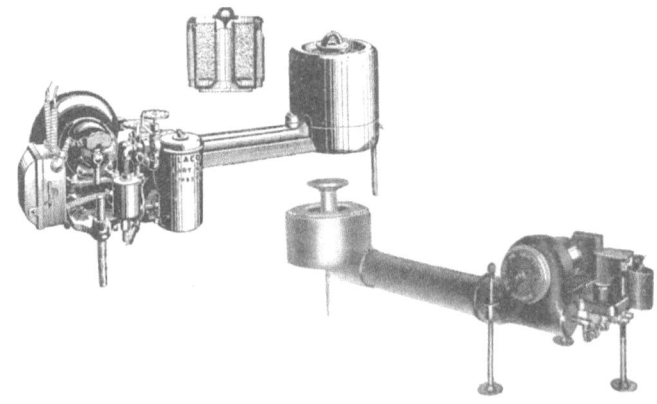

Fig. 57. LACO burner.     Fig. 58. BOCK burner.
(mechanically assisted evaporation).

data given in *The Fuel Oil Journal* and which may give an idea of the enormous variety of applications of comparatively few basic principles.

The work of big associations like the *A.O.B.A.* (American Oil Burner Ass.) the *A.S.T.M.* (American Society for Testing Materials), the *A.P.I.* (Am. Petr. Inst.) and *A.S.H.V.E.* (American Society of Heating and Ventilating Engineers) and of the Fuel Oil Department of the board of Fire Underwriters contributed largely to the stabilisation of the domestic oil burner business, and especially of late years a mutual understanding between

Fig 59. NORTHERN vaporizing pot type (natural draught).

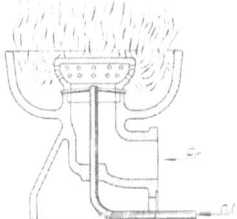

Fig. 60. HAAG vaporizing pot type.

oil burner manufacturers and oil producers has been cultivated to the benefit of both.

A few names of fully automatic mechanically atomizing American oil burners well-known also in Europe are: ABC, CHALMERS, GILBERT & BARKER (fig. 63), HEIL-COMBUSTION (fig. 64), JOHNSON, QUIET-MAY (fig. 65), PETRO-NOKOL (a phonetic advice to use petroleum and no coal!), PROGRESSIVE (fig. 66) (in Holland known as NITEK), RAY (fig. 67), SILENT-GLOW, WAYNE, WILLIAMS' OIL-O-MATIC, NU-WAY and YORK.

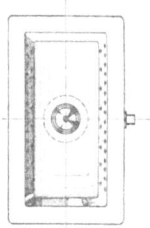

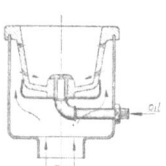

Fig. 61. PRIOR vaporizing pot type.

Fig. 62. IMPERIAL trip bucket for safety against leakage

About 1930, after the American people had got to appreciate the comfort of oil burning and discarded coal, there was a second phase in which they were gradually "educated" to the comforts of *automatic room temperature control*. This evolution took place much more easily in America than it is doing at present in Europe, because domestic labour is more expensive in America and where many European families keep a servant who

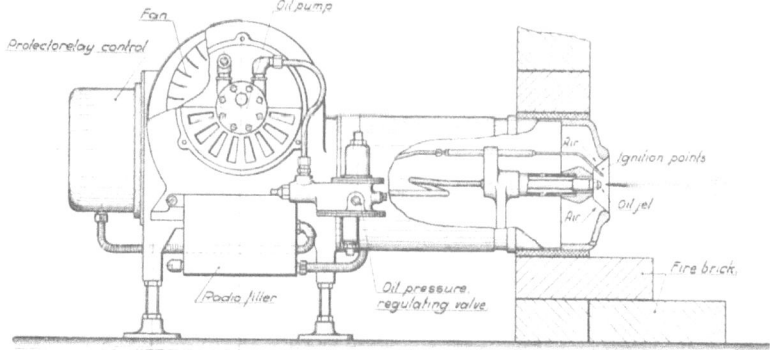

Fig. 63. GILBERT and BARKER press. atom. burner.

attends now and then to the central heating plant such could not be afforded under similar American conditions, so that in the latter country the benefits of oil are felt more directly than in Europe.

The above-mentioned "education" in connection with automatically controlled

The Baby-type HEIL COMBUSTION.    The large capacity HEIL COMBUSTION.
Fig. 64. HEIL COMBUSTION domestic oil burners.

oil heating is more or less similar to that in the automobile industry: about 1925 motorists had come to appreciate the easier performance of a 6-cylinder motor-car, and in 1932, notwithstanding the higher costs, more complications and higher gasoline

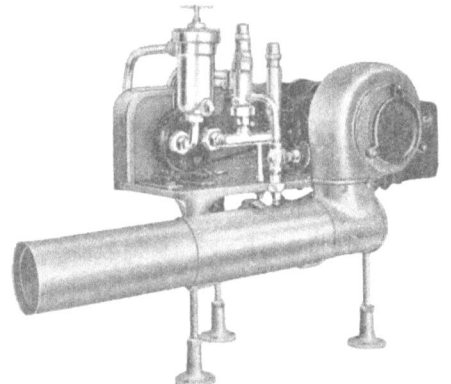

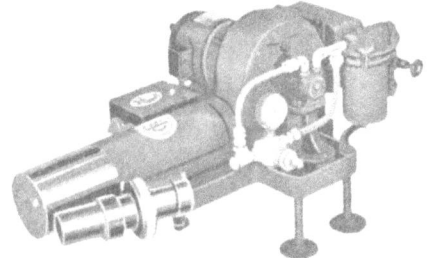

Fig. 65. QUIET MAY press. atom. burner.        Fig. 66. PROGRESSIVE press. atom. burner.

consumption of the 6-cylinder engine, a 4-cylinder car could scarcely be sold at all in America; the second stage was entered upon about 1931, when the 8-cylinder motor-car came into favour, although this is still more expensive. If once the public has become familiar with some form of comfort it will not easily part with it again.

It may seem somewhat beside the point to make these remarks here, but it should be borne in mind that the development of domestic oil burners is governed by rules quite different from those governing the industrial oil burner. Whereas the

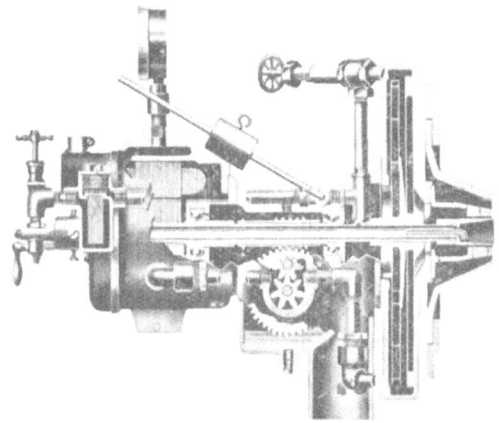

Fig. 67. RAY hor. rotary domestic burner.

Fig. 68. CHALMERS gun type oil burner.

prospective purchaser of an *industrial* oil burner only asks for an efficient and reliable instrument with which to cut down his costs of manufacture or to improve his products, *the buyer of a domestic oil burner asks for a certain form of luxury and comfort in his home*, demands which are similar to those considered when buying a car.

It is not surprising, therefore, that also the *outer appearance* of the domestic oil burner is getting more and more important and this, for instance, accounts for the "stream line" having been applied even to the domestic oil burner, an example of which is shown on fig. 68, "CHALMERS' *gun type*" oil burner (also: "clean-lined").

Another feature is the recent development of the *burner boiler units* (GENERAL ELECTRIC, YORK, HEIL COMBUSTION, JOHNSON's Oil heat Servant, see fig. 69) of which the basic idea was that oil burner and boiler

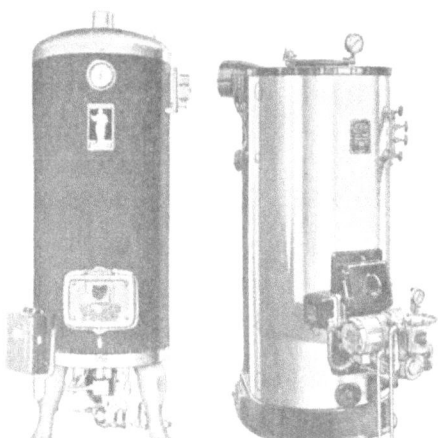

Fig. 69. Some burner boiler units.

should form one harmonious whole (fig. 70), whilst only too often a good oil burner gives bad results owing to its being put into the wrong boiler, or into the right boiler but in a wrong way. Although the principle of this is quite correct, it has its drawbacks because, in the case of conversion jobs when the old boiler becomes useless, the cost is higher.

Another new type of burner, although its principle has been known from 1916 onwards (e.g. Br. Patents 159,467; 188,871; 247,833), that appeared on the American oil burner market some years ago (about 1927, Lynn, Silent Glow) and that is now being applied more and more (in 1932 about 65 makes), is the so-called *range-burner* (see fig. 71) consisting of a cast-iron base with grooves from which the oil evaporates by radiation of heat from a number of red-hot perforated shells placed on the base as concentric rings. The air is sucked into the holes by natural draught and some kind of surface combustion takes place inside the shells, with a *blue flame*. These range burners have become very popular, mainly because of their being easy to install, absolutely noise-

"Putting a modern oil burner in an old-fashioned boiler is like putting a modern motor in an old buggy...."

Fig. 70. Striking advertisement of a burner-boiler unit (American Radiator Comp. Fuel Oil Journal, June 1934).

less and needing very little draught. Although this burner is sometimes claimed to be suitable for gas oil, an absolutely reliable performance can only be obtained by

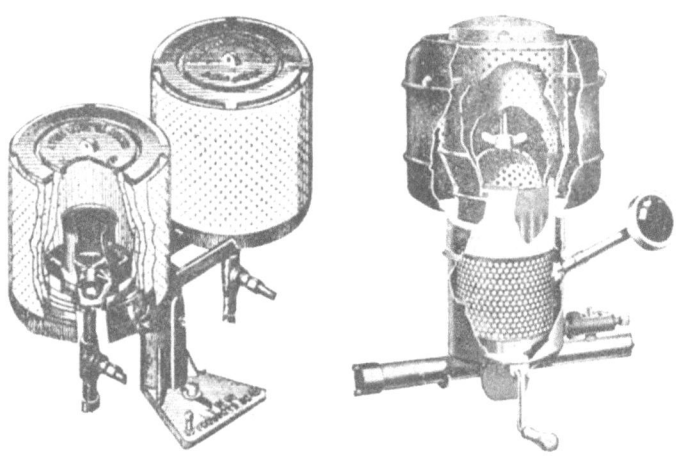

Fig. 71. Range burners.

using kerosene as fuel. In America a special brand of oil is now being distributed for this purpose, known as *range oil*, which is a heavy kerosene.

B. *The Development of Domestic Oil burning in Europe*

Whereas domestic oil heating was started in America in about 1920 and came to a rapid development there in a few years' time, in Europe the application of oil burning for domestic use prior to 1925 was hardly worth mentioning.

In general it may be said that, whilst America took and maintained the lead, Europe is still some 3 to 15 years behind, more in some countries than others; *Switzerland, Austria* and *Holland* may be regarded as forerunners in Europe with respect to domestic oil burning, with *Germany* coming in at the end. Curiously enough the conditions for favourable development of this branch of oil burning have been quite different in all the three first-mentioned countries.

Thus *Switzerland* has been very successful in domestic oil burning because of its being a wealthy country and very popular for winter-sports. Tourists demand luxurious comfort and are willing to pay well for it. Moreover the country itself produces neither coal nor oil and consequently there are no reasons of mercantile policy for putting higher import duties on either of them. Then again it is of the utmost importance for such a mountainous country that the transportation cost of one heat-unit (Cal. or B.T.U.) 1000 metres above sea-level is considerably less for oil than it is for coal (not only because of the different heat value per kilogram but even more so because of the higher thermal boiler efficiencies obtainable with oil).

*Austria*, which is a relatively poor country, has nevertheless a population which was originally accustomed to luxury and is used to taking life easy. Various loans have been negotiated in exchange for low import duties on foreign products and thus oil prices are comparatively low, which is the same with oil burner equipment. This must be one of the main reasons why oil burning has found such a rapid application in that country.

*Holland* is a rather wealthy country. It has world-wide trade connections and its people readily adopt new foreign inventions. The population preferably live in houses of their own and consequently there was a demand for small convenient oil burning equipments. Low fuel oil prices and mild winters, resulting in comparatively low heating cost per season, have been factors largely contributing to the growth of domestic oil heating.

In *England* and *France*, although both rather wealthy countries too, domestic oil burning is seriously handicapped by very high fuel oil prices, due to high import duties, which makes it extremely difficult for oil to compete with coal.

In *Germany*, which is not a wealthy country, the price of oil is so high and the average living conditions of the inhabitants so poor that domestic oil burning has found hardly any application there.

The *Scandinavian countries* are comparatively wealthy but their high fuel oil prices and long severe winters make it difficult for oil to compete with coal and the national fuels, wood and charcoal.

*Italy* and *Spain* are relatively poor countries having high fuel oil prices, but they enjoy extremely mild and short winters and consequently domestic oil burning has found some application only in the northern regions of these countries.

In general it may be remarked that in most European countries at least three-fourths of the burners used are of American origin, with the exception of England, in which country a national domestic oil burner industry has sprung to life under the influence of the *"Buy British"* propaganda; some *British makers* are: PARWINAC, (abbreviation of: Parker, Winder & Achurch), HOPE, CLYDE, AUTOMESTIC, COMBUSTIONS LTD, HOMATRA, ELECTROMATIC, LAIDLAW DREW, FILMA, URQUHART, BRITISH OIL BURNER, KALOROIL, VICTORY. — *France*, too, has its own domestic oil burners, such as: C.A.T., S.I.A.M., LOY et AUBÉ, MAZAL, LINKÉ, CAUVET-LAMBERT, AUTOPROGRESSIF (Boutillon) and FLAMME BLEUE. — *Switzerland* has its CUÉNOD, HAAG, PRIOR, BECKER, LIEBER and ELKA, and *Austria* the COMBINA, SIMPLEX, O.F.A., DANUBIA, GARVENS and PRODOMO, whilst *Dutch* makes are the N. F. burner and the REINHART-LANG burner; moreover the Swiss burners PRIOR and HAAG are manufactured in Holland under licence.

This Section may be concluded with one more general remark concerning the differences between Europe and America, viz. that the domestic fuels distributed in Europe are as a rule much heavier and more expensive than those in America, a fact which has sometimes led to disappointments encountered with imported oil burner equipments needing a lighter grade of fuel oil than the average products distributed as domestic fuel oils in the country in question.

In this connection it is interesting to note that during the last few years in the West European countries, especially in Great Britain, a very pronounced and fruitful research has taken place aiming at the development of domestic oil burners capable of burning the *heavier grades* of fuel oil. For this purpose a new element was added to the equipment, viz. a *preheating device*, which automatically keeps the oil consistently at the necessary temperature for a suitable viscosity and efficient pulverization. Further details will be discussed in the Fourth Section dealing with the Future of Oil burning.

This historical survey, then, may be concluded with the remark that, if we consider the various stages of development up to the present day, it is scarcely subject to doubt that not only is oil burning in internal combustion engines one of the predominant factors governing the world, but also its importance for industrial and domestic heating is growing rapidly and in fact may in the near future prove to be "a burning question" indeed.

# SECTION II



## BASIC PRINCIPLES OF OIL BURNING

———————

# CHAPTER I

## INTRODUCTION

Liquid fuel is a collective noun for several products obtained from crude oil, which latter as it occurs in the natural state may be considered to belong, roughly speaking, to one of three classes according to the *base of the oil*, viz.:

1) Paraffin Base,
2) Asphaltic Base, or
3) Naphthenic Base,

although in fact the constituents of every crude oil belong more or less to all of these three classes.

The American Pennsylvanian oils and the Sumatra oils from the Dutch East Indies (D.E.I.) are examples of paraffin-base crudes; Venezuelan and Mexican oils and Java (D.E.I.) oils of the asphaltic-base crudes; Californian, Russian, Borneo (D.E.I.) crudes are examples of naphthenic-base crude oils. — How little, however, the origin of the oil tells about its properties may be illustrated by the fact that *Louise*, one of the Borneo fields, produces crude oils of all three classes, sometimes from wells practically on the same spot.

In regard to their use as combustibles, there is a considerable difference between the fuels obtained from these classes of crude oil, especially with respect to *the tendency to soot or to form oil-coke*. In general it may be said that the paraffin-base fuels can be burnt easiest without smoke, whereas asphaltic-base fuels give most trouble in trying to get clean combustion; the naphthenic-base fuels lie in between.

*Crude oil* as it comes from the earth is not suitable for use as fuel oil, because it is a *highly inflammable* mixture of several kinds of oil, from the very light and volatile naphthas or gasolines to the very heavy non-volatile waxy or asphaltic residues. As only oil vapour and not the liquid itself is inflammable when mixed with air, it follows that the more volatile the oil, the more inflammable it is and the more dangerous to handle.

According to their volatilities, oil products may be classified as follows:

*a) Gasoline or naphta.*

On the Continent of Europe these oils are known as: *"Benzin(e)"* (not to be confounded with *"benzene"* or *"benzol"*, which is $C_6H_6$), *"spirit"*, *"essence"* or *"petrol"* (the latter not to be confounded with *"petroleum"*, which is *"lamp oil"*). These oils contain the most volatile liquid constituents of crude oil and since their vaporization needs no special arrangements they are eminently suitable for use in automobile engines. On the other hand they are dangerous to handle and for their storage stringent precautions have to be taken against fire.

*b) Kerosene or lamp oil.*

The name *"kerosene"* was originally used by an American firm as a trade mark for its heavier oils, but it has been gradually applied to *illuminating oils* in general. In Europe the name *"petroleum"* is often heard for this product, which is a component of crude oil next heavier to gasoline. Being less volatile it is not so dangerous to handle and precautions against fire are less severe. On the other hand its clean combustion is less easily obtained than that of gasoline.

*c) Gas oil.*

The name *"gas oil"* was originally given to the oils next heavier than kerosene because they were used to produce oil gas by cracking according to processes such as evolved by YOUNG, PINTSCH and others. It is usually a yellowish-brown oil, which contains so small a percentage of volatile fractions that it is a very safe liquid indeed. A burning match will be extinguished when thrown into it. Its clean combustion, however, requires special arrangements.

*d) Residue.*

This oil is a collective noun for *all that is left* from crude oil after the more valuable constituents have been distilled off from it. In general, residues may be *waxy* or *asphaltic*, depending on the origin of the crude. Other European names are *"astatki"*, *"pacura"* and *"mazouth"*. Residues are the safest fuel oils imaginable, but on the other hand their efficient and clean combustion requires special arrangements, and the heavier the residue, the more complicated such arrangements are.

Of course, there are many oils on the market which do not exactly belong to one of the above classes; the classification is only intended as a rough guide.

In order to avoid confusion it is well to remember that engines and oil burners are often advertised as being able to work on *"crude oil"* (*"Rohöl-Motoren"*) whereas in fact it is *gas oil* that is meant, viz. *"crude"* taken as the opposite of *"refined"*. From the above it will be clear that the use of actual *"crude oils"* for engines and burners would by no means be recommendable from the point of view of safety. Again, one may often read advertisements in French papers for *"brûleurs au mazout"* (oil burners on mazouth), when it is not the heavy residue, but a *gas oil* which is meant.

*Fuel oil* in the sense in which it will be used here is to be understood as being a *"safe" oil having a flash point of at least 65° C. (150° F.)*, which is the legal limitation in most countries for an oil which is stored and handled without extraordinary precautions against fire. The *flash point* is the lowest temperature to which the oil must be heated to make it possible to ignite the oil vapours above its surface *just for a moment*, for instance with a small gas flame, burning match, electric spark or the like. After this initial ignition the flame of oil vapour will soon die out, and this is the reason why that temperature is called the *"flash" point*. To obtain a *longer-lasting* flame it is necessary to heat the oil to a still higher temperature (for mineral fuel oils generally some 30° C. above the flash point), and when the flame continues to burn for 5 seconds this is called the *"fire" point*. The temperature at which the oil, when mixed with air, ignites spontaneously — *"spontaneous ignition temperature"* or S.I.T. — lies still higher, generally between 300° and 500° C.

Because of its international importance with regard to the fuel oil market, this flash point test (generally determined by the PENSKY-MARTENS method) and several other tests, which will be discussed later on, were explicitly prescribed years ago by Boards such as the *American Society for Testing Materials* (A.S.T.M.) and the *Bureau of Mines*, which regulations have now been generally adopted all over the world.

From the above it will be clear that safety against fire in storage and handling is one of the main considerations for the use of fuel oil, but then if the oil answers the above requirements it is necessary to prepare the oil and air in a suitable way for combustion, as otherwise it is impossible to burn it efficiently.

The *process of oil burning* in general may be roughly divided into 7 stages:
1) *Preparing the oil* for combustion.
2) *Preparing the air* for combustion.
3) *Vaporizing* or *gasifying* the oil (if not already done in the first stage).
4) *Mixing* the oil, oil-gas and air (so far as not already done in the first or third stage).
5) *Igniting* the mixture and *burning* it to water and carbon dioxide.
6) *Transferring* the heat from combustion products to the materials to be heated.
7) *Carrying away* the flue gases.

Strictly speaking, the denomination *"oil burning"* is wrong, because actually it is not the oil as a liquid but its gas that burns. Likewise the name *"oil burner"* is not quite correct, because the apparatus designated by that name is only a contrivance to prepare the oil in such a way that *its gas* can be burnt, which process takes place in another device comprising a combustion chamber.

Before entering further into the details of these 7 stages of oil burning, it is advisable first to deal with the question: What is oil really, physically and chemically speaking, and how does its combustion take place?

# CHAPTER II

## THE PHYSICAL PROPERTIES OF FUEL OIL

### A. *Distillates versus Residues*

Crude oil as collected from the wells is generally split up by distillation into several *"fractions"* or *"cuts"* before it is used. The diagram of fig. 72 may serve as a simple scheme for the distillation of crude oil. The products framed by thick lines are known as *fuel oils;* kerosene, being a *refined* product, which is also sometimes used as a fuel for heating purposes, represents a class of its own.

Fuel oil for burning purposes may be divided into two classes:
1) pure distillates, and
2) residue-containing fuel oils,
which show a marked difference in behaviour when burnt.

Usually distillates are produced by distillation under atmospheric pressure, a physical process, which may be repeated in an oil burner, as contrasted with residues which are the bottom products left behind in a distillation under atmospheric pressure. The reason why such residues cannot be distilled is that the boiling points of their constituents lie above the temperature at which the reaction of thermal decomposition starts, a process known as *"cracking"*. Distillation of these fractions can only be rendered possible to a certain extent by artificially lowering the boiling points by reducing the pressure, such as is done when producing lubricating oils by means of a high vacuum distillation plant.

It is clear that *residue-containing fuels cannot be completely evaporated at atmospheric pressure*, so that it is impossible to burn them in burners based upon the principle of previous evaporation of the oil, without leaving behind unburnt heavy ends. Nevertheless it happens from time to time that new oil burners are invented based upon that same principle of previously evaporating the oil, e.g. it being even claimed to burn the *crank case drainings* (the residual lubricating oils) of automobile engines to heat the gasoline service stations along the roads! The rough sketch in fig. 73 shows how the curve of boiling points of most hydrocarbons intersects that of thermal decomposition at about 350° C, and therefore lubricating oils having atmospheric boiling points generally above their decomposition temperatures would be decomposed first before being evaporated. Now this would not be so serious if the cracking process were not generally accompanied by the formation of oil coke, which very soon interferes with the regular operation of the burners.

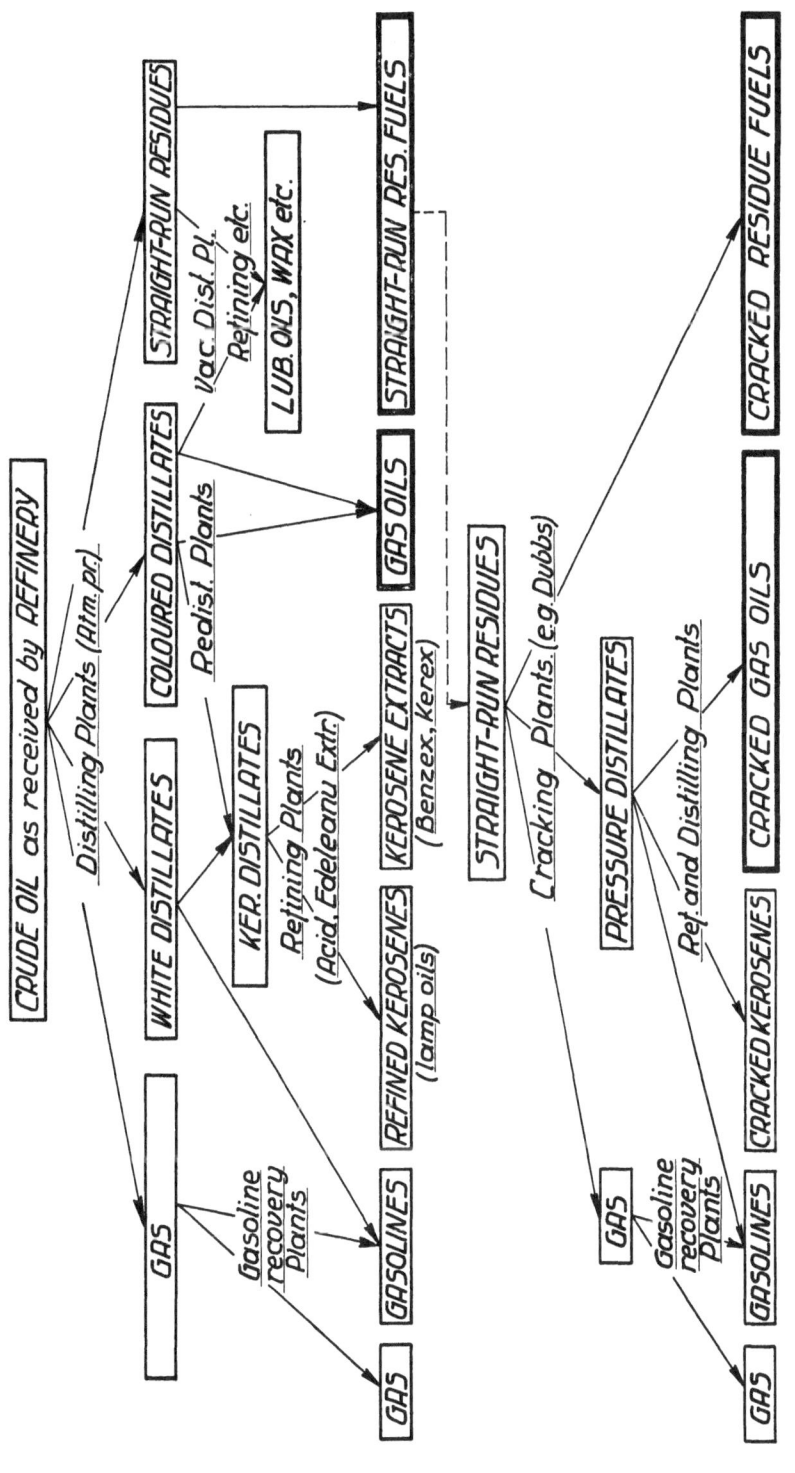

Fig. 72.
Scheme of Crude Oil Products.

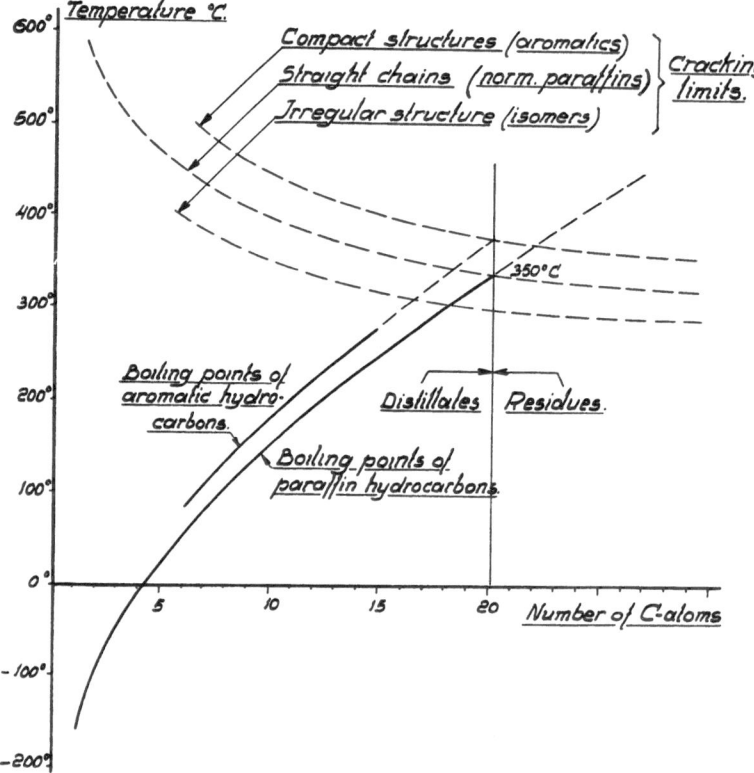

Fig. 73. Boiling points and Cracking limits of hydrocarbons.

## B. *Boiling range*

The above shows the importance of the so-called *"boiling range"* of a fuel oil, which may be determined by a distillation test as prescribed, for instance, by the American Society of Testing Materials and which is often known as the *"A.S.T.M.-analysis"*. This test consists in the determination of the temperature at which distillation under atmospheric pressure starts (*"initial boiling point"* = I.B.P.) and at which it stops (*"final boiling point"* = F.B.P.) and, moreover, of the temperatures at which 10, 20, 30 etc. per cent by volume of the oil is distilled over under standard conditions.

If the fuel is a *pure distillate* it will be possible to determine the A.S.T.M. curve *completely*, because what is left behind is only a negligible quantity, in contrast to a *residue-containing fuel*, of which the distillation is generally *stopped at 350° C.*, whatever the rest may be, as cracking will make further distillation useless. Therefore it is clear that the distillation test does not say so much for residue-containing fuels as it does for distillates.

## C. *Dew Point*

A very important physical property of a distillate oil is its *"dew point"*, which is the lowest temperature at which the oil as a liquid is *in equilibrium with its own complete vapour*, or in other words the temperature at which the oil may be completely evaporated under atmospheric pressure, or at which the first trace of condensation (fog) appears in the complete oil vapour when slowly cooled down. It is necessary to speak of *"complete"* *vapour* because at lower temperatures the oil may also give off a *lighter* vapour consisting of relatively more volatile constituents, which could be denoted as *"incomplete"* vapour. This shows that it is far more difficult to determine the dew point of fuel oil, as a mixture of numerous hydrocarbons, than of a single substance.

From its definition it is clear that only the dew point of distillates can be determined and not that of residues. Furthermore it is obvious that an oil burner based upon the principle of previous evaporation of the oil must have an evaporation device which has a temperature equal to. or higher than the dew point of the oil used.

The apparatus for determining the dew point is rather complicated and is not standardized. Generally the dew point of a distillate lies in the neighbourhood of the temperature at which about 65 to 70% is distilled over in the distillation test.

## D. *Viscosity and Surface Tension*

Another important physical property of fuel oil is its *"viscosity"*, this being a measure for the *internal friction* of the liquid and therefore of primary importance if the fuel is to be pulverized before being burnt. The *viscosity*, together with the *capillary constant of surface tension*, of the oil determines the magnitude of the oil drops when pulverized by pressure or centrifugal force. It is therefore obvious that the viscosity is of minor importance if the oil is not pulverized, as for instance is the case with distillate burners based upon the principle of previous vaporization of the oil. In that case, however, viscosity may still have some importance if wicks are used.

To determine the viscosity several methods are adopted in various countries, such as the VOGEL-OSSAG (fig. 74) and ENGLER tests on the Continent of Europe, the REDWOOD test in England, and the SAYBOLT test in America, and as none of these has yet been adopted as universal standard there is sometimes much trouble and confusion, the more so because viscosity is very dependent on temperature, and temperatures of various tests are different.

The *physical* measure of viscosity or the so-called *absolute viscosity* is the force in dynes necessary to give two oil surfaces of 1 cm² at a distance of 1 cm. a relative velocity of 1 cm./sec. One dyne-sec./cm² is called *one poise*. Water at 20° C. possesses exactly 0.01 poise or one *centi-poise* (c.p.), which is the unit of absolute viscosity generally adopted by physicists. The ratio of absolute viscosity to density is known as the *kinematic* viscosity, the density, of course, being expressed in units compatible with the system of units employed. The dimensions of kinematic viscosity are $L^2/T$ and its units are called *stokes* expressed in cm²/sec., or *centi-stokes* ($= 0,01$ stoke).

The different values of viscosity may be converted by means of the table of fig.

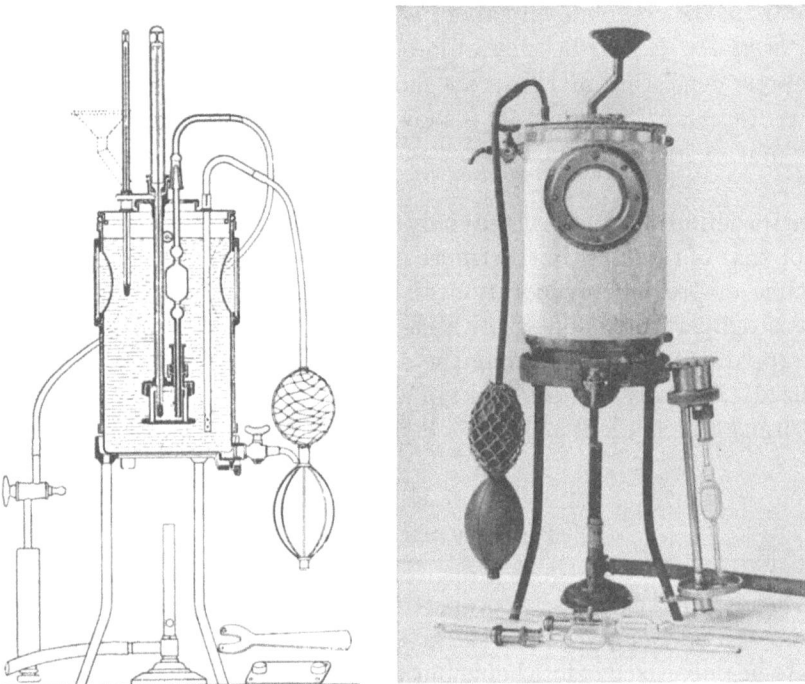

Fig. 74. The VOGEL-OSSAG viscosimeter.

75 in which $V_k$ denotes the kinematic viscosity. The SAYBOLT test for lighter fuels is known as "*Saybolt-Universal at 100° F.*", whilst for heavier fuels there is another test at 122° F. (boiling point of Furol) known as the "*Saybolt-Furol test*".

The surface tension is measured by means of the so-called *stalagmometer* (TRAU-BE), based upon the determination of the magnitude of drops falling from a polished horizontal plane. Viscosity and surface tension are independent of each other, but both decrease with increasing temperature, till about zero is reached at the critical temperature of the hydrocarbon.

## VISCOSITY CONVERSION TABLE

| $V_k$ = Abs. Kinematic Visc. (centi-stokes) | | | | S.U. = Seconds SAYBOLT-UNIVERSAL | | | |
|---|---|---|---|---|---|---|---|
| E = Degrees ENGLER | | | | R.I. = Seconds REDWOOD I | | | |

| $V_k$ | E | R.I. | S.U. | $V_k$ | E | R.I. | S.U. | |
|---|---|---|---|---|---|---|---|---|
| **1,0** | **1,00** | **28½** | **32** | **20,0** | **2,9** | **87** | **97** | |
| 1,5 | 1,06 | 30 | 33 | 20,5 | 2,95 | 88 | 99 | |
| 2,0 | 1,12 | 31 | 34 | 21,0 | 3,0 | 90 | 101 | For higher figures use: |
| 2,5 | 1,17 | 32 | 35 | 21,5 | 3,05 | 91 | 104 | |
| 3,0 | 1,22 | 33 | 36½ | 22,0 | 3,1 | 94 | 106 | |
| 3,5 | 1,26 | 34 | 38 | 22,5 | 3,15 | 96 | 108 | $V_k = 7,58$ E |
| 4,0 | 1,30 | 35½ | 39½ | 23,0 | 3,2 | 98 | 110 | |
| 4,5 | 1,35 | 37 | 41 | 23,5 | 3,3 | 100 | 112 | $V_k = 0,244$ R.I. |
| **5,0** | **1,40** | **38** | **42½** | 24,0 | 3,35 | 101 | 114 | |
| 5,5 | 1,44 | 39½ | 44 | 24,5 | 3,4 | 103 | 117 | $V_k = 0,218$ S.U. |
| 6,0 | 1,48 | 41 | 45½ | **25** | **3,45** | **105** | **119** | |
| 6,5 | 1,52 | 42½ | 47 | 26 | 3,6 | 109 | 123 | |
| 7,0 | 1,56 | 44 | 48½ | 27 | 3,7 | 113 | 128 | or |
| 7,5 | 1,60 | 45 | 50½ | 28 | 3,85 | 117 | 132 | |
| 8,0 | 1,65 | 46½ | 52 | 29 | 3,95 | 121 | 136 | |
| 8,5 | 1,70 | 48 | 54 | **30** | **4,1** | **125** | **141** | $E = 0,132\ V_k$ |
| 9,0 | 1,75 | 49½ | 55½ | 31 | 4,2 | 129 | 145 | |
| 9,5 | 1,79 | 51 | 57 | 32 | 4,35 | 133 | 150 | $E = 0,0322$ R.I. |
| **10,0** | **1,83** | **52½** | **59** | 33 | 4,45 | 137 | 154 | |
| 10,2 | 1,85 | 53 | 60 | 34 | 4,6 | 141 | 158 | $E = 0,0287$ S.U. |
| 10,4 | 1,87 | 53½ | 60½ | **35** | **4,7** | **144** | **163** | |
| 10,6 | 1,89 | 54 | 61 | 36 | 4,85 | 148 | 167 | |
| 10,8 | 1,91 | 55 | 62 | 37 | 4,95 | 152 | 172 | |
| 11,0 | 1,93 | 55½ | 63 | 38 | 5,1 | 156 | 176 | or |
| 11,4 | 1,97 | 56½ | 64 | 39 | 5,2 | 160 | 181 | |
| 12,2 | 2,04 | 59 | 67 | **40** | **5,35** | **164** | **185** | |
| 12,6 | 2,08 | 60½ | 68 | 41 | 5,45 | 169 | 190 | R.I. = 4,10 $V_k$ |
| 13,0 | 2,12 | 62 | 70 | 42 | 5,6 | 173 | 194 | |
| 13,5 | 2,17 | 63½ | 72 | 43 | 5,75 | 177 | 199 | R.I. = 31,1 E |
| 14,0 | 2,22 | 65 | 74 | 44 | 5,85 | 181 | 203 | |
| 14,5 | 2,27 | 66½ | 76 | **45** | **6,0** | **185** | **207** | R.I. = 0,894 S. U. |
| **15,0** | **2,32** | **68½** | **77** | 46 | 6,1 | 189 | 212 | |
| 15,5 | 2,38 | 70 | 79 | 47 | 6,25 | 193 | 216 | |
| 16,0 | 2,43 | 72 | 81 | 48 | 6,45 | 197 | 221 | or |
| 16,5 | 2,50 | 74 | 83 | 49 | 6,5 | 201 | 225 | |
| 17,0 | 2,55 | 75 | 85 | **50** | **6,65** | **205** | **230** | |
| 17,5 | 2,6 | 77 | 87 | 52 | 6,9 | 213 | 239 | S.U. = 4, 58 $V_k$ |
| 18,0 | 2,65 | 79 | 89 | 54 | 7,1 | 221 | 248 | |
| 18,5 | 2,7 | 81 | 91 | 56 | 7,4 | 229 | 257 | S.U. = 34,81 E |
| 19,0 | 2,75 | 83 | 93 | 58 | 7,65 | 237 | 266 | |
| 19,5 | 2,8 | 85 | 95 | **60** | **7,9** | **245** | **275** | S.U. = 1,12 R.I. |
| **20,0** | **2,9** | **87** | **97** | | | | | |

Fig. 75. Viscosity Conversion Table.

### E. *Pour point*

A physical property of fuel oil intimately related to viscosity is the *pour point*, which is the lowest temperature at which the oil will pour or flow when chilled without disturbance under certain specified conditions. This pour point is very important for fuel oils carried in the bunkers of ocean-going vessels, because it would cause difficulties to draw the oil up by pumps if its pour point should happen to be higher than the sea-water temperature.

Fuel oils of paraffin base are generally characterized by high pour points and viscosities, as against the naphthenic oils which have low pour points and viscosities and therefore are often considered as superior fuels in this respect.

### F. *Specific Gravity*

Another rather important physical property of fuel oil is its *specific gravity*, generally determined with a hydrometer at 15° C. or 60° F. As fuel oils are bought by volume and their heat content per kilogram is about equal to 10,000 cals., the specific gravity roughly fixes the price of one heat-unit, and for this reason the determination of the specific gravity is of some importance, although it says very little about the burning qualities as a fuel.

The specific gravity is usually expressed as the ratio between the weight of a given volume of oil to an equal volume of water at the same temperature (usually 15° C. or 60° F.) or in kilogrammes per litre. As the specific volume of oil is very dependent on the temperature it is necessary to state the temperature at which the measurement was carried out, and if not stated it is taken to be 15° C. or 60° F.

In English speaking countries specific gravity is often expressed in BEAUMÉ units according to the formula:

$$\text{BEAUMÉ gravity} = \frac{140}{\text{Spec. Gravity}} - 130,$$ which is the original scale, whereas in America another BEAUMÉ scale is used, known as:

$$\text{A.P.I. gravity} = \frac{141.5}{\text{Spec. Gravity}} - 131.5.$$

For water at 60° F. both BEAUMÉ scales correspond to the value 10. The Table below gives a comparison between the original BEAUMÉ scale and the Specific Gravity scale.

It may be of interest to point out here that the denomination *"heavy fuels"* generally relates to *high viscosity* and *high boiling points* rather than to a high specific gravity. In fact there are certain fuels (especially those containing paraffin hydrocarbons) having a very high viscosity or low fluidity and a lower specific gravity than other fuels containing more naphthenes and aromatics, which are decidedly more fluid at the same temperatures.

CONVERSION TABLE FOR SPECIFIC GRAVITIES

| BEAUMÉ | Spec. Gravity | Lbs/Imp. Gallon | Imp. Gallons/ton |
|---|---|---|---|
| 10 | 1.0000 | 10.000 | 224.00 |
| 11 | 0.9929 | 9.929 | 225.55 |
| 12 | 0.9859 | 9.859 | 227.13 |
| 13 | 0.9790 | 9.790 | 228.80 |
| 14 | 0.9722 | 9.722 | 230.49 |
| 15 | 0.9655 | 9.655 | 231.98 |
| 16 | 0.9589 | 9.589 | 233.42 |
| 17 | 9.9523 | 9.523 | 235.11 |
| 18 | 0.9459 | 9.459 | 236.66 |
| 19 | 0.9395 | 9.395 | 238.30 |
| 20 | 0.9333 | 9.333 | 239.82 |
| 21 | 0.9271 | 9.271 | 241.34 |
| 22 | 0.9210 | 9.210 | 242.95 |
| 23 | 0.9150 | 9.150 | 244.40 |
| 24 | 0.9090 | 9.090 | 246.01 |
| 25 | 0.9032 | 9.032 | 247.64 |
| 26 | 0.8974 | 8.974 | 249.15 |
| 27 | 0.8917 | 8.917 | 250.84 |
| 28 | 0.8860 | 8.860 | 252.53 |
| 29 | 0.8805 | 8.805 | 254.00 |
| 30 | 0.8750 | 8.750 | 255.85 |
| 31 | 0.8695 | 8.695 | 257.47 |
| 32 | 0.8641 | 8.641 | 258.94 |
| 33 | 0.8588 | 8.588 | 260.61 |
| 34 | 0.8536 | 8.536 | 262.14 |
| 35 | 0.8484 | 8.484 | 263.83 |
| 35 | 0.8433 | 8.433 | 265.40 |
| 37 | 0.8383 | 8.383 | 266.82 |
| 38 | 0.8333 | 8.333 | 268.42 |

(A.P.I. figures are slightly higher than the original BEAUMÉ figures, viz., 40.3 for 40 Bé; 30.2 for 30 Bé; 20.1 for 20 Bé and equal for 10 Bé.)

(Lbs/Imp. Gallon = 10 × Spec. Gravity)

(For conversion to Am. Gallons/ton multiply by 1.20)

## G. *Specific heat*

There are some other physical properties of fuel oil that are of minor importance, e.g. the *specific heat* of the liquid, a property only necessary to know if the heat required to preheat the oil has to be calculated, and its *latent heat of vaporization*, which is of more importance for designers of distilling plants than it is for oil burning and therefore will be passed over here.

The determination of the physical properties referred to above does not give so very much trouble and the only complication is the fact that *fuel oil is a mixture of several constituents* of different physical properties, so that it is only possible to speak of *average* values of specific gravity, boiling point, dew point, etc. We will now see what the chemical analysis can tell us.

---

# CHAPTER III

## CHEMICAL PROPERTIES OF FUEL OIL

### A. *Classification of hydrocarbons*

The so-called *chemical elementary analysis* shows that generally fuel oils consist of about 85 to 90% of carbon, 13 to 8% of hydrogen and a small percentage of sulphur, oxygen and nitrogen, but two fuel oils of the same elementary analysis may behave in quite a different manner in burning practice, so that this chemical analysis is really of very little value. This may be made clear with the aid of Table I of the appendix: "CLASSIFICATION OF HYDROCARBONS".

Hydrocarbon molecules are supposed to consist of a skeleton of carbon (C) atoms linked together by one or more of their valencies, and of hydrogen (H) atoms compensating their other valencies. Every individual hydrocarbon may be denoted by a formula $C_nH_{2n+m}$ with suitable values for n and m as *whole* numbers. The number of n may be odd or even, whereas for m only the values $+2$, zero, $-2$, $-4$, $-6$ etc., are allowable, because odd numbers for n would make it necessary to suppose *three* valencies for one of the carbon atoms, which is contradicted by many other statements proving that a carbon atom always has *four* valencies. For a *mixture* of several hydrocarbons n and m represent average values and may be *fractions* too.

In Table I the hydrocarbons are arranged horizontally according to their value of n (number of carbon atoms), and vertically according to their value of m, each horizontal line thus representing a *homologous series*, so called because there is a constant difference $(CH_2)$ between any one compound and the next higher or lower member, so that all the compounds of the series appear to be proportional.

With the exception of the first series $(C_nH_{2n+2})$ there are two or more series for every value of m, namely one or more series with *open chains* (without rings) or so-called *acyclic series*, and one or more series with *closed chains* (with rings) or so-called *iso-cyclic* or *cyclic series*.

Thus there are:

A. $C_nH_{2n+2}$: **Paraffin series** or *saturated hydrocarbons*, so-called because it is impossible to combine more hydrogen to the molecules. The name *"paraffin"*, which means *"against affinity"*, originates from the fact that these hydrocarbons are very stable and resistant to nearly all chemical action except burning. In the paraffin series the carbon atoms are supposed to form open chains.

B. $C_nH_{2n}$ *as acyclic compounds:* **Olefine series** or *unsaturated hydrocarbons*, so called because the arrangements of the atoms make it necessary to suppose *two* affinity units between two carbon atoms (double link), thus leaving the possibility open to apply more hydrogen or other elements (chlorine, bromine, sulphur, etc.) to the molecules, which in fact may often readily combine. These hydrocarbons do not occur in crude oil in its natural state, but they do frequently if the oil has been exposed to high temperatures, as may happen for instance during a distilling or a cracking process.

C. $C_nH_{2n}$ *as isocyclic compounds:* The **Naphthene series** or *cyclanes*, which, although being of the same chemical composition as the olefine series, has entirely different physical and chemical properties, owing to the arrangement of the atoms being different. We suppose the carbon atoms in the hydrocarbons of the naphthene series to form *closed chains* (rings), a structure different from that of the paraffin —and olefine — series. The *cyclic formations* make it possible to avoid a double link between carbon atoms and therefore the naphthene series must be considered as being to a certain extent *saturated hydrocarbons* too. Consequently they are rather resistant to chemical actions. They occur, for instance, in Caucasian and Borneo oils in the natural state, and even cyclopropane, cyclobutane, etc., are present in considerable quantities in the natural gases of these fields.

$D_1$. $C_nH_{2n-2}$ *as acyclic compounds:* This is the **Acetylene series,** which comprises highly unsaturated hydrocarbons, the acyclic arrangement of the molecules making it necessary to suppose *three affinity units between two carbon atoms* (threefold link), which must be regarded as a very *unstable* forced situation that may be readily changed if the opportunity presents itself to combine with other elements. These hydrocarbons of the acetylene series are characterized by their *strong affinity for oxygen*, resulting in a *very high burning velocity*, which in turn may result in high flame temperatures. Moreover there is an *extra amount of latent heat* stored in the threefold link between the carbon atoms. They do not occur in crude oil in its natural state, but may be formed during distilling or cracking processes.

$D_2$. $C_nH_{2n-2}$ *as another series of acyclic compounds:* The **Diolefine series,** in which the presence of two affinity units between two carbon atoms (typical for the olefine series) occurs *twice* in the same hydrocarbon molecule which is expressed in the name: *di*-olefine. In general the hydrocarbons of this series bear a close resemblance to the olefines, except in the important case of the *conjugated double bonds:*

$$\ldots\ldots -\overset{|}{C} = C - \overset{|}{C} = C - \ldots., \text{ the hydrocarbons of this typical structure being}$$

very liable to gumming, polymerising and oxidation reactions.

E. $C_nH_{2n-2}$ *as isocyclic compounds:* The **Cyclene series** *or unsaturated naphthenes* are of the same chemical composition as the acetylene series but have entirely different properties. The cyclic formation makes it possible to reduce the threefold link between two carbon atoms to a double one. It is obvious that the hydrocarbons of this series have less affinity to oxygen than the former (acetylene series) and therefore the flame temperature of an iso-cyclic representative usually will be lower than that of the corresponding member of the acyclic series. They do not occur in natural crude oil in appreciable quantities, but may occur after a distillation and certainly do occur in large quantities after cracking.

F. $C_nH_{2n-4}$ *as acyclic and cyclic compounds:* The **Terpene series,** so called because the principal constituent of *turpentine* $C_{10}H_{16}$ belongs to this series. Though rather varying in structure and nature of the links between their carbon atoms, the terpenes are surprisingly analogous in their properties. They occur mostly in essential oils, not in crude mineral oil, but may occur in its products.

G. $C_nH_{2n-4}$ *as cyclic compounds:* The **Cyclopentadiene series.** The ring formations make it possible to arrange the atoms only with double links between carbon atoms. They do not occur in crude oil, but may occur in its products.

H. $C_nH_{2n-6}$ *as acyclic compounds:* The **Diacetylene series** comprises highly unsaturated hydrocarbons easily decomposed when heated. They do not occur in crude oil, but may occur in its products.

I. $C_nH_{2n-6}$ *as cyclic compounds:* The **Benzene series.** Its first member, *benzene* ($C_6H_6$ = "benzol") consists of a closed *six-carbon ring,* typical for the so-called *aromatic* compounds, which by its double symmetrical arrangement is, chemically speaking, so stable that it may be regarded as a *self-contained unit* from which several other compounds may be derived. Thus benzene is the starting point for numerous aromatic hydrocarbons, as is methane for paraffin hydrocarbons. It is curious that benzene, though being the *lowest* member of its family, shows one of the *highest* specific gravities of its series, the higher boiling members mostly having lower specific gravities contrasted with the paraffin series. Further it is of interest to note that there are even four substances of the formula $C_6H_6$ other than benzene, with entirely different properties: one, *fulvene*, is also a cyclic compound with a *five-carbon ring* which makes it far less stable than benzene. Aromatic hydrocarbons do occur frequently in crude oil and its products.

J. $C_nH_{2n-8}$ *as acyclic compounds* are so unstable, because of several threefold links between carbon atoms, that their occurrence may be ignored in practice.

K. $C_nH_{2n-8}$ *as cyclic compounds* offer better possibilities because of a benzene ring being used as a building stone, which makes it possible to avoid three-fold links between carbon atoms. The third member is well known as *tetralin*. They are comparatively stable, and occur in crude oil and its products.

L. $C_nH_{2n-10}$, which is only possible *as cyclic compounds*, is in general less stable than $C_nH_{2n-8}$, especially when threefolds are present in the side chains. Contrasted with the above there is *indene*, $C_9H_8$, which has a *bicyclic* structure related to naphthalene (see below), a rather stable hydrocarbon, though evidently unsaturated.

M. $C_nH_{2n-12}$ *as cyclic compounds* offer new possibilities, namely provide a new keystone consisting of a *bicyclic* structure built up by *two benzene rings having two carbon atoms in common*, thus forming *naphthalene*, which in turn is a *very stable compound* acting as a self-contained unit, with the aid of which a large number of other compounds may be built up.

N. $C_nH_{2n-14}$ *as cyclic compounds* offer a similar possibility, namely by connecting *two complete benzene rings* by one of their valencies, thus forming *diphenyl*, which in turn is a *very stable compound* too, acting as a self-contained unit.

How far this principle of building up new units from benzene rings can be carried is shown by the schemes of the first members of the series $C_nH_{2n-56}$ and $C_nH_{2n-78}$. $C_{42}H_{28}$, *tetra-α-naphthyl-ethylene*, in particular is a remarkable substance, as it shows reversible oxidation (see Annales de Chimie, 10ième série 1931, MONDAIN-MONVAL).

As far as *mineral oil* is concerned, it may be stated that:

I. *Crude oil* in its natural state may contain *a*) paraffin, *b*) naphthene and *c*) aromatic hydrocarbons.

II. *Products obtained from crude oil* by means of pure *physical distillation* (viz. at sufficiently low temperatures) contain the same hydrocarbons as I.

III. *Products obtained from crude oil* by means of so-called *destructive distillation* (viz. constituents partially "cracked" at high temperatures) may contain all kinds of hydrocarbons, generally from $C_nH_{2n+2}$ up to $C_nH_{2n-14}$ or sometimes even higher.

This survey may give the reader some idea of the number of hydrocarbons present in fuel oil and, if he is fully aware of the fact that all of these hydrocarbons have their own peculiar properties, he may understand too that when comparing two fuel oils it may be found that, although having the same chemical composition, they may show an enormous difference in combustion performance.


## B. *Isomerism*


Having obtained an idea of the great number of series of hydrocarbons which may occur in the products of crude oil, we shall now see how many members there may be of *one* series, for instance that of the Paraffins.

It is assumed that the reader will know that hydrocarbons having the same chemical formula and belonging to the same series may have quite different properties because of a *different arrangement of atoms in the molecule*. We have already encountered such a case with *olefines* and *naphthenes*, both of the formula $C_nH_{2n}$, but the former having an *acyclic* against the latter a *cyclic* arrangement, which makes such a great difference in properties that these substances are considered as belonging to two different series.

If this phenomenon occurs with members of *one* series it is called *isomerism*, and the first member of the paraffin series showing this property is *butane*. Butane may be arranged as a straight chain (normal butane, boiling at —1° C.) or as a "T" (isobutane, which boils at —11° C.), both being of the formula $C_4H_{10}$.

With *pentane* there are *three* possible arrangements, one as a *straight chain*, one as a **T** and one as a *cross*, each having its own boiling point. The number of isomeric possibilities increases with increasing molecular weight; with hexane there are 5 isomers, heptane 9, octane 18, and so on. Of 18 possible octanes $C_8H_{18}$ several boiling points are known; we give the following examples:

| Name of isomer | Approx. boiling point |
|---|---|
| 1) *normal octane* . . . . . . . . . . . . . . | 125° C. |
| 2) 2-methyl heptane . . . . . . . . . . . | 117° C. |
| 3) 3-  „        „      . . . . . . . . . . . | 118° C. |
| 4) 4-  „        „      . . . . . . . . . . . | 118° C. |
| 5) 2-2-dimethyl hexane . . . . . . . . . . | 106° C. |
| 6) 2-3-   „         „     . . . . . . . . . . | 114° C. |
| 7) 2-4-   „         „     . . . . . . . . . . | 110° C. |
| 8) 2-5-   „         „     . . . . . . . . . . | 109° C. |
| 9) 3-3-   „         „     . . . . . . . . . . | 111° C. |
| 10) 3-4-   „         „     . . . . . . . . . . | 117° C. |
| 11) 2-2-3-trimethyl pentane . . . . . . . . | 111° C. |

| NAME OF ISOMER | APPROX. BOILING POINT |
|---|---|
| 12) 2-2-4-trimethyl pentane  (= „iso-octane") | 99° C. |
| 13) 2-3-3-  „  „  . . . . . . . . | 114° C. |
| 14) 2-3-4-  „  „  . . . . . . . . | 113° C. |
| 15) 3-ethyl hexane  . . . . . . . . . . . | 118° C. |
| 16) 2-methyl ethyl pentane 3 . . . . . . . . | 114° C. |
| 17) 3-methyl ethyl pentane 3 . . . . . . . . | 119° C. |
| 18) 2-2-3-3-tetra methyl butane  . . . . . . | 106° C. |

Now these differences may not seem so important at first sight, but it must be realized that other properties, especially those of *combustion performance, may show enormous differences;* for instance *normal octane* is a substance which *"knocks" very heavily in gasoline engines,* whereas so-called *iso-octane* (2-2-4-trimethyl pentane) is used as the *"non-knocking reference fuel"* in knock-testing.

CAYLEY calculated the numbers of possible isomers for the paraffin series up to $C_{13}H_{28}$ and found 802. From his data the rule may be deduced that the increase of the number of isomers follows approximately a *parabola on a logarithmic scale* according to the formula $(\log_{10}i)^{0.567} = 0.139$ n, where i = number of possible isomers and n = number of carbon atoms.

If the reader now imagines a gas oil exclusively consisting of pure paraffin hydrocarbons with a boiling range of 214°C  $(C_{12}H_{26})$ up to 381°C. $(C_{22}H_{46})$, from the above equation it follows that *the number of possible isomers for this gas oil is approx. 20,000,000!!* Even if only $1^0/_{00}$ of the theoretically possible isomers was really present, there would still be 20,000 isomers left, and if one takes into consideration that every isomer has its own properties, it is clear that the properties of such an extremely simple hypothetical gas oil, even though consisting of only paraffin hydrocarbons, may show an almost infinite number of variations in combustion performance.

And what about the isomers of the other series which may occur in actual gas oil, and what about *residues!*

From the above it may be clear that it is practically impossible at present to predict the burning qualities of a fuel oil merely from data on the chemical composition and consequently a number of *empirical tests* have been developed in course of time to meet the requirements of the trade. This way has proved to be reasonably successful notwithstanding the great number of variations possible. One might compare the constituents of the oil with the inhabitants of a country for, although its inhabitants show an enormous number of variations in individual behaviour, one can certainly speak of a pronounced general character of such a country and more or less predict their behaviour in reactions as a whole.

## C. *Flash point*

A physico-chemical test on fuel oil of great practical importance has already been mentioned, viz. the *determination of the flash point* for which the method evolved by Pensky-Martens (A.S.T.M.-D 93–22) has been adopted all over the world (see

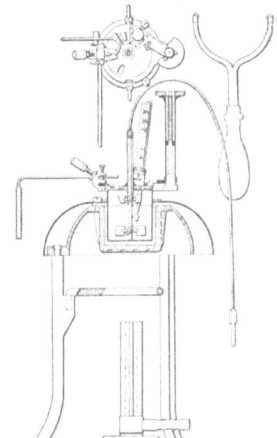

fig. 76). Of course it will make a great difference whether the oil vapours are collected in a *closed* cup or are allowed to diffuse freely into the atmosphere. What is known as the *"closed-cup test"* is generally taken and gives the lowest value for the flash point.

Whereas a *minimum* flash point is required for safety, a *maximum* value may be demanded for the easy starting of fully automatic oil burner equipment.

Other chemical tests are:

## D. *Water and Sediment*

Fig. 76. Pensky-Martens flash point tester.

A combined determination by means of a centrifuge is prescribed in the A.S.T.M. test No. D 96–23 or by distillation and extraction in the Bureau of Mines Techn. Paper No. 323 B. The sediment percentage is the quantity of the oil (not being water) insoluble in hot benzene.

## E. *Sulphur*

Sulphur is determined by means of a *sulphur lamp* (Richardson, Heslinga and others) for distillates up to gas oil, or by means of the *oxygen bomb* for heavier fuel oils. These methods are based on the combustion of a suitable known weight of the oil and absorption of the sulphur dioxide formed in a diluted solution of hydrogen peroxide, by which it is converted into sulphur tri-oxide and sulphuric acid, which can be easily determined with sodium carbonate.

In former times the importance of a low sulphur content for fuel oil was very much exaggerated, but nowadays it is generally agreed that *sulphur is not dangerous as a corrosive agent as long as the flue gas is not cooled below its dew point* (generally not higher than 55° C.), because sulphur dioxide gas in the dry state has no corrosive properties. Whereas in some cases, e.g. when nickel or nickel alloys or certain sensitive glasses and enamels have to be in contact with the flame gas and flue gas, it may be necessary to limit the percentage of sulphur in the oil to 0.5%, it is interesting to note that brass, copper, cast iron and most of the bronzes can be melted without any difficulty in the atmosphere of the open oil flame with fuel containing up to 2% sulphur.

## F. *Ash*

A suitable quantity of oil is burnt in a crucible and the residue glowed, until the weight is constant, precautions being taken not to volatilize any chlorides present. The ash of fuel oil very often consists of mud from the wells, but its quantity is almost negligible.

## G. *Heat Value*

The heat value may be determined by means of a *bomb calorimeter* (MAHLER, PARR), in which a carefully weighed quantity of oil is burnt with enough oxygen to ensure complete combustion. The result is calculated from the increase in temperature imparted to a weighed quantity of water in which the bomb is immersed during the experiments.

The calorific value, which is a predominant factor for solid combustibles — for example, it may vary from 4500 cal/kg for very low grade coal to 8000 cal/kg for high grade coal — loses much of its importance for liquid fuel, *the heat content for all kinds of fuel oil being about 10,000 cal./kg.* (heavy oils perhaps 300 cal. less and light oils 500 cal. more). As a rule, therefore, it is not considered worth while taking the trouble to determine this property and sometimes a fuel oil with lower calorific value may even be considered superior by the buyer if it has qualities more conducive to clean and efficient combustion. The question is not only: How much heat is there in the fuel, but also how much can be got out of it in practice?

There is, however, one important thing to be observed in this connection. There are *two different heat values* of hydrocarbons, one high and the other low, according to whether the water of the flue gases is condensed or not. The difference between these two values (approx. 600 cal./kg. fuel) is equal to the heat of vaporization of the water formed by combustion of the fuel (about 1 kg./kg. fuel). It is obvious that the efficiencies of an oil-burning equipment may also be expressed in two ways, viz. whether the net heat production is compared with the *highest* heat value (giving a *lower efficiency*) or with the *lowest* value (resulting in a *higher efficiency*).

In *English-speaking countries the upper heat value* (or *gross* calorific value) is generally adopted, whereas on the *Continent of Europe the lower value* (or *net* calorific value) is accepted, because it is considered more logical there to subtract from the heat value the latent heat of the water vapour, which is always lost in practice too; in fact this latent heat could only be turned to advantage by cooling the flue gas to below its dew point, which introduces a serious danger of damage by corrosion and is therefore never applied in practice.

Although, of course, it makes no fundamental difference, it is well to remember that in English or American literature on oil burning the boiler efficiencies are expressed in values generally 7 to 9% lower than those in European Continental literature, merely because they are based on different heat values.

## H. *Aromatics, Naphthenes, etc.*

There are several tests for fuel oil to determine its composition in 1) Paraffins, 2) Aromatics, 3) Naphthenes and 4) Olefines, but they are all very rough and unreliable.

"Aromatics" are generally extracted with sulphuric acid of 98% or 100%, but also olefines are removed in this way. The *bromine value* (MCILHINEY method) should be a measure for unsaturated hydrocarbons, but if the test is not carried out quickly enough, other hydrocarbons are attacked too.

"Naphthenes" are still more difficult to determine, even for gasolines (e.g. Aniline Point Method).

In general it can be stated that the difficulties with these determinations increase with the molecular weight and that *these tests are unreliable for gas oil and worthless for residues or residue-containing fuels.*

## I. *"Asphaltene" content*

A test of more practical importance is the determination of the *"asphaltene" content*, sometimes called the *"hard asphalt" content*. This test determines the percentage of the oil which is insoluble in a special spirit (such as di-ethyl ether) but which, at the same time, is soluble in pure benzene ($C_6H_6$).

The asphaltene test forms a *practical demarcation between distillate and residual fuels*, because the former generally show negligible asphaltene figures. A rough indication of the percentage of "asphaltenes" (which is a collective noun for highly complex black or dark brown hydrocarbon compounds of low H to C ratio, usually containing some combined oxygen, having a tarry, asphaltic character) may be obtained by drying a few drops of the oil on a piece of blotting paper and then washing out the oil with petroleum ether.

## J. *Carbon residue test*

There is still another practical test to be mentioned, namely the *Carbon residue test*, also called: CONRADSON *coking test*, which is described as A.S.T.M. method No.

D 189–28. This method is a means of determining the amount of carbon residue left on evaporating an oil under specified conditions, and it is intended to throw some light on the relative *carbon-forming propensity* of the oil.

A sample of the oil is heated for about half an hour in a crucible under restricted air supply (fig. 77) and the amount of carbon left is weighed. The CONRADSON *figure* or *"coking value"* denotes this weight of carbon expressed as a percentage of the weight of the original sample. To avoid very small figures for distillates (gas oils) it is sometimes recommended to carry out the test not on the original oil but on the 10% bottom that is left by a partial A.S.T.M. distillation test. The CONRADSON figures are thus magnified about 10 times and are indicated as: "CONRADSON on *10% bottoms"*.

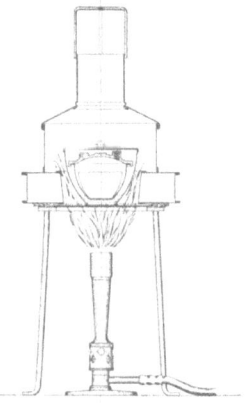

Fig. 77. CONRADSON carbon residue test.

Special attention is drawn to the fact that the carbon as found by the Conradson test is *not present in the oil as such* but is *formed by the cracking* of the oil during heating. Thus it may very well be possible to burn a gas oil of 0.05% carbon residue in a burner based on the principle of previous vaporization of the oil without any carbon being left, if only the conditions in the burner are less favourable for cracking and carbonizing than they are during the Conradson tests.

---

## CHAPTER IV

### COMMERCIAL GRADES

Fuel oils are marketed under many commercial designations which vary considerably in different countries. Broadly, however, they are classified according to the following groups:

A. *Distillates.*
1. Gas oil.
2. Light Diesel Oil.
3. Light domestic fuel oil.

B. *Residue-containing fuels.*
1. Diesel oil or heavy domestic fuel oil.
2. Light industrial fuel oil.
3. Heavy industrial fuel oil.

## THE A.O.B.A. SPECIFICATIONS FOR FUEL OILS

### Old Specifications (CS12-33)

| GRADE OF OIL | Flash Point, deg. Fahr. Min. | Max. | Pour Point, deg. Fahr. Max. *a* | Water and Sediment, per cent Max. | Distillation Temperatures, deg. Fahr. Max. 10% | Max. 90% | Max. End | Viscosity sec. Max. |
|---|---|---|---|---|---|---|---|---|
| No. 1—A distillate oil for use in burners requiring a volatile fuel | 110 or legal | 165 | 15 | 0.05 | 420 | | 600 | |
| No. 2—A distillate oil for use in burners requiring a moderately volatile fuel | 125 or legal | 190 | 15 | 0.05 | 440 | 620 | | |
| No. 3—A distillate oil for use in burners requiring a low-viscosity fuel | 150 or legal | 200 | 15 | 0.1 | 460 | 675 | | 55 S.U. @100 |
| No. 4—An oil for use in burners requiring low-viscosity fuel | 150 | b | c | 1.0 | | | | 125 S.U. @100 |
| No. 5—An oil for use in burners *equipped with preheaters* permitting a medium-viscosity fuel | 150 | | | 1.0 | | | | 100 S.F. @122 |
| No. 6—An oil for use in burners *equipped with preheaters* permitting a high-viscosity fuel | 150 | | | Water 1.75 sed. 0 25 | | | | 300 S.F. @122 |

### New Specifications (CS12-35)

| Flash Point, deg. Fahr. Min. | Max. | Pour Point, deg. Fahr. Max. | Water and Sediment, per cent Max. | Carbon Residue, per cent Max. | Ash, per cent *a* Max. | Distillation Temperatures, deg. Fahr. 10% Max. | 90% Min. | 90% Max. | End Point Min. | End Point Max. | Saybolt Universal at 100° F Min. | Max. | Saybolt Furol at 122° F Min. | Max. |
|---|---|---|---|---|---|---|---|---|---|---|---|---|---|---|
| 100 or legal | 150 | 15b | 0.05 | 0.02 | | 420 | | | | 600 | | | | |
| 110 or legal | 190 | 15b | 0.05 | 0.05 | | 440 | | 620 | 600 | | | | | |
| 110 or legal | 200 | 15b | 0.1 | 0.15 | | | 620 *d* | | | | | 70 | | |
| 150 | | c | 1.0 | | 0.1 | | | | | | 70 *e* | 500 | | |
| 150 | | | 1.0 | | 0.15 | | | | | | | | 25 | 100 |
| 150 | | | 2.0 *f* | | | | | | | | | | 100 *g* | 300 |

NOTES:

Descriptions of the various grades of fuel oil given in this column conform to the *new specifications*. Descriptions for the old specifications are practically the same as these except *equipped with preheaters* was not mentioned for Nos. 5 and 6 oils, and No. 5 oil was described as being the same as Bunker B oil, No. 6 as being the same as Bunker C oil.

*a.* Lower or higher pour points may be specified whenever required by conditions of storage and use. However, these specifications shall not require a pour point less than 0° F, under any conditions.

*b.* Whenever required, as for example in burners with automatic ignition, a maximum flash point may be specified. However, these specifications shall not require a flash point less than 250° F, under any conditions.

*c.* Pour point may be specified whenever required by conditions of storage and use. However, these specifications shall not require a pour point less than 15° F, under any conditions.

*a.* Recognizing the necessity for low sulfur fuel oils used in connection with heat-treatment, non-ferrous-metal, glass and ceramic furnaces and other special uses, a sulfur requirement may be specified in accordance with the following table:

| Grade of Fuel Oil | Sulfur, Max., Per Cent |
|---|---|
| No. 1 | 0.5 |
| No. 2 | 0.5 |
| No. 3 | 0.75 |
| No. 4 | 1.25 |
| No. 5 | no limit |
| No. 6 | no limit |

Other sulfur limits may be specified only by mutual agreement between the buyer and seller.

*b.* Lower or higher pour points may be specified whenever required by conditions of storage or use. However, these specifi-

cations shall not require a pour point lower than 0° F, under any conditions.

*c.* Pour point may be specified whenever required by conditions of storage or use. However, these specifications shall not require a pour point lower than 15° F, under any conditions.

*d.* This requirement shall be waived when the carbon residue is more than 0.07 per cent and less than 0.15 per cent.

*e.* This requirement shall be waived when the carbon residue is more than 1.0 per cent.

*f.* A deduction in quantity shall be made for all water and sediment in excess of 1.0 per cent.

*g.* This requirement shall be waived when the carbon residue is 4 per cent or more.

Fig. 78. The A.O.B.A. Specifications for Fuel Oils.

The nomenclature is anything but uniform and this has caused very much confusion, which was the reason why the American Oil Burner Association fixed some years ago what are known as the *"A.O.B.A. specifications"*, which are now generally adopted in the United States. According to these specifications, which have been recently revised, there are *six grades of fuel oil*, as shown in the table of fig. 78.

As already mentioned, the original meaning of *"mazouth"* is a thick paraffin residue, whereas *"les brûleurs à mazout"* in France really burn a domestic fuel oil equal to gas oil. Now, for instance, the distillate known as *"gas oil"* in England may be termed *"stove oil"* in the United States of America, while the expression *"solar oil"* is used for a similar product in many countries in the Far East and in Scandinavia because of its having been used in former times as a lighting fuel in what were termed solar lamps.

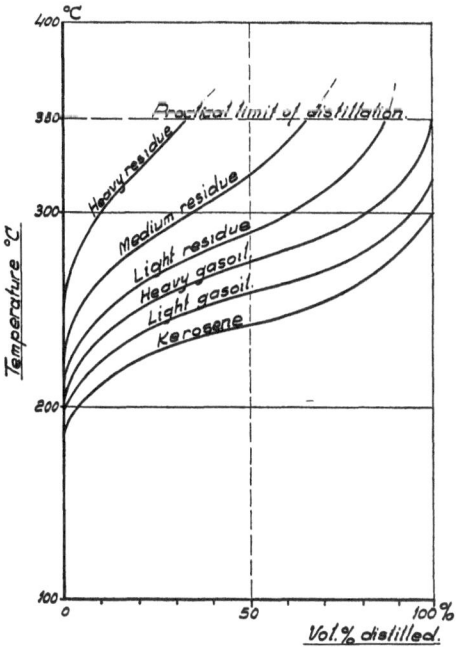

Fig. 79. Some typical distillation curves of fuel oils.

*Gas oil* received its name from its application in oil-gas production (YOUNG, LOWE, PINTSCH), and the bluish colour of the Roumanian gas oil known as "Blau-Oel" gave rise to the name of "Blau-gas" or "blue-gas" for the oil gas obtained from it. In the U.S.A. the term *"furnace oil"* is usually applied to a very light distillate more in the range of a heavy kerosene than a gas oil, but in most other countries the name "Furnace oil" is used for the heaviest grade of residual fuel. "Domestic fuel oil" may be a gas oil distillate or a light fuel oil, according to the exigencies of any particular market.

Some typical A.S.T.M. curves of fuel oils are given in the graph of fig. 79.

## CHAPTER V

### THEORIES OF COMBUSTION OF FUEL OIL

As previously pointed out, fuel oil is composed of hydrocarbons except for a small percentage of oxygen, nitrogen and sulphur compounds. When fuel oil is burnt, its hydrocarbon molecules are not oxidized instantaneously to carbon dioxide and water, but several decomposition products and (or) incompletely oxidized products will be formed as intermediates. Besides these intermediate combustion reactions the combustion of fuel oil or liquid hydrocarbons is substantially the combustion of:

A) Carbon,
B) Gaseous hydrocarbons,
C) Hydrogen,
D) Formaldehyde,
E) Carbon monoxide.

The combustion of carbon will be discussed first.


### A. *Combustion of Carbon*

The historical development of the theory of the combustion of carbon may be divided into stages, as described in a very instructive way in HASLAM and RUSSELL's book "Fuels and their Combustion" (1926).

1) *Carbon dioxide theory.*

Up to about 1870 this theory was generally accepted. *Carbon dioxide* was thought to be the *primary* product of combustion, whilst any *carbon monoxide* formed was supposed to result as a *secondary* product from the reaction of $CO_2$ with carbon (reduction). This theory was mainly based on the fact that a diamond (pure carbon) burns to $CO_2$ *without any visible flame*, for if CO had been formed first and thereafter burnt to $CO_2$ a flame having the blue colour of a CO flame would have been observed. Moreover SMITH (1863) made an experiment in which oxygen was absorbed by carbon (charcoal) at —12° C. and on the carbon then being heated to 100° C., $CO_2$ was evolved and no CO.

2) *Carbon monoxide theory.*

About 1870 BELL, however, proved that gas samples drawn from the middle of the gas flames of blast furnaces might contain much more CO than $CO_2$, which gradually led to the idea that *CO* was the *primary combustion product* and not $CO_2$.

In 1887 BAKER (C.J.) attacked the main pillar of the $CO_2$ theory, SMITH's experiment, as he found that although $CO_2$ is in fact evolved at 100° C. if *moist* oxygen is absorbed by carbon at —12° C., no gas at all is released at 100° C. if carefully *dried* oxygen is absorbed by *dry* carbon at —12° C. In this latter case it proved to be necessary to heat even as high as 450° C. to evolve gases (mainly consisting of CO) form the charcoal.

It was clear that moisture acted as a catalytic agent with regard to the oxidation of carbon to $CO_2$, a fact which was further supported by the empirical observation that coal heaps are especially apt to get hot by slow combustion if they are moist. Therefore a new theory was developed on the idea that $CO_2$ was the primary product of the *catalytic* combustion, whereas CO was the initial product of the *real* combustion of carbon. Moreover, it was found in 1888 by BAKER (H.B.) that dry carbon dioxide brought into contact with dry carbon at red-hot temperatures was not reduced, and therefore $CO_2$ was considered to be a final product.

3) *Carbon Complex Theory.*

About 1912 RHEAD and WHEELER published several objections against both the $CO_2$ and the CO theory, and on the basis of their experiments they developed the so-called *"carbon complex theory"*.

Briefly, this theory states that the first step in the combustion of carbon is the formation of a loosely bound *physico-chemical compound of carbon and oxygen* of the general formula $C_xO_y$. This unstable compound is supposed to break up later on into CO and $CO_2$ in various proportions, depending on temperature and other conditions.

It is called a *"physico-chemical"* compound because it is not a purely chemical bond but something between this and a physical adsorption. On the one hand, the ratio x to y of $C_xO_y$ is not a fixed one; on the other, the oxygen cannot be released by purely physical actions, such as the application of vacuum. When the compound is heated in vacuo the oxygen comes off together with a mixture of CO and $CO_2$.

According to our modern views we would express it as follows: The oxygen is fixed by the *residual affinity* or remaining valencies of the carbon atoms, resulting from their molecular arrangement in the solid.

Besides the formation of the carbon-oxygen complex *direct reactions* are possible, forming $CO_2$ and CO, but it is supposed that the major part of these constituents is formed by decomposition of the complex. In a next stage CO may be oxidized further to $CO_2$ if an excess of oxygen is present, or $CO_2$ may be reduced to CO if an excess of carbon is present.

This theory has received strong support from the research work of LANGMUIR, who studied the oxidation of a carbon filament of an electric lamp in which he admitted oxygen of extremely low pressures. To prevent any secondary action between the carbon filament and any initially formed $CO_2$, which might produce CO as a secondary product and thus spoil the study of initial combustion, LANGMUIR immersed the bulb in liquid air, by means of which any $CO_2$ initially formed was immediately frozen out on the cold walls of the lamp bulb.

These experiments have been further developed by MEYER, who published in

1932 his results in "Zeitschr. f. phys. Chemie", Abt. B, Bnd. 17, Heft 6, April 1932, page 385. He proved that LANGMUIR, when making *static* experiments, would never be able to avoid the secondary actions completely and therefore he improved LANG-MUIR's method by blowing a strong current of oxygen along the filament in such a manner that any initial product formed on the carbon surface was immediately carried away and had no opportunity of reacting for a second time with carbon. He had to overcome several difficulties before his method proved successful, and although these are very interesting to read they will be passed over here and only some con-clusions mentioned. He used for his experiments a carbon filament of *graphite obtained by the thermal decomposition of methane.*

Briefly, these experiments made it probable that there are *two entirely different mechanisms in the combustion of carbon.* The first, so-called *"absorptive oxidation"*, takes place from room temperature up to about 1200° C. For this type of combustion MEYER gives the following explanation: The oxygen penetrates into the hexagonal arrangement of the atoms of solid carbon and is fixed by the remaining valencies. This is the same as was meant before by the $C_xO_y$ complex. As soon as every free place behind the outer layer of carbon atoms is occupied by oxygen molecules, that layer is broken off from the piece of solid carbon and its carbon atoms combine with oxygen according to the equation: $4 C + 3 O_2 = 2 CO_2 + 2 CO$ (Ratio $CO_2/CO = 1$).

It is curious to note that MEYER was able to deduce that, of the 3 oxygen molecules necessary for this reaction, two were originally absorbed and one came direct from the gas space, but it is still more interesting to read how he was able to explain that, in a $CO_2$-molecule formed, two oxygen atoms originated from different oxygen molecules, namely from an initially absorbed one and from a free one of the gas space. Fig. 80 A shows diagrammatically how this reaction is supposed to take place.

The second combustion process of carbon may be called the *"true combustion"*, which starts from about 1500° C. This reaction consists in the carbon atoms' being attacked only by oxygen molecules from the gas space without any previous ab-sorption, such according to the equation $3 C + 2 O_2 = CO_2 + 2 CO$ (ratio: $CO_2/CO = \frac{1}{2}$) (see fig. 80 B).

The carbon atoms at the *edges* are more readily oxidized because they are more exposed to the bombardment of oxygen molecules than are the other atoms. The difference between these two forms of combustion is clearly visible on the surface of the carbon filament, because the absorptive oxidation attacks layers and leaves edges, whereas the true oxidation burns away the edges first. Moreover the difference may be observed distinctly by the sudden change of the ratio of $CO_2/CO$ from 1 to $^1/_2$, which is in accordance with the above equations. This discovery is very important with regard to the luminosity and tendency to soot of oil flames, and therefore one more detail will be given.

The *reaction velocity* of the *"absorptive combustion"* first increases with the temperature as an ordinary chemical reaction would do, but attains a maximum at about 1100° C. and decreases from that point to practically zero at about 1300° C. This curious behaviour is due to the *decreasing solubility of oxygen* in carbon with increasing temperature.

As, however, the second form of combustion — *"true oxidation"* — only starts at about 1500° C., it follows that there is *no active form of combustion for carbon between*

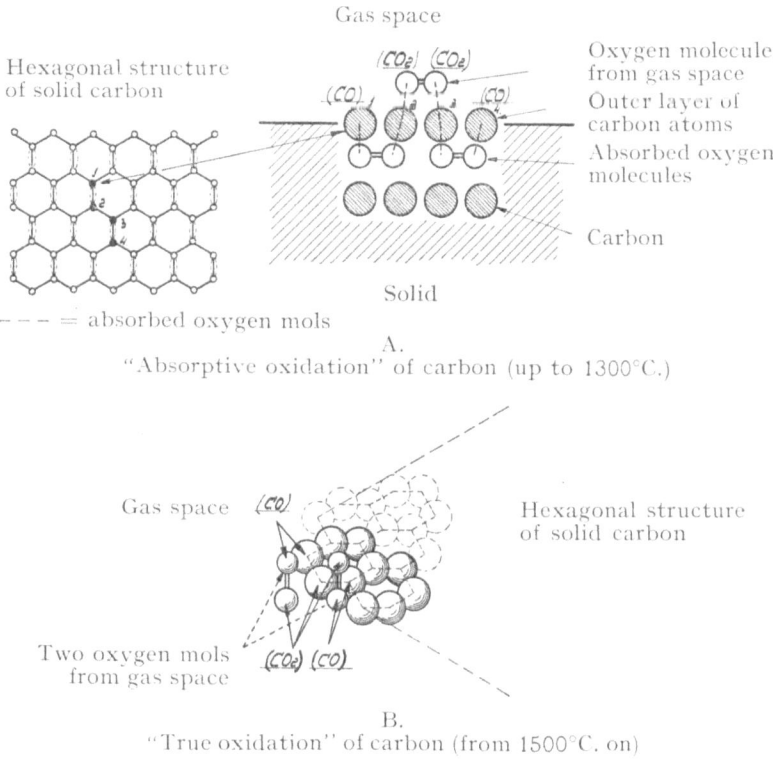

Fig. 80. Two combustion processes of carbon (MEYER, Z. f. ph. Chemie.)

*1300° and 1500° C.*, and therefore carbon (graphite) burns very badly between these temperatures. This may be *the reason why oil flames are often luminous in consequence of glowing carbon particles, which are not burnt notwithstanding the high temperature of 1300–1400° C. to which they are exposed.*

MEYER states, for instance, that the reaction velocity of the combustion of carbon (viz. graphite obtained by thermal decomposition of methane) between 1200 and 1500° C. and 0.05 mm. $O_2$ pressure amounts to only $\frac{1}{40}$ of that at bright red heat.

### B. *Combustion of Gaseous Hydrocarbons*

We will now pass on to the historical development of the combustion of gaseous hydrocarbons. It is advisable first to discuss the phenomenon for hydrocarbons in *gas form* because this problem is less complicated than that for hydrocarbons in the *liquid state*, such as with pulverized fuel oil.

A certain analogy may be observed between the stages of the development of the combustion of carbon and that of the combustion of gaseous hydrocarbons, for this also shows 3 stages of development, as clearly demonstrated in HASLAM and RUSSELL's above-mentioned book.

1) *Theory of preferential combustion of hydrogen.*

Until about 1892 the basic principles of FARADAY were generally accepted, according to which the *hydrogen* of a hydrocarbon molecule was supposed to be *ignited first*, this being a very vividly burning gas and liberating the carbon, which might or might not burn, according to whether conditions were favourable or not. The hydrogen was regarded as being, so to speak, *"singed off" the molecule.*

2) *Theory of preferential combustion of carbon.*

In 1892 DIXON, however, made a very remarkable discovery, which cast some doubt on the accuracy of the idea of the preferential combustion of hydrogen. He found that a mixture of equal volumes of ethylene and oxygen yields on detonation almost twice its volume of hydrogen and CO, according to the equation $C_2H_4 + O_2 = 2H_2 + 2CO$, and therefore the oxygen, although present in the right proportion to hydrogen to burn to water, is as it were *"snatched away by the carbon from the very nose of the hydrogen"*. Thus carbon proved to be *"not so lazy"* as FARADAY supposed it to be!

A similar experiment was made by mixing equal parts of methane and "detonating gas" consisting of $H_2$ and $O_2$ in the right proportion to form water. The combustion takes place according to the equation:
$$CH_4 + 2H_2 + O_2 = CO + H_2 + H_2O + 2H_2$$
which shows that even the well-known violent detonation of $(2H_2 + O_2)$ does not take place, but instead the carbon of methane is burnt completely against only half of its hydrogen.

In this connection we may recall the fact that the well-known *"water gas"* (mixture of CO and hydrogen, a substitute for coal gas in town gas) is made by the action of steam on glowing coke. Here too the carbon is burnt and hydrogen liberated.

This led to the new supposition that *carbon is burnt preferentially*, a theory that had already been suggested by KERSTEN as early as 1861. Flame experiments by SMITHELS and INGLE showed hydrogen present in the centre of hydrocarbon flames, which was also regarded as a proof of this theory.

In this connection it should be pointed out, however, that some of the above experiments and arguments apply to cases where either an extremely high temperature, or a comparatively long time is available for combustion, as in such cases the equilibrium $C + H_2O = CO + H_2$ may be reached which, at the high temperatures concerned, lies much on the CO-side.

In flames the *speed of combustion* plays a very important rôle, as this determines to a large extent the degree of equilibrium which may be reached. From this point of view DIXON's experiment is not a conclusive proof in favour of the theory of the preferential combustion of carbon.

3) *Theory of "Hydroxylation"*.

About 1903 the famous English chemist and physicist, BONE, began to publish a series of exhaustive studies on the problems of the combustion of hydrocarbons. From his experiments, consisting in 1) static bomb tests, 2) dynamic flow tests and 3) explosion tests, carried out on numerous hydrocarbons such as paraffin hydrocarbons, olefines, aromatics, acetylene, kerosene, etc., he had become convinced that there was no question of any preferential combustion either of hydrogen or of carbon, but that *several intermediate products* are formed, such as *alcohols and aldehydes* and especially *formaldehyde*, which are burnt to CO, $CO_2$ and $H_2O$ according to the conditions prevailing.

Thus BONE established the so-called *"hydroxylation theory"*, which was subsequently completed by the work of CALLENDAR and his collaborators and by BLAIR and WHEELER. Briefly, this theory assumes that oxygen is attached to the hydrocarbon molecule by replacement of one of the H atoms by a hydroxyl group (—OH), resulting in the formation of alcohols or, in a second step, to aldehydes (= O). The name *"hydroxylation"* is given to it because this *hydroxyl-group attachment* is an essential part of the theory.

4) *Theory of Peroxides*.

In February 1927 CALLENDAR started publishing a series of articles in "ENGINEERING" in which he further developed the hydroxylation theory of BONE to explain the phenomenon of knocking in gasoline engines.

Among the *"oxygenated" products* which are formed during the combustion processes of such engines, and which are found when, for example, hexane and oxygen react at comparatively low temperatures, aldehydes were found to be very abundant, alcohols on the contrary being practically absent. CALLENDAR concluded that the aldehydes are not formed via alcohols — as was assumed in BONE's original theory — but via a type of compounds of hydrocarbons with oxygen, which are called *peroxides*. These peroxides are supposed to be formed during the compression stroke of the engine, especially on the surface of small liquid drops of gasoline and, being rather unstable, are thought to cause detonation of the mixture in the cylinder after being ignited by the spark.

It is interesting to note a certain similarity between this supposition of peroxides being an intermediate stage for the oxidation of hydrocarbons and that of the physico-chemical compound $C_xO_y$ as an intermediate of the combustion of carbon.

Now this *"peroxide theory"* of knocking has given rise to many objections from several quarters, mainly because the occurrence of these peroxides could not be proved incontestably by experiments. Anyhow, whatever the value of CALLENDAR's peroxide theory of engine knocking may be, it is certain that his work contributed largely to the knowledge of the combustion of hydrocarbons. As a very important feature must be mentioned his discovery of a considerable difference between what is usually called *"spontaneous ignition temperatures"* (S.I.T.) of mixtures of air with hydrocarbons and the temperatures at which the first signs of oxidation become apparent. (CALLENDAR: *"temperatures of initial combustion"* or T.I.C.).

In his experiments mixtures of the inflammable vapours and air were passed at very low rates of flow (e.g., 1 litre of air per hour through a tube of 12 mm. diameter) through heated glass tubes. Tests on various reaction products were carried out and it was found that in several cases a loss of oxygen occurred at temperatures about 100 to 200° C. below the so-called *"spontaneous ignition temperatures"*. The *"temperature of initial combustion"* depends on the mixture strength and passes through a minimum. The minimum values which CALLENDAR observed are given in the following table and compared with *"spontaneous ignition temperatures"* measured by other investigators.

| Substance | Temp. Initial Comb. (T.I.C.) | Spontaneous Ignition Temp. (S.I.T.) |
|---|---|---|
| Norm. pentane . . . | 295° C. | 476° C. |
| Iso pentane . . . . . | 297° C. | ? |
| Hexane . . . . . . | 266° C. | 487° C. (Tapp) |
| Paraffin . . . . . . | 210° C. | 374° C.; 432° C. |
| Ether . . . . . . . | 145° C. | 400° C.; 347° C. |
| Benzene . . . . . . | 670° C. | 740° C. (Spiers) |
| Toluene . . . . . . | 550° C. | 810° C. ( „ ) |

Particular attention is drawn to the extremely high T.I.C. value shown by benzene, which is only about 70° C. below its S.I.T. In this connection it may be remembered that aromatic hydrocarbons have a *high anti-knock value* in gasoline engines, a point which will be discussed further in a separate chapter on discontinuous oil flames.

These very important experiments have been completed by several famous French chemists and physicists such as MOUREU, DUFRAISSE, MONDAIN-MONVAL, PRETTRE and DUMANOIS. An intelligent summary of this French development may be found in "Annales de Chimie, Tome XV, April 1931 by MONDAIN-MONVAL.

5) *Theory of Destructive Combustion.*

Now these versions of the hydroxylative combustion of hydrocarbons — with or without peroxides as intermediates —, which may be regarded as being typically

English and French, were not so readily accepted by German scientists, whose main objection was that flames of hydrocarbons apparently contained carbon particles, since they radiated a *continuous spectrum*, which in fact is a characteristic of glowing solid materials. Consequently it could not be doubted that *carbon* was an intermediate product of combustion of hydrocarbons, apparently formed by *"cracking"* of hydrocarbons.

Among the German supporters of this theory of *thermal destructive combustion* may be mentioned: AUFHÄUSER ("Brennstoff und Verbrennung"), WILKE ("II. Kongress für Heizung und Lüftung, Dortmund 1930) and WIRTH ("Brennstoff-Chemie"). According to this theory hydrocarbons are "cracked" (thermally decomposed) first from higher to lower hydrocarbons and ultimately even down to *methane, hydrogen and carbon*, after which cracking process oxidation is supposed to start.

*Cracking* is a well-known method for converting heavy residues

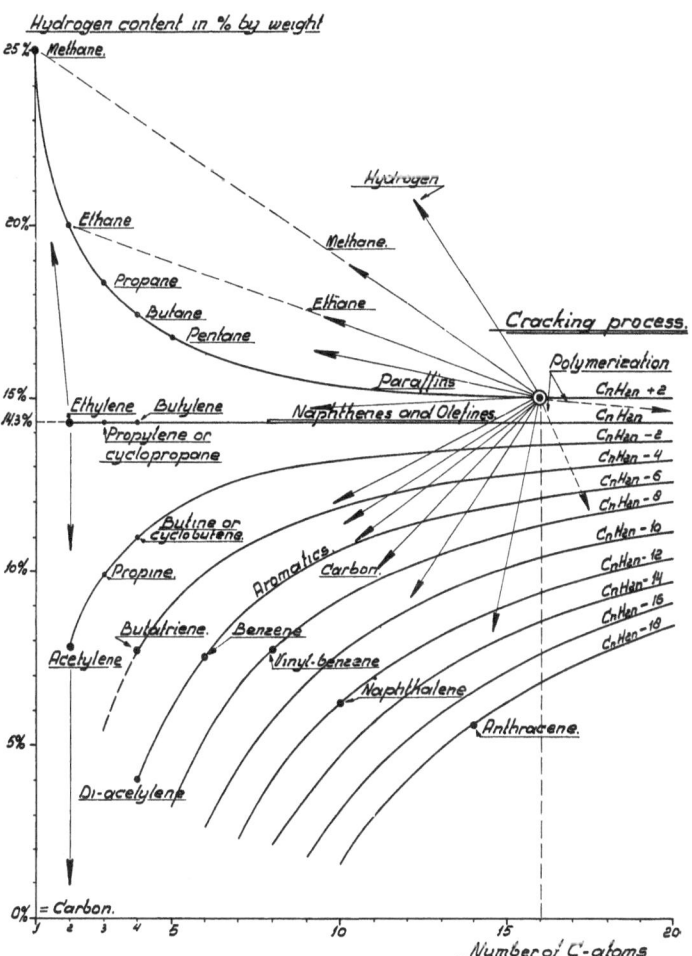

Fig. 81. Hydrogen content of various hydrocarbons.

into lighter and more valuable oils. In its oldest form it consists in a *destructive distillation* at 350 to 500° C. and pressures up to 12 atm., which method, however, is characterized by comparatively long times of reaction necessitating the building of "reaction chambers" in which the oil and its vapours are given time to decompose.

In the flame process only a very short time is available for destruction but it must not be forgotten that in general cracking is strongly influenced by temperature, time and the density of the material and, roughly speaking, it is found that high temperatures favour the formation of light products, gaseous hydrocarbons and finally carbon and hydrogen, when comparing processes of the same yield.

At about 900° C. and under atmospheric pressure residues are almost completely converted into gas and carbon; for instance, the oil gas processes of YOUNG, DAYTON, PINTSCH and others involve temperatures as high as that. At oil flame temperatures (1200 to 1500° C.) the cracking conditions may be so favourable that, notwithstanding the short time for reaction, even methane will for the greater part be cracked into carbon and hydrogen.

In the graph of fig. 81 the percentages of hydrogen contained in the various series of hydrocarbons are plotted against the number of carbon atoms, from which it is clearly seen that a complete conversion of a paraffin hydrocarbon of say 16 carbon atoms into methane and ethane would be absolutely impossible because of a *shortage of hydrogen*. Consequently *hydrocarbons of lower hydrogen content*, such as belong to the acetylene and aromatic series, must inevitably be formed, or even, in an extreme case, carbon itself.

AUFHÄUSER represents his view of the burning process of hydrocarbons in the way shown by fig. 82. The area of the largest isosceles triangle represents a certain weight of the original oil, e.g. one gram. The ratio of height to base has been chosen equal to the ratio of hydrogen to carbon in the oil. Whereas the largest triangle represents the quantity of oil as it starts burning, the smaller ones show the different periods of combustion in such a way that the carbon content of the remaining oil increases

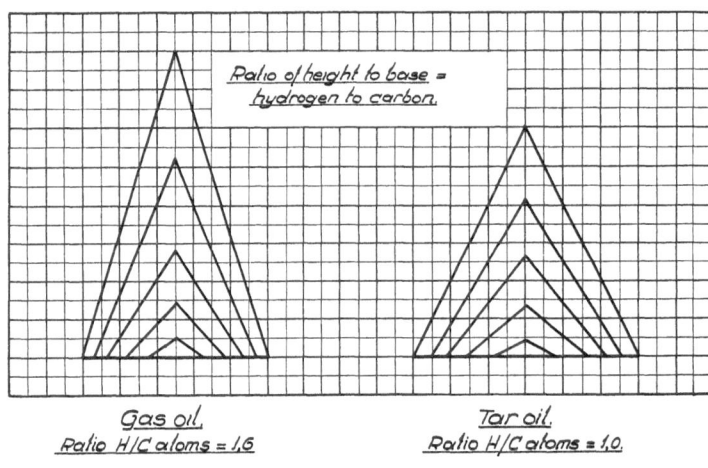

[Fig. 82. Stages of combustion of hydrocarbons. (AUFHÄUSER).

until finally only carbon will be left, because gaseous hydrocarbons such as methane and ethane, formed by cracking, will carry away more hydrogen than corresponds to the hydrogen content of the original oil.

In fig. 82 AUFHÄUSER compares a gas oil of natural origin (steep triangle) with a tar oil from coal distillation (flat triangle) and if in both cases equal quantities of methane, ethane and hydrogen are supposed to be formed by initial cracking, it is possible that the "carbon limit" due to lack of hydrogen may be reached sooner with tar oil than with gas oil.

## CHAPTER VI

### COMBUSTION OF HYDROCARBONS TAKEN AS A PROCESS COMPOSED OF TWO EXTREMES: VIZ. ALDEHYDEOUS AND CARBONIC COMBUSTION

#### A. *Blue and Yellow Flame Combustion*

Experiments seem to justify the view that both versions, the hydroxylative theory and the destructive one, may be taken as two extremes occurring simultaneously in every hydrocarbon flame.

This idea has been indicated by HASLAM and RUSSELL in their book "Fuels and their Combustion" (1926) but it will be worked out here more concretely.

In Chapter IX (page 185) one may read as follows:

"The heavier hydrocarbons usually undergo reactions other than simple hydroxylation, and the same is true of the lighter hydrocarbons if conditions are unfavourable for the formation of hydroxylated compounds. The heavy hydrocarbons may be "cracked" to give saturated and unsaturated lighter hydrocarbons, or they may be decomposed completely into carbon and hydrogen." and further:

"Thus in the ordinary combustion of hydrocarbons, there is a race between thermal decomposition and hydroxylation. If the conditions favour hydroxylation (such as preheating the hydrocarbons and air, and allowing time for the entrance of oxygen into

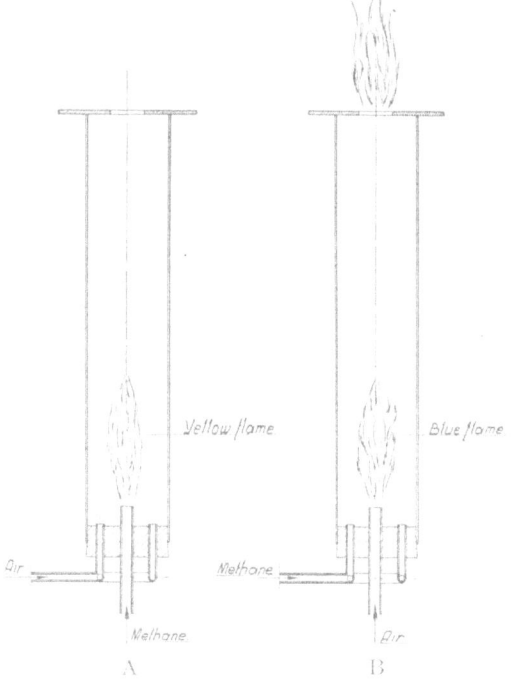

Fig. 83. The reserve flame principle.

the hydrocarbon molecule), there will be no soot. If, however, conditions favour cracking, as, for example, if the hydrocarbons and oxygen from the air are not thoroughly mixed together, the heat from the combustion of part of the hydrocarbon decomposes or cracks the remainder."

The two types of combustion, which stand at antipodes, are:

1) The *blue flame combustion* (hydroxylative, or better, **aldehydeous** combustion).

2) The *yellow flame combustion* (destructive, or better, **carbonic** combustion).

They may be realized in a simple way for the combustion of *methane* as follows: If methane burns in an *atmosphere of air* (see fig. 83 A) its flame will be *yellow*, but if one uses the lamp-glass device of fig. 83 B to *burn air in an atmosphere of methane* ("reverse flame"), the flame will be *blue*. A represents the carbonic combustion and B the aldehydeous combustion of methane.

The reason for this different behaviour is the following: The centre of the flame in both cases is exposed to a strong radiation of heat, which causes in the first case A a cracking of methane before oxidation takes place. Carbon particles are liberated and set glowing, causing the emission of a continuous spectrum which is characteristic for solid materials. In the second case, however, the strongly heated centre of the flame consists of non-decomposable air and the methane being on the outside of the flame is only moderately heated as the radiation dissipates to all sides. On the other side the air which is fed to the flame is strongly preheated before it comes into contact with methane, a condition which is very favourable to the so-called "pre-oxygenation" of the hydrocarbon, resulting in aldehydeous combustion.

This so-called *"reverse flame principle"* has found widespread application in the "range-burners" of the perforated shell type (see fig. 84). With this type of burners the air is sucked by natural draught into the holes of the perforated shells, which form a space filled with oil vapour evolved

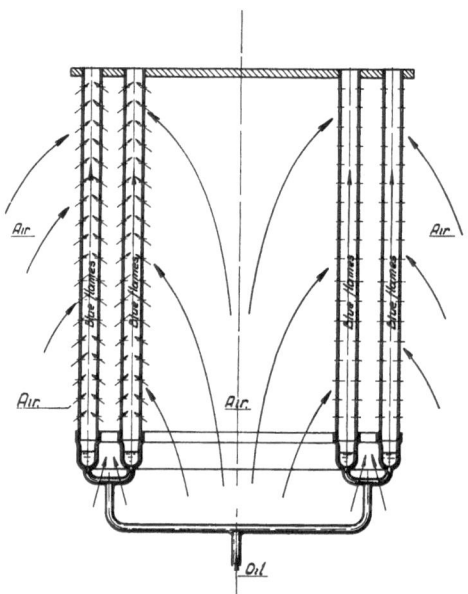

Fig. 84. The range burner principle.

by vaporization of the oil which is admitted in the base by some constant level device. Thus every little hole will form a small perfectly blue *flame of air burning in hydrocarbon gas* in accordance with the aldehydeous combustion process, these flames making the shells red hot.

A curious property of this kind of burners is that *incomplete combustion*, as caused for instance by poor draught when the shells are too cold, is not characterized by the formation of soot (product of destruction), but by the occurrence of *invisible acrid vapours containing aldehydes*, which in fact are the typical intermediate products of the blue flame combustion.

## B. *Simple Schemes of Combustion*

The scheme for the **carbonic combustion of methane** looks as follows:

First stage:                              Second stage:
*"thermal destruction"*                   *"oxidation"*

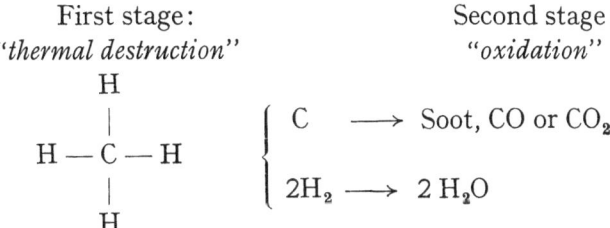

This reaction practically only takes place above 650° C., the decomposition of methane being too slow below that temperature. As a matter of fact, methane is much more stable than its homologues from the point of view of equilibrium and of the rate of decomposition, and this is the reason why it occurs so often as an intermediate product of combustion from hydrocarbons.

If a flame of hydrocarbons is rapidly cooled down to below about 650° C., e.g., by contact with comparatively cold walls of boilers etc., methane even may be found as a flue-gas constituent, which very often may escape the determination, it being very difficult to burn the last traces of methane of a flue-gas sample even by leading it over red hot copper oxide.

The scheme for the **aldehydeous combustion of methane** looks like this:

First stage:                                          Second stage:
*"hydroxylation or peroxidation"*                     *"combustion of formaldehyde"*
(often called "pre-oxygenation")                      (incl. its thermal decomposition)

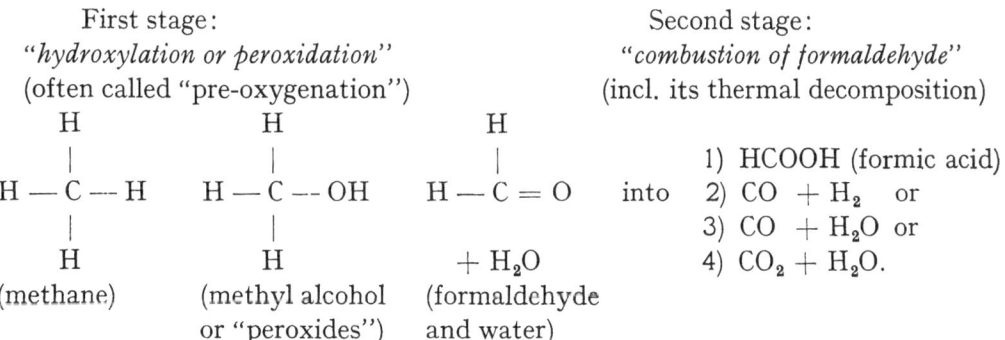

The *formation of formaldehyde is certain* and may be easily determined with imperfect burning blue flame devices, such as the above-mentioned "range burners", which then give off acrid, invisible, tear-drawing vapours, without, however, the slightest trace of soot. The *occurrence of alcohols is doubtful* and it remains an open question whether the stage between hydrocarbon and formaldehyde is formed by some kind of peroxides instead of alcohols.

For this reason it is better to speak of *"aldehydeous"* combustion than of *"hydro-*

*xylative"* combustion, as this latter name lays too much stress upon the formation of alcohols, which is anything but certain. ("Hydroxylation" means: "substitution of H-atoms by hydroxyl-groups"). *Formaldehyde* may be called the *key-product of blue flame combustion*, it being an inevitable link in the process, and in the same way *carbon* may be regarded as the *key-product of the yellow flame combustion.*

One scheme out of several for the first stage of **the carbonic combustion of a long chain hydrocarbon** of the paraffin series may look like this:

a)  −C−C−C−C−C−C−C−C−C−C−    (norm. decane)

(For the sake of simplicity the hydrogen symbols are not indicated)

b)  −C−C−C−    C=C−C−    C=C−C−C−    (propane)    (propene)    (butene)

c)  −C−    C=C    C≡C    −C−    C=C    C=C    (methane) (ethylene) (acetylene) (methane) (ethylene) (ethylene)

d)  CH$_4$    2C+2H$_2$    2C+H$_2$    CH$_4$    2C+2H$_2$    2C+2H$_2$

First stage: (Thermal destruction)

After this first stage of thermal destruction, which of course may take place in several other ways, the decomposition products, mainly consisting of carbon, hydrogen and methane, are oxidized:

a) Carbon to CO or CO$_2$ or left unburnt as soot.

b) Hydrogen to H$_2$O.

c) Methane burnt more or less completely, or left unburnt or further decomposed.

One scheme possible for the **aldehydeous combustion** of the same long chain hydrocarbon of the paraffin series may look like this:

a)  −C−C−C−C−C−C−C−C−C−C−H    (norm. decane)

$$b) \quad -C-C-C-C-C-C-C-C-\overset{\overset{H}{|}}{C}-\overset{\overset{H}{|}}{C}-OH \quad \text{(alcohol)}$$

$$c) \quad -C-C-C-C-C-C-C-C-\overset{\overset{H}{|}}{C}-\overset{\overset{H}{|}}{C}=O \quad \text{(aldehyde)}$$

$$d) \quad -C-C-C-C-C-C-C-\overset{\overset{H}{|}}{C}-\overset{\overset{H}{|}}{C}-\overset{\overset{H}{|}}{C}=O$$

$$e) \quad -C-C-C-C-C-C-C-\overset{\overset{H}{|}}{C}-\overset{\overset{H}{|}}{C}=O \quad + \quad H-\overset{\overset{H}{|}}{C}=O$$

(formaldehyde split off, after which it is burnt or decomposed and burnt)

$$f) \quad -C-C-C-C-C-C-C-\overset{\overset{H}{|}}{C}-\overset{\overset{H}{|}}{C}=O$$

$$g) \quad -C-C-C-C-C-C-\overset{\overset{H}{|}}{C}-\overset{\overset{H}{|}}{C}=O \quad + \quad H-\overset{\overset{H}{|}}{C}=O$$

(formaldehyde split off, etc.)

The long chain hydrocarbon thus *burns gradually like a fuse*, forming first higher alcohols, fatty or other organic acids and aldehydes (the possibility of the formation of peroxides instead of alcohols must be reckoned with), which are gradually transformed into lower alcohols, organic acids and aldehydes, at the same time *splitting off formaldehyde molecules*, which are eventually burnt separately either to $CO_2$ and $H_2O$ or to CO and $H_2O$ or broken down thermally to CO and $H_2$.

Of course this scheme represents a theoretical case just to show the principle of the thing. Thus the same process may of course start from the other side too, or even somewhere in the middle of the chain. This latter case, especially, may occur if the hydrocarbon has *branches* or *side-chains*, $CH_3$ groups being more favourable to "hydroxylation" or "oxygenation" than $CH_2$ groups. Thus the *presence of $CH_3$ groups* may be regarded as an indication of *preference for aldehydeous combustion*, a point which will be discussed in greater detail further on, when dealing with the oil properties.

## C. *Thermal Destruction*

Yet another *variant of the aldehydeous combustion* includes some form of a *moderate thermal decomposition* either of the hydrocarbon itself or of "pre-oxygenated" hydroxylated or oxidized products which are formed as intermediates, *provided no carbon is formed*, after which the resulting products are oxidized further in accordance with the aldehydeous scheme.

This form of combustion may be taken as a rule for those higher hydrocarbons which have comparatively *low thermal stability*. Thus a mixture may be formed which has enough thermal stability to continue the combustion process according to the aldehydeous scheme, the lower hydrocarbons usually being more stable than their higher homologues.

Special attention must be drawn to *acetylene* which may be easily decomposed into carbon and hydrogen, and for this reason a thermal destruction of hydrocarbons, which yields acetylene, will be followed certainly by *carbonic combustion*. Acetylene can probable be burnt only by the *carbonic or destructive* scheme of combustion. It is famous for its extremely high rate of combustion and this seems to be the reason why the maximum rate of combustion obtainable with the carbonic process is far higher than that obtainable under most favourable conditions by the aldehydeous process. This point will be dealt with further when discussing discontinuous oil flames in engines.

Finally, it may be observed that, although acetylene will in most cases be an intermediate product of carbonic combustion, this is no general rule as, for example, may be demonstrated by the above-mentioned scheme of destructive combustion of methane. For this reason it is not right to speak of "acetyleneous combustion" instead of "carbonic combustion".

The name "destructive combustion" is not quite correct either, because, as explained above, *moderate* destruction (no carbon formed) belongs to the opposite form, viz., aldehydeous combustion.

### D. *The Technical Side of the Problem*

From a technical standpoint it is important to make a clear demarcation between the yellow flame combustion and the blue flame combustion for the following reasons:

1. Blue flame combustion is characterized by a *discontinuous spectrum of bands* due to the emission of light from *glowing gases*. The *heat radiation* of the flame is *far less* than that of the yellow flame under similar combustion conditions of temperature, surface etc., because the yellow flame emits a *continuous spectrum* due to the presence of *solid carbon particles*, which serve as an excellent medium for heat radiation.

2. Yellow flame combustion is characterized by the *presence of solid carbon particles* in the flame, which involves a certain risk of the occurrence of a *smoky flame*. Soot is not only a waste of fuel and a nuisance to neighbours but also reduces the heat transfer efficiency through boiler walls etc.

3. Blue flame combustion usually takes place with *long flames* due to a *low rate of combustion*, which entails a risk of impingement of the flames upon the walls or boiler tubes and consequent damage to them. Yellow flame combustion may take place with *short, hot flames* which may damage the refractory bricks of the combustion chamber.

The technical policy of oil burning, therefore, must be focussed on choosing the happy medium between the two extreme forms of combustion, as neither of them is ideal from the technical standpoint.

### E. *Factors governing both Forms of Combustion*

We will now pass on to the discussion of the *factors which govern both forms of combustion*, and to the combustion conditions which prevail with oil flames in practice.

A flame consists of a flow of combustible gases in which liquid and solid particles may be suspended and the temperature of which is so high that combustion takes place. Flames may be *continuous* or *discontinuous*, the latter being used in internal combustion engines. A flame will be continuous if the heat developed by exothermic reactions is sufficient to maintain combustion of the incoming fresh materials and if an equilibrium can be established between the speed of propagation of the flame front and the velocity of flow of the fresh mixture.

Although thousands of reactions take place in an oil flame, it is possible to classify the flames roughly according to what might be called: *the orientation of the combustion between the aldehydeous and the carbonic prototypes.*

If the conditions are *favourable to primary oxidation*, such as may be brought about by:

1) *giving the oil time and opportunity for complete, previous evaporation,*
2) *thoroughly mixing the air and oil vapour before combustion,*
3) *suitably preheating the air or the mixture in order to favour the formation of oxygen compounds ("pre-oxygenation"),*
4) *exposing the mixture to radiation of heat gradually, if at all,*

then the oil will for the greater part be burnt according to the *aldehydeous* combustion scheme, resulting in a *predominantly blue flame without any tendency to soot*, but liable to produce acrid flue gases as soon as incomplete combustion occurs.

Such conditions are fulfilled with blue burners of the "reverse flame type" or those of the "pre-oxygenation type", which will be discussed further in Section III, and, moreover, in gasoline engines.

If the conditions are more *favourable to thermal destruction*, such as may be brought about by:

1) *suddenly exposing the oil to intensive heat without giving it time and opportunity for previous evaporation*, which may occur if the oil is not, or only coarsely pulverized, ("liquid phase cracking"),
2) *badly mixing the oil and air*, which causes cracking of the oil gas by its being evolved without enough oxygen available at the same spot to burn it, ("vapour phase cracking"),
3) *omitting preheating of the air or mixture,*
4) *exposing the mixture of combustibles suddenly to strong radiation of heat,*

then the combustion will mainly proceed according to the *carbonic* process, resulting in a *yellow flame with a strong tendency to soot*, and producing flue gases with a fatty smell typical of oil soot.

Some examples of such predominantly carbonic combustion processes are: Ordinary wax candle flame, that of a kerosene lamp, most oil flames for industrial purposes under boilers etc., and, moreover, the discontinuous oil flames in Diesel engines.

### F. *Internal Factors governing both Forms of Combustion*

The above-mentioned conditions may be called the *external factors* influencing the orientation of the combustion. The chemical and physical properties of the fuel oil constituents also have very important influence on the orientation towards aldehydeous or towards carbonic combustion, which influences might be called *internal factors*. Certain hydrocarbons show a propensity for primary oxidation and others for primary destruction.

On pages 75 to 77 we discussed the possible reaction schemes for the oxidation of hydrocarbon chains by both processes. There is little doubt but that they both really do occur and CALLENDAR's experiments, for example, prove that hydroxylation plays an important rôle in the so-called "low temperature oxidation" or "sly oxidation" processes.

*a) For Aldehydeous Combustion*

With regard to the predestination for *aldehydeous* combustion $CH_3$-*groups* seem to have a favourable influence, obviously because H-atoms of such groups may be easily substituted by —OH or =O. If one compares the homologues of the paraffin series it is clear that, if the above rule holds good, *ethane* with two $CH_3$-groups and 2 C-atoms and consequently *one $CH_3$-group per C-atom* will be in a more favourable condition than the higher normal paraffin hydrocarbons, as shown in the following table:

RATIO OF NUMBER OF $CH_3$-GROUPS TO NUMBER OF C-ATOMS

| Formula | Name | Normal chain | Isomers with | | |
|---|---|---|---|---|---|
| | | | 1 side-chain | 2 side-chains | 3 side-chains |
| $CH_4$ | methane | | | | |
| $C_2H_6$ | ethane | 1.— | | | |
| $C_3H_8$ | propane | 0.67 | | | |
| $C_4H_{10}$ | butane | 0.50 | 0.75 | | |
| $C_5H_{12}$ | pentane | 0.40 | 0.60 | 0.80 | |
| $C_6H_{14}$ | hexane | 0.33 | 0.50 | 0.67 | |
| $C_7H_{16}$ | heptane | 0.29 | 0.43 | 0.57 | 0.71 |
| $C_8H_{18}$ | octane | 0.25 | 0.38 | 0.50 | 0.63 |

From this table it may be seen that the isomeric modifications, especially those with many side-chains, are probably more adapted to aldehydeous combustion than their normal representatives. This point acquires considerable interest in connection with the problem of knocking gasoline engines, as the normal combustion seems to be mainly aldehydeous, whereas if knocking occurs, the combustion probably moves steadily towards the destructive side (*blue* combustion changes into *yellow* combustion). This point will be further discussed when dealing with discontinuous flames in gasoline engines.

The reason why $CH_3$-groups seem to lend themselves well to aldehydeous combustion is that each group provides a *starting point for primary oxidation*. The more side-chains there are, the more chance there is that the aldehydeous combustion will proceed satisfactorily to the end before cracking of the remaining part of the molecule becomes inevitable.

*b) For Carbonic Combustion*

As regards the *propensity of hydrocarbons for carbonic combustion* it seems that *aromatic rings (benzene rings)* are specially favourable for this type of combustion. This is probably due to the very stable arrangement of such rings. Thus the temperatures of initial combustion (T.I.C.) observed by CALLENDAR are exceedingly high for benzene (670° C.) and toluene (550° C.) and differ little from the spontaneous ignition temperatures (see table on page 70).

Although hydroxylation of benzene is possible under special conditions with special catalysts (resulting in phenol and polyhydroxyl benzenes, for example), such delicate conditions do not exist in an ordinary flame of benzene and consequently *benzene burns destructively.*

Thus naphthalene, diphenyl and other hydrocarbons, which are built up of aromatic rings, show a preference for carbonic combustion too, yielding carbon, hydrogen, acetylene and methane, for example, as intermediate products.

If the *cyclic hydrocarbon contains a straight side-chain* this may be hydroxylized, but, as soon as the aromatic ring is reached by the oxidation process, the combustion will change into a destructive one. An example of such combustion may be given diagrammatically for *propylene benzene* ($C_9H_{10}$) as follows:

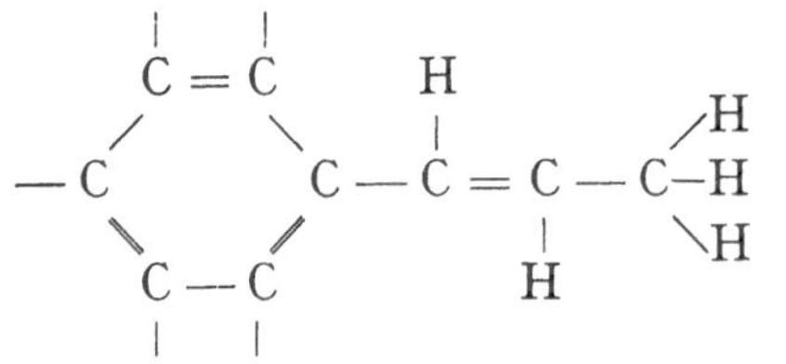

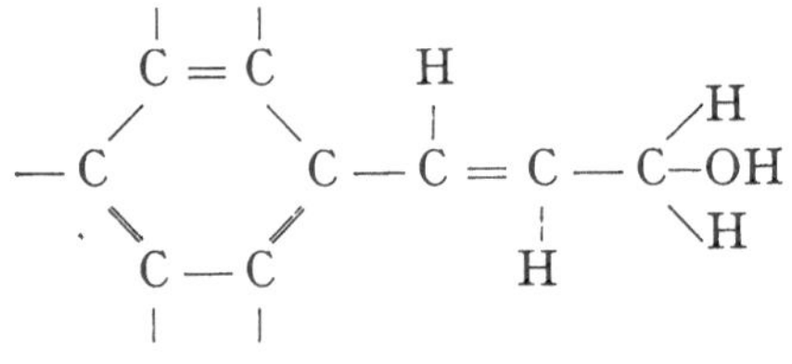

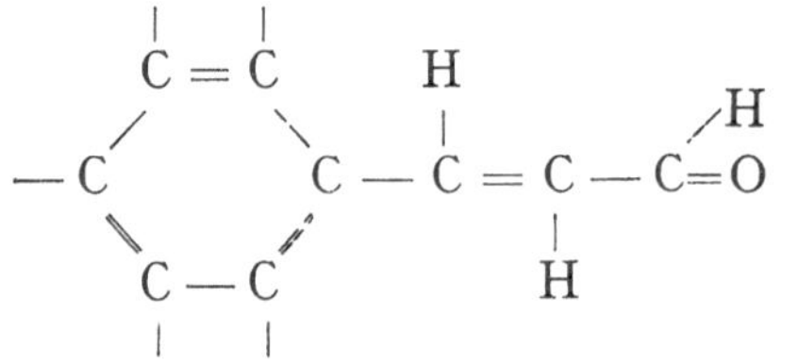

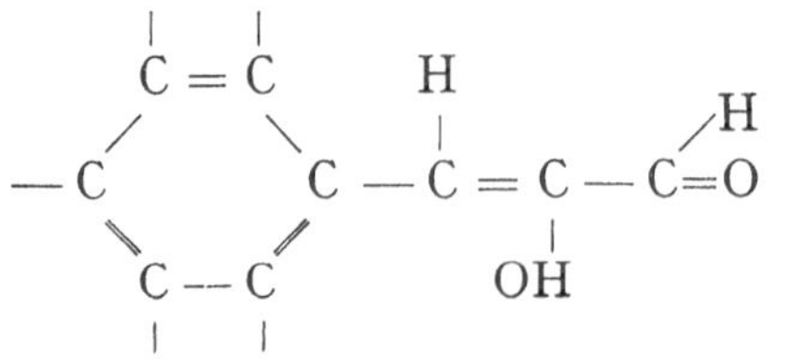

e)

$$
\underset{\substack{| \quad\ |\\ \mathrm{C-C}\\ / \quad\ \backslash}}{\overset{\substack{|\quad\ |\\ \mathrm{C=C}\\ \backslash\quad /}}{-\mathrm{C}\quad\quad\mathrm{C}}}-\mathrm{C=C=O} \quad + \quad \overset{\mathrm{H}}{\underset{|}{\mathrm{H-C=O}}}
$$

(formaldehyde)

f)

$$
\underset{\substack{| \quad\ |\\ \mathrm{C-C}\\ / \quad\ \backslash}}{\overset{\substack{|\quad\ |\\ \mathrm{C=C}\\ \backslash\quad /}}{-\mathrm{C}\quad\quad\mathrm{C}}}-\mathrm{OH} + \mathrm{C} \quad + \quad \mathrm{CO} \quad + \quad \overset{\mathrm{H}}{\underset{|}{\mathrm{H-C=O}}}
$$

(phenol)   +   (carbon)  + (carbon monoxide)   + (formaldehyde)

---

# CHAPTER VII

## DISCUSSION OF VARIOUS OIL FLAME TYPES OCCURRING IN PRACTICE

Let us now consider oil flames as they occur in practice, viz., from *mixtures* of hydrocarbons.

A flame as produced by a *wax candle* or *kerosene lamp* is an example of a hydrocarbon flame *mainly* consisting of *carbonic combustion*. The centre of the flame consists of practically pure hydrocarbon vapour exposed to strong radiation of heat from the surrounding flame fronts, resulting in decomposition of the oil gas in the centre before it has had an opportunity of combining with oxygen. The carbon particles thus formed are set glowing as they reach the flame front (inner side) and are burnt as they gradually traverse the flame.

The explanation of why the carbon particles are not burnt as soon as they are formed might be found in the above-mentioned studies of MEYER, who showed that two forms of combustion of carbon, viz., "absorptive" and "true" combustion, do not overlap each other, with the result that there is a *region from about 1300 to 1500° C. in which carbon* (as obtained by the thermal decomposition of methane) *burns rather badly*.

It is clear that such flames used for illumination must be of the carbonic type, as the aldehydeous combustion would give but a faint bluish light emission unsuitable for the purpose. The technique of making a good candle or a good kerosene is thus mainly based upon three factors:

1) Production of a *suitable quantity of carbon particles*, which must serve as a radiation medium.

2) Production of a *high flame temperature* in order to obtain an intensive light emission by the solid particles.

3) An *efficient way of burning the carbon particles again* after they have done their work as a light radiant, in order to leave no soot.

With reference to point 1) it may be stated that a luminous oil flame is in fact a gaseous body with suspended solid particles. The emission of light depends not only on the temperature, but also on the radiating surface, i.e., the sum of the area of all carbon particles which are visible from the outside of the flame. From this it follows that the emission of light depends strongly upon the *"carbon density of the flame"* i.e., upon the quantity of free carbon per unit of volume.

Each carbon particle practically radiates as a "black body", i.e., it absorbs nearly all radiant energy received or, in other words, it has very little reflective power. The laws governing this emission will be discussed further on, but from the above it will be clear that, as the gases between the carbon particles have very little absorptive power, the flame taken as a whole may not be considered as a "black body". It is usually taken to be a *"grey body"*, *the reflective power of which depends entirely upon the quantity of free carbon per unit of volume.*

A body is called *"grey"* if it shows the same shape of radiation curves as a *"black body"* (see fig. 107) but requires higher temperatures to radiate a given quantity of energy than a "black body" would require. In the graph of fig. 107 the curves must be lowered if applying to "grey body radiation".

The same effect of *"reduced black body radiation"* will be obtained if the radiating surface (flame surface) is not continuous but *discontinuous*, viz., consisting in a number of small surfaces (carbon particles). The quantity of energy radiated per unit of surface (flame surface) will be decreased according to the ratio: actual surface to flame surface.

Only the destructive process is able to furnish carbon; therefore, flames for lighting purpose must have a considerable orientation in the direction of the carbonic combustion process.

The *flame temperature* depends upon several factors, such as the quantity of excess air, the influence of which will be discussed later on, and particularly upon the rate of combustion; a high rate of combustion means a short flame in which the latent heat of the combustibles is set free in such a concentrated form that a high flame temperature is produced.

The *liquidation* ,as it might be called, of the carbon particles after having done their work as a radiant medium depends upon the combustion conditions in the last part of the flame. If these conditions are favourable to good combustion, such as enough oxygen left and thorough mixing by turbulency, there will be no soot. If not, a smoky flame will be produced.

Yet another very important factor in this connection is the *quantity of water vapour* in that part of the flame, because, provided the temperature is high enough, water may be able to convert carbon forming carbon monoxide and hydrogen according to the so-called "water-gas reaction" and these gases burn with non-luminous, smokeless flames. The water necessary for this soot-preventing action may derive not only from the humidity of the air or from steam, if used for pulverisation, but also from burnt hydrogen of the fuel itself and therefore *the ratio of carbon to hydrogen in a fuel is of the utmost importance with regard to its tendency to soot.*

This influence may be demonstrated by the experiment of fig. 85, consisting of six ordinary oil lamps filled with the following hydrocarbons:

1) *Tetralin* (tetra-hydro-naphthalene, $C_{10}H_{12}$),
2) *Mesitylene* (aromatic series, $C_9H_{12}$),
3) *Benzex* (aromatic extract from an unrefined kerosene distillate),
4) *Refined kerosene* (mainly consisting of paraffin hydrocarbons),
5) *Cetene* (olefine series, $C_{16}H_{32}$),
6) *Cetane* (paraffin series, $C_{16}H_{34}$).

Before the photograph was taken the wicks of all lamps had been adjusted to the so-called *"smoke point"*, which is the limit to which the wicks may be turned up without causing the flames to smoke. Thus the *flame length is a rough measure for the tendency to soot or "TTS" of the fuels* (a better measure would be a comparison of the weight of oil burnt at smoke-point adjustment per unit of time).

It may be clearly seen that from left to right the tendency to soot decreases with increasing ratio hydrogen to carbon. On the other hand this effect is not only caused by the higher water content of the right-hand flames, but also by the fact that the latter, being mainly paraffin and olefine hydrocarbons, are more adapted to aldehyde-ous i.e. soot-free combustion than the former containing aromatic rings and being predestined to the carbonic combustion process.

The effect of water vapour may be demonstrated in the following way, which, however, is not visible from the photograph: If the wicks of the left-hand lamps are turned up to about the same extent as the right-hand lamps, very heavy smoking will ensue. If, however, the air around the smoking lamps is saturated by water-vapour, c.g., by means of the spout of a boiling water kettle, the smoke will disappear immediately and a clear flame and a clean combustion will be produced.

Details of the effect of water vapour on the combustion process will be discussed further in Section IV.

From the foregoing it is clear that for *oil lamps* we shall have to make a kind of compromise: On the one hand there should be no soot, i.e., the tendency towards carbonic combustion should not be too strong and carbon should be effectively removed in the outer part of the flame; on the other hand, we want carbonic combustion in order to furnish carbon particles for the emission of light.

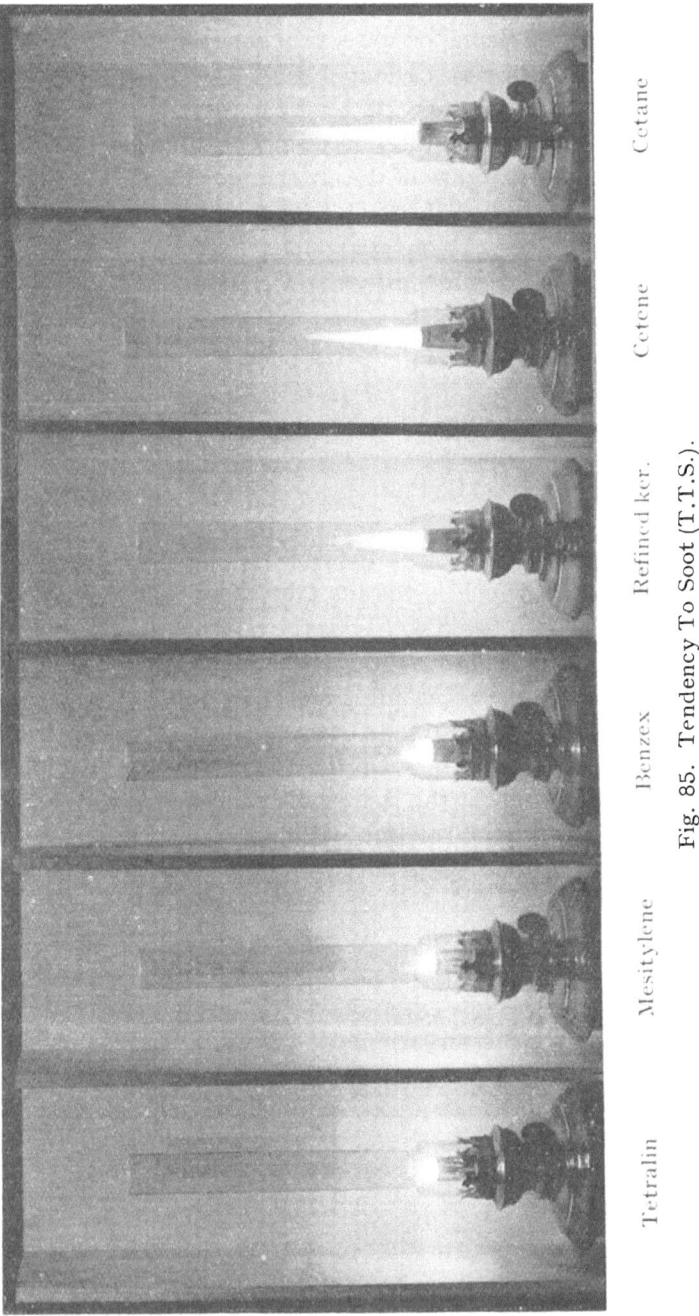

Fig. 85.  Tendency To Soot (T.T.S.).

This problem is technically solved by taking a fuel with a high hydrogen to carbon ratio (saturated hydrocarbons), as such a fuel has a sufficient tendency towards aldehydeous combustion and enough water vapour in its flame to guarantee the absence of soot. The construction of the lamp then forces the fuel locally to carbonic combustion by not supplying enough oxygen in the centre of the flame. So we can use for candles: paraffin wax, and for lamps: kerosene from which the aromatic hydrocarbons (if present) have been partly removed, for instance by the EDELEANU Process (extraction by liquid sulphur dioxide).

Another reason why a very aromatic fuel is unsuitable for lighting purposes is the following: If there are too many carbon particles, they may radiate so much energy that (except in cases where the rate of combustion is artificially raised by, say, preheating the air or by using oxygen instead of air, as with blow pipes) the flame temperature drops too much and a dull, reddish, smoky flame is produced.

With respect to the *constructive development of kerosene lamps* which, as a matter of fact, are a simple but delicate form of "fuel oil burners", the following remarks may be of interest.

Nowadays people very often speak disparagingly of oil lamps, which are considered to be ridiculously primitive devices, but it is forgotten that they are the result of very elaborate and painstaking research done half a century ago. Readers who are interested in this matter are recommended to read STEPANOFF's book *"Grundlagen der Lampentheorie"* (1894) for which he was awarded the NOBEL prize.

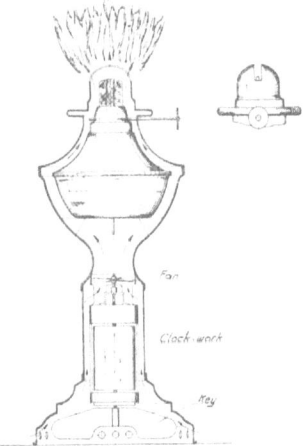

Kerosene lamps were originally made with *massive wicks*, like those of candles, but very soon *annular wicks* with air admittance in the centre proved to reduce the tendency to soot considerably, obviously because the centre, being exposed to a strong radiation of heat, then consisted of thermally stable air instead of unstable hydrocarbons. This constructive change therefore moved the orientation of the burning process somewhat in the direction of aldehydeous combustion.

After a contrivance (see fig. 86) by which air was directed into the flame by means of a propeller driven by clockwork, in fact a *motor-driven oil burner of about 1880!*, had proved to be unsatisfactory, *natural draught* was applied for this purpose by means of a glass chimney,

Fig. 86. HITCHCOCK
Clock-operated oil lamp.

resulting in a more vivid combustion and a higher temperature, which in its turn meant more light and less soot. According to the researches of STEPANOFF it proved to be a most important point to establish a suitable distribution of air for the inside of the hollow wick as compared with the outside of the flame. This controlled the dimensions of the cone and — as one would express it now — controlled the orientation of the combustion between the aldehydeous and carbonic process.

The further development of oil lamps to blue burning devices for heating will be discussed in Section III.

---

## CHAPTER VIII

### DISCONTINUOUS OIL FLAMES IN ENGINES

A. *Two types of combustion for two types of engines*

The same difference between "blue" and "yellow" combustion may be met with in engines and although it may not seem entirely to the point to discuss it here, some space will be devoted to the subject because, as a matter of fact, the difference between *continuous oil flames in furnaces* and *intermittent oil flames in engines* is not as great as it is usually thought to be.

The main difference between the two applications of oil burning is the great importance of *combustion speed* in the case of oil engines i.e., the fact that for every single combustion process in the engine the flame must be ignited, and in the second place that combustion occurs at *high pressure*. Pressure and combustion speed are intimately related, as the very use of this high pressure is one factor in promoting combustion speed, since the heat transfer to neighbouring combustibles necessary for propagating the flame is increased by the higher density of the working medium.

There are mainly two different classes of internal combustion engines, viz.,

1) The *carburettor type* (engines with spark ignition, gasoline or petrol engines), where the fuel is mixed with the air in a more or less completely evaporated state before being introduced into the cylinder and being compressed, and

2) The *injection type* (compression ignition engines of the Diesel type) where fuel is introduced in a liquid, more or less pulverized state into the combustion chamber filled with hot compressed air.

The *combustion in the carburettor engine is mainly blue* as is shown, for instance, by the spectroscopic studies of WITHROW and RASSWEILER (See fig. 87 taken from their publication in Ind. and Eng. Chem., July 1931, page 679), which seems to point to a *combustion of predominantly the aldehydeous type*.

This is not astonishing considering the fact that with this type of engines several conditions for this kind of combustion are present such as:

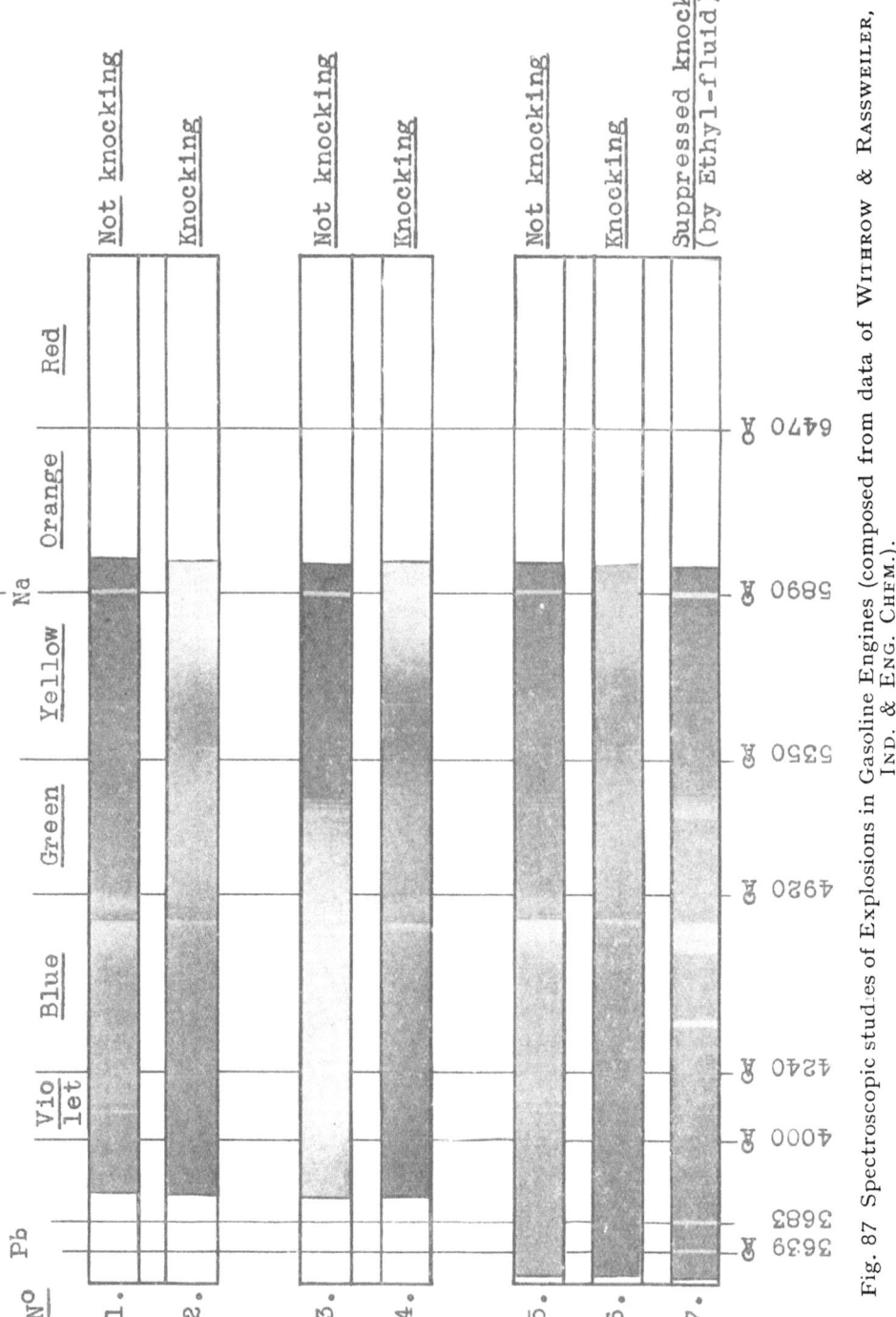

Fig. 87 Spectroscopic studies of Explosions in Gasoline Engines (composed from data of WITHROW & RASSWEILER, IND. & ENG. CHEM.).

1) The oil (i.e., gasoline = volatile fractions) has been given time and opportunity for a more or less complete *previous evaporation*.

2) The oil vapour and air are thoroughly *mixed before combustion*.

3) The *mixture is preheated* not only before entering into the cylinder but even more so by compression and by contact with hot cylinder walls.

Indeed observations with engines equipped with quartz windows in the cylinder head have shown that when using a suitably heated carburettor the flame is blue and the window remains clear, whereas if no heat is applied to the carburettor, the resulting inequality of the mixture and the presence of fine drops of unvaporized gasoline will lead immediately to reddish-yellow colouring of the flame and to a gradual deposition of soot on the window.

The *combustion in a compression ignition engine (Diesel type) is yellow*, which seems to point to *combustion of the carbonic type*. This is obvious because of the conditions, which are the following:

1) The oil is *not previously evaporated* but is injected into the combustion chamber in a liquid state, though finely pulverized.

2) The combustion is started by initial reactions of the vapours released from the fine oil drops with oxygen, after which the drops are *suddenly exposed to the intensive heat* of the initial flame thus formed around, and during the resulting intensive evaporation *cracking of oil vapours and liquid rests* takes place before they have had time to find oxygen.

The question as to how and by what kind of reactions the oil in the compression ignition engine comes to ignite spontaneously is a problem in itself, which is outside the scope of the present work. It seems, however, as if this very first stage of combustion must be of the *aldehydeous* type, as the conditions are conducive to that form.

Of course, neither the spark ignition engine (S.I. engines) nor the compression ignition engine (C.I. engine) has a purely aldehydeous, and a purely carbonic combustion respectively. Thus partial carbonic combustion may be obtained with the S.I. engine by over-rich setting of the carburettor or by detonation of a part of the charge (knocking engines).

## B. *Compression Ignition Engines*

As to the compression ignition engine, the conditions for aldehydeous combustion due to intimate mixing before combustion may be more rarely met with; *at low loads*, however, and with *long ignition lags* conditions may be sufficiently favourable and it is noticeable that under such conditions the exhaust gases have often a very offensive acrid smell due to the presence of aldehydes, which points to *aldehydeous* combustion.

The most vital problem connected with the combustion in compression ignition engines is to obtain proper ignition, but the next important point is to establish *clean, soot-free combustion,* which in fact is scarcely a problem at all with the S.I. engines. This contrast is due to the different combustion processes in both types.

Measures to ensure soot-free carbonic combustion will be discussed further in section III, but one typical means which has found considerable appreciation for compression ignition engines may be mentioned here, viz. *by bringing about intensive turbulency during combustion.* This may be done by imparting a swirling motion to the compressed air, such as for instance with the RICARDO-DORMAN engine of fig. 88.

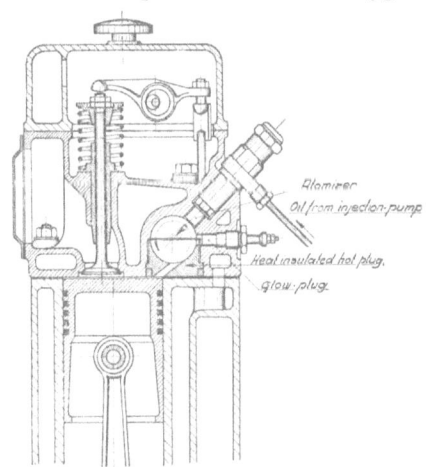

Fig. 88. COMET-head of the DORMAN-RICARDO Diesel engine.

## C. *Some Remarks on Spark Ignition Engines*

The author will close this chapter on discontinuous oil flames in engines with some remarks on the phenomenon of *knocking engines* of the carburettor type, most of which remarks, through lack of available data, must be taken merely as suggestions mainly given in order to draw the attention of those concerned with research work on oil firing to the results of studies of the combustion processes in engines, likely to be to the benefit of both.

From the above-mentioned experiments of RASSWEILER and WITHROW (see fig. 87) it follows that, whereas the combustion of a spark ignition engine under *normal* conditions is mainly *blue,* the *occurrence of knocking* may be characterized by the *colour of the explosion turning from blue to yellow,* which seems to justify the idea that the mainly *aldehydeous* combustion changes into a mainly *carbonic* one.

As will be discussed further on (page 97) when dealing with rates of combustion, the maximum obtainable combustion speed seems to be considerably faster with the carbonic type than that maximum obtainable with the aldehydeous type of combustion. When burning equal quantities of gaseous hydrocarbons per unit of time, those which burn according to the *carbonic* scheme, e.g. acetylene, will burn with *shorter flames* than those which have greater preference for aldehydeous combustion, such as methane, ethane, etc. Consequently it is not astonishing that *knocking,* a phenomenon which is caused by too rapidly increasing pressure on the piston, *seems to be accompanied by the occurrence of carbonic combustion.*

WITHROW and RASSWEILER, moreover, showed that if the knocking was suppressed by admixing "ethyl fluid" (a lead-containing compound soluble in gasoline, see SPIERS, "Technical Data on Fuel", 1935 page 278) with the gasoline, the colour of the explosion changed from yellow to blue again. Apparently this anti-knock dope had a favourable influence on the aldehydeous type of combustion, or an unfavourable influence on the carbonic type.

The mechanism of the "knock" may be understood as follows: The combustion is started by the spark under conditions favourable to aldehydeous combustion (see arguments mentioned before), and consequently the first part of the explosion wave is mainly aldehydeous. The heat radiation of this aldehydeous start, combined with a temperature rise by rapidly increasing pressure may, however, cause a *thermal decomposition of the remaining unburnt mixture* (in literature often called "end-gas") resulting in a *very rapid spontaneous carbonic combustion* of the same, the almost instantly increasing pressure causing the well-known pinking sound. This "knock" should not be confounded with that occurring when the compression ratio is raised to such an extent that the *whole* charge of combustibles, instead of being ignited by the spark, ignites spontaneously ("Diesel-knock" of S.I.-engines).

The suppression of the knock is called *"anti-knock effect"*, which may be obtained by *all conditions favourable to aldehydeous combustion*, provided this form of combustion be rapid enough for the purpose under consideration.

Thus anti-knock effects may be obtained by:

1) *Suitable combustion conditions*, or
2) *Suitable fuel properties*.

As regards point 1), knocking may be suppressed by *more complete evaporation*, viz., by *preheating* the mixture (provided at not too high a temperature, as otherwise there will be self-ignition by the heat of compression, "Diesel-knock") or by *pulverization*, by *better mixing* the gasoline vapours with air before entering into the cylinder and, last but not least, by imparting *turbulency* to the flame gases. This last method in particular has found widespread application, as in the course of years the compression ratio has been raised considerably thanks to improved design of combustion chambers.

Regarding the influence of *fuel properties* on knocking, it should follow from the above-mentioned theory that hydrocarbons burning preferably according to the aldehydeous scheme would have higher anti-knock values than those which prefer to burn after the carbonic scheme.

### D. *Anti-knock values of various hydrocarbons*

First of all considering the *hydrocarbons of the paraffin series*, it is a well-known fact that the *normal representatives show a decreasing anti-knock value with increasing molecular weight* (see fig. 89). This accords with the theory of aldehydeous combustion, as the tendency towards primary oxidation also decreases with higher molecular

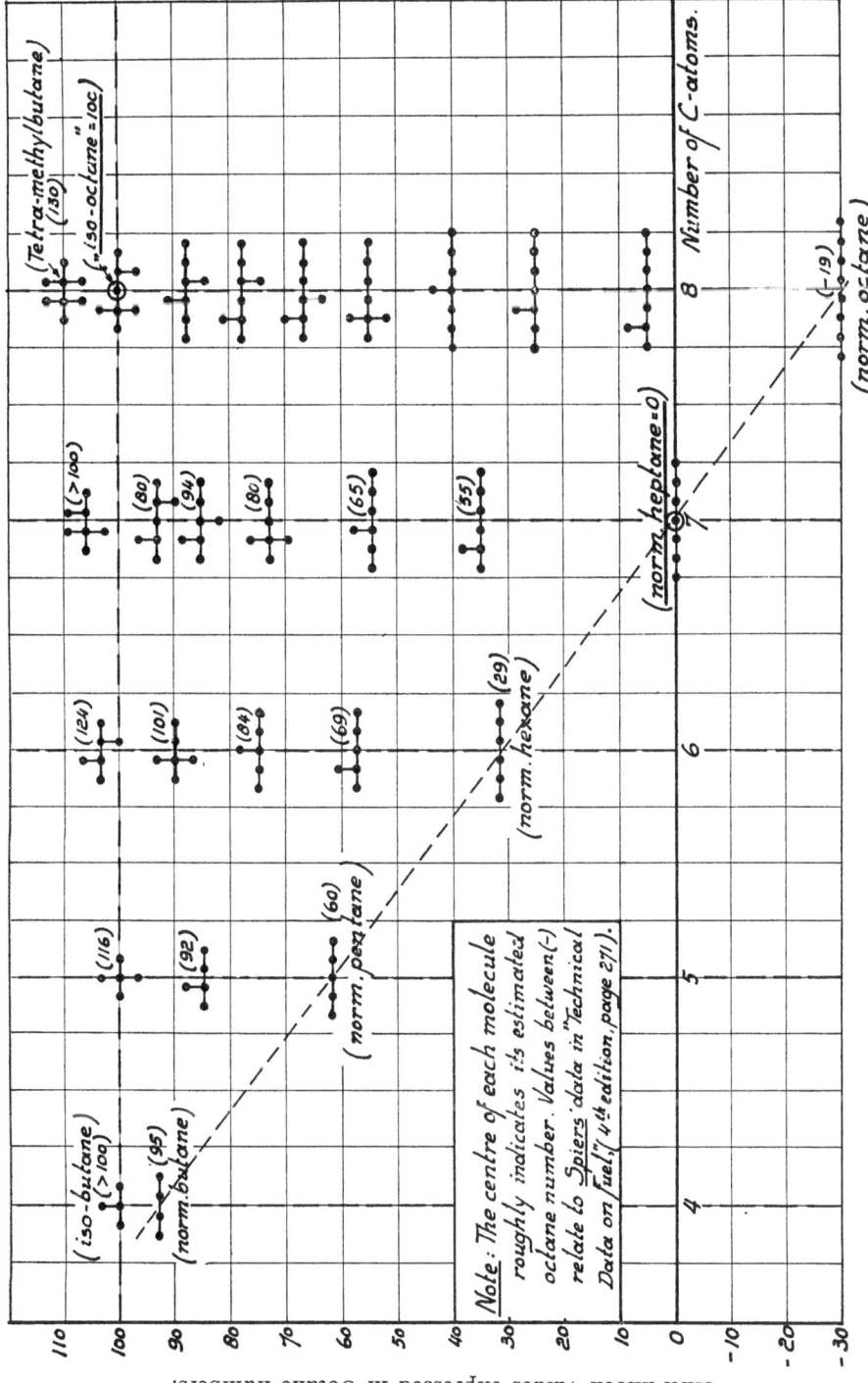

Fig. 89. Influence of Side-chains on Anti-knock values.

weights, owing to a decrease in their $CH_3$ to C ratio, which runs from 1 for ethane down to 0.25 for normal octane (see page 000).

On the other hand the presence of *side-chains (isomers) increases the tendency towards primary oxidation*, which tallies with the fact that the *isomeric modifications of the paraffin series have higher anti-knock values* than the corresponding *normal* representatives. Thus iso-octane (2.2.4-tri-methyl-pentane), which is used as the "non-knocking" reference fuel for knock testing, has a ratio of 0.63 as compared with normal octane having only 0.25 and normal heptane, which is the "knocking" reference fuel with 0.29. Fig. 89 may be given as a suggestion for the probable influence of side-chains on the anti-knock values of paraffin hydrocarbons, which graph tallies rather well with the meagre data available on the subject.

If a hydrocarbon molecule (e.g. normal octane) has too little preference for primary oxidation (long straight chain with low $CH_3/C$ ratio), its *aldehydeous* combustion may be *too slow* for the short time available in the engine, as in the meantime temperature and pressure may be increased to such an extent that the unburnt remainder of the molecule will be cracked and burnt according to the *carbonic* scheme. Thus there seems to exist a parallel between the destructive detonation of *a part of the combustibles* in a knocking engine cylinder, and the partial detonation of *one separate molecule*. Such *"partial molecular detonation"* may shorten the time of combustion considerably and, therefore, by increasing both the temperature and pressure, may favour the detonation of the end-gas.

Another important factor of the anti-knock problem is the *thermal instability of hydrocarbons*, and the high anti-knock values of isomeric paraffins might be explained too by their being rather unstable formations which are *easily cracked into lower hydrocarbons of high anti-knock values* (methane, ethane, ethylene, propane, etc.) which occurs at *moderate temperatures without free carbon* as yet being formed. These products of moderate cracking, being easily burnt according to the *aldehydeous* scheme, may thus contribute to the high anti-knock value of the original hydrocarbon. This would imply what might be called a *"reverse molecular detonation"*, viz., consisting of a destructive start followed by aldehydeous combustion of the decomposition products, the ordinary detonation being taken as an aldehydeous start followed by a destructive combustion of the intermediate products.

In this connection *benzene* takes up a peculiar position in the problem, for, as has been mentioned before, benzene shows a *strong tendency towards destructive combustion* and, regarded superficially, that might be expected to entail a *low* anti-knock value. This, however, is not so. On the contrary, benzene, although being adapted to *carbonic* combustion, is still *more resistant to destruction than*, say, *normal heptane*, and for this reason benzene shows a *very high* anti-knock value notwithstanding its carbonic burning.

## E. *Final Observation*

For the sake of brevity the carbonic combustion has been denoted as *"yellow* combustion" as contrasted with the aldehydeous combustion which we call *"blue* combustion". For normal oil firing at atmospheric pressure and when using air which is not artificially enriched with oxygen, the above holds good, but it should be borne in mind that under certain conditions, e.g. by using higher pressures or enriching with oxygen, the colour of the carbonic combustion may become *white* too, as the colour of the light emitted by carbon particles only depends on the temperature.

It is therefore a better definition to characterize yellow combustion by its emitting a *continuous* spectrum as contrasted with blue combustion showing a *spectrum of bands*.

This remark applies especially to the combustion of *benzene* in carburettor engines, which shows a bright flame of almost white colour and which might wrongly give the impression of a blue aldehydeous combustion. Another example is the flame of an oxy-acetylene welding torch, which is bluish white, though acetylene doubtlessly burns according to the carbonic scheme.

---

## CHAPTER IX

### CONTINUOUS OIL FLAME

## A. *Stable and Unstable Flame Fronts*

After being ignited, a continuous stream of oil and air may produce a continuous oil flame if ignition conditions of the mixture of oil and air are sufficiently good to support combustion. Experiments have shown that one of the most important conditions is that *sufficient oil vapours* be present in order that an initial flame may deliver sufficient heat to evaporate the rest of the oil. For this reason it has been found necessary to pulverize the oil before combustion and the more so the heavier oil is.

If ignition conditions are not good enough, the oil flame will be blown out again as soon as the source of ignition, flame, spark or the like, is extinguished or moved away. Whether the flame will be blown out or not depends on whether or not the combustion gets an opportunity of *establishing an equilibrium between the velocity of flow and the combustion speed*, the latter being directly related to the ignition conditions.

If the oil and air mixture flows through a *tube with parallel walls* (constant cross-section, e.g. cylindrical tube), its velocity of flow will be approximately the same all over the tube and an equilibrium within the tube will only be possible if the *rate of combustion is exactly equal to the velocity of flow.* If, however, the ignition conditions are slightly improved, the flame front will move upstream and leave the tube, and if the ignition quality is a little worse, the flame front will go downstream and leave the tube on the other side.

If the air and oil mixture flows through a *divergent tube* (= increasing area in the direction of flow, whatever shape the cross-section may be) the velocity of flow decreases gradually in the direction of flow and consequently *an equilibrium will always be established*, whatever the combustion speed and ignition quality may be (fig. 90).

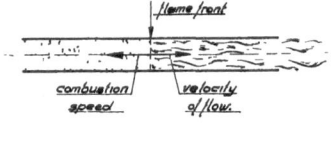

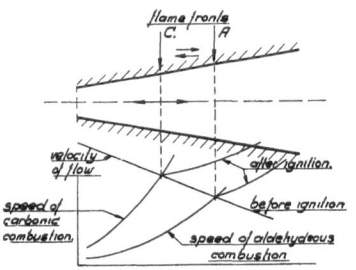

Fig. 90. Stability of flame fronts in tubes.

If, however, the mixture flows through a *convergent tube*, the velocity of flow increases in the direction of the stream and therefore the flame, if ignited outside, will only be able to enter the tube by itself if the ignition quality permits of a combustion speed *higher* than the velocity of flow at the narrow mouth of the tube. In that case the flame front will run rapidly through the length of the tube, as *no equilibrium is possible* within it. If the ignition quality is made worse, the flame front will travel through the tube again as soon as the combustion speed is lower than the velocity of flow at the wide mouth of the tube, and, as again no equilibrium can be established, the flame will leave the tube downstream.

The above shows that a flame front which is stable under all combustion conditions can only be obtained by means of a *divergent* flow, which offers the opportunity of establishing an equilibrium of speed whatever the ignition condition may be. In fact, it is not strictly necessary to use a divergent *tube*, as the *free outflow* of gas from an orifice shows a *natural divergency*, which may serve for this purpose too. *Convergent or cylindrical tubes, however, are not suitable forms for combustion chambers.*

Furthermore it must be borne in mind that flames are glowing gases which are governed by the laws of friction and resistance just as well as any other gas flow of atmospheric temperature. Gas resistance is caused by velocity and is therefore mainly dependent on the *volume* and less on the *weight* of the gas. A temperature rise from 100° to 1500° C. increases the volume of non-reacting nitrogen from the air in an oil flame by about *five times;* moreover, when considering the increase of volume caused by chemical reactions and dissociation, it will be clear why flames may often exert a back pressure soon after ignition. This phenomenon is due to a combined action of *resistance* and *acceleration of the flame gases* in the combustion chamber.

A divergent cone of some 70 degrees has proved to be conducive to stable oil flames.

## B. *Vibrating Flame Fronts and Roaring Flames*

If a steady oil flame is observed more closely it will be seen that although the flame as a whole has relatively a stable form, the *flame front*, which is the separation between the burning and non-burning part of the gas flow, is continuously in motion, due to a rapid jumping to and fro.

This phenomenon may probably be taken as a *hesitation*, so to speak, of the flame between its two forms of combustion, the *aldehydeous* and the *carbonic*, and, as it has been explained before that the latter probably has considerably higher rate of combustion than the former, each form will have its own flame front fixed by the equilibrium of combustion speed and velocity of flow. *The flame front of rapid carbonic combustion lies nearer to the burner tip than that of slower aldehydeous combustion.*

Remembering the above-mentioned remarks (see Chapter VIII) on knocking in gasoline engines, it is curious to observe that this *jumping to and fro of the flame fronts is a phenomenon which is perhaps very similar to that of knocking combustion*, both consisting in a start of aldehydeous combustion which suddenly changes into a very rapid carbonic one.

It is not so easy to give a sharp definition of the "flame front" as in fact a combustion process of gases becomes *visible* as a flame as soon as the oxidation has liberated enough heat to cause an emission of light by the combustibles and combustion products, and, remembering the studies of CALLENDAR on low temperature oxidation (T.I.C.), this point of the flame is probably not identical with that at which oxidation or combustion really starts. If, for instance, a flame is studied in a dark room, it will be seen that very often a pale bluish light is emitted by the gas stream before reaching the flame front where it is seen in daylight, which obviously is caused by previous oxidations. A photograph taken on a film which is sensitive to infra-red rays will show a flame front lying even still nearer to the burner tip.

From this it will be clear that *"flame front" or "flame beginning" is a very vague indication, which is entirely dependent on the conditions of observation.*

PRETTRE, DUMANOIS and LAFITTE (Comptes Rendus de l'Académie des Sciences 191, 1930, pp. 329 and 414) made experiments with a pentane and air mixture flowing through a tube and discovered two kinds of spontaneous ignition, one between 260 and 300° C. and the other between 660 and 670° C. Between 220 and 260° C. a silent luminous combustion was observed. The addition of anti-knock dopes proved to be able to suppress this first oxidation considerably. Other hydrocarbons showed a similar behaviour.

The flame front emits an intensive radiation of heat towards the oncoming air and oil mixture, and apparently preheats it in such a way that some kinds of primary oxidation at comparatively low temperatures take place, which may cause a faint emission of blue light. The length of the zone of this pre-combustion depends on the velocity of flame propagation. If this velocity is high, this length may be shortened to say 1 cm.; if the velocity is low and the conditions of pre-combustion are extremely favourable (fine pulverization, good mixing and strong preheating ),this zone may be considerably longer e.g. 10 cm. or more. If, however, the preheating of the zone of pre-combustion by radiation of the flame front gets too strong, this primary oxidation process (aldehydeous type) may be changed suddenly into a rapid carbonic combustion, at the same time causing the flame front to jump up stream, i.e. towards the burner tip. As, however, the combustion conditions of the mixture near the tip are less favourable than those at some distance (less preheating, less evaporation and less good mixing) the *carbonic combustion cannot maintain its front near the burner tip* and consequently will be blown back again by the gas stream. A new start of aldehydeous pre-combustion will be established until again it changes into the carbonic form, and thus a *rapid, periodical movement of the flame front* may be produced, resulting in the well-known *roaring oil flame* (see fig. 90).

To show the *analogy of roaring oil flames with knocking engines* it may be mentioned that the noise of such flames may be reduced considerably by influences which are known to have a certain anti-knock value in engines, such as:

1) *Admixing intermediate or final combustion products with fuel or air* (alcohols, aldehydes, water or steam, carbon dioxide, or recirculating flame or exhaust gases),

2) *Reducing the amount of oxygen* by reduction of its pressure or admixing an inert gas.

Such silencing influences fix the flame front at the *largest* distance from the tip according to the equilibrium of *aldehydeous* combustion and change the coarse, sharp and short "detonation flame" into a softer, milder and longer flame, which is more of the aldehydeous type.

---

# CHAPTER X

### SEVEN STAGES OF THE OIL BURNING PROCESS

As already mentioned, the technical oil burning process may be roughly divided into seven stages, viz.:

1) *Preparing the oil for combustion.*

2) *Preparing the air for combustion.*

3) *Vaporizing or gasifying the oil* (in so far as not already done in the first stage).

4) *Mixing of oil or oil gas with air* (in so far as not done in the first stage).

5) *Igniting the mixture and burning it* to carbon dioxide and water.

6) *Transferring the heat* to the materials to be heated.

7) *Carrying away the flue gases,*

which stages will be discussed in succession.

### A. *First stage: Preparing the oil for combustion*

It is necessary to prepare the oil for combustion in some way or other because the unprepared fuel oil will either burn with a thick smoky flame or not burn at all.

The object of this preparation is:

*a)* To remove *foreign matter*.

*b)* To measure or regulate the *quantity of oil* delivered to the flame.

*c)* To give the oil a *suitable surface* per unit of weight.

*d)* To give it a *suitable temperature or fluidity*.

*e)* To give it a *suitable motion*, i.e. velocity and direction.

These points will be discussed briefly:

*a) Preparing the oil in order to remove foreign matter.*

This is done by means of sieves or filters, the former being generally placed in the suction line of oil pumps in order to protect the valves, and the latter in the pressure line of the pumps before the burners in order to avoid clogging of the small holes in the burner tips. With the early steam-atomizing burners of the drooling type with large oil passages there was no need of filtering the oil and, as a matter of fact, they are even used at the present day for burning thick dirty fuels.

Although filtering seems to be a very simple matter it involves several problems, the discussion of which will be omitted here, with the exception of a typical difficulty occurring *when certain fuels are*

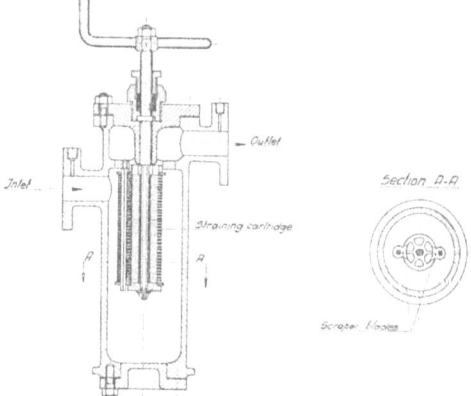

Fig. 91. Self-cleansing filter WALLSEND-HOWDEN (Patent of AUTO-KLEAN STRAINERS Ltd.).

*mixed*, none of which fuels would have given any difficulty at all if filtered *separately*. This difficulty arises if so-called *sludge* is formed after mixing, which may occur especially when "cracked fuels" containing very complicated unsaturated compounds are mixed with "straight-run oils." Certain reactions then liberate a precipitate which may entirely clog up the filters.

Filters are often made of the so-called *"self-cleansing"* type, but in most cases it is the operator who has to clean them, though sometimes merely by turning a spindle, (see fig. 91) an operation which may be done by an electric motor.

b) *Preparing the oil in order to measure or regulate its quantity.*

As compared with solid fuels, the use of fuel oil introduced many new problems, one of which being that it was found to be absolutely necessary for good combustion *to maintain continuously the right ratio of fuel to air.* When burning coal this was not so urgent, the fuel being shovelled on to the grates from time to time and the combustion process being controlled by regulating the quantity of air admitted under the grates by means of an adjustable damper. Thus with *coal burning*, being a process operating with a certain *stock of fuel in the combustion chamber*, a narrow regulation of the fuel-air ratio was not so imperative: A wrong adjustment of the air quantity admitted to the fire would cause lower efficiencies but would not prevent the coal-fire from burning.

On the other hand, *oil burning*, being a process operating practically *without a stock of fuel in the combustion chamber*, depends every moment entirely on the correct adjustment of fuel and air, the limits of which ratio are rather narrow: An oil flame may eventually be blown out by applying excess air or suffocated on the instant by shortage of air.

Some 25 years ago the designers of automobile engines were faced with the same problems of *correct fuel to air ratio adjustment* and an enormous amount of research work has been necessary to develop the present *carburettors*. Without doubt oil burning research is somewhat behindhand regarding this problem, but certainly there is no reason why less attention should be paid to a shilling's worth of wasted fuel oil than to a shilling's worth of wasted gasoline. The reason why this problem has been somewhat neglected in connection with fuel oil may be that incorrect adjustment of a carburettor seriously influences the performance of the engine, whereas incorrect adjustment of oil fires is less easily detected.

The first oil burning installations were exclusively operated by manual control, the fireman looked at the flame and at the chimney and adjusted accordingly, a practice that even now is very common for industrial oil burning. The development of small domestic oil burning units for automatic central heating has aroused interest in *automatic adjustment of the oil to air ratio*, the more so because a correct adjustment, being necessary for clean and safe combustion, was an indispensable safety factor for this domestic application. This in turn has had a good influence on the development of automatic industrial oil burning and on the knowledge of the advantages of such control, which details will be discussed later on.

For *pressure atomizing burners* the quantity of oil burnt is generally regulated by means of the *pressure at the burner tip*. A pressure-regulating, spring-loaded valve returns a variable surplus of the pumped oil back to the reservoir in order to keep the pressure constant. A serious objection to this method of regulating the capacity of pressure atomizing burners is that a reduced pressure means less perfect atomization. Means to avoid this drawback (PEABODY-FISHER, PILLARD D X, BARGEBOER's burner) will be discussed later on in Section III.

For *atmospheric burners* (= burners to which the oil is fed under atmospheric

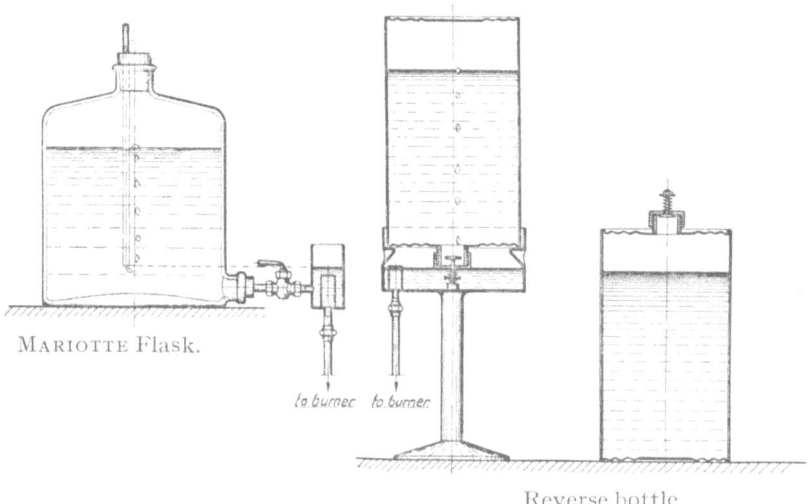

MARIOTTE Flask.

to burner   to burner

Reverse bottle.

Fig. 92. Constant Level Devices.

pressure) another method is used, which is very similar to the regulation of an oil flame by turning a wick up and down, viz., the so-called *open level control*, which is very commonly applied for centrifugal burners and for those of the vaporizing type (range-burners). A constant level of oil is maintained in the vaporizer of the burner by means of a MARIOTTE flask generally consisting of a *reverse bottle* in which the air is gradually admitted by the oil level itself (see fig. 92). The capacity of the flame is then increased by raising the oil level, thus increasing the quantity of oil exposed to evaporation.

### c) *Preparing the oil in order to give it a suitable surface.*

One of the reasons why the early burners consisting of heated containers with oil (NOBEL's trough, ASTRAKAN oil pans, BIDLE's oil fire, etc.) failed to produce a smoke-less combustion was that the oil had only a *very small surface per unit of weight*. It was gradually recognized that the only way to obtain a good combustion with less vola-tile oil was to give it a larger surface, viz., by spreading it out in a film, pulverizing it into drops or by evaporating it, thus giving the oil an opportunity of collecting sufficient radiant heat from the flame to evaporate and to be thoroughly mixed with air thereafter.

With respect to the magnitude of oil drops required for good combustion the following consideration is very instructive: *One kilogram of an ordinary fuel oil* will on an average need for its complete combustion about *15 cub. metres of air* (20° C. and atm. pressure), assuming 25% of excess air, which is a reasonably good figure for practical conditions. Taking the flame temperature at 1400° C. this quantity will correspond to 85.5 cub. metres of 1400° C. and atmospheric pressure or, if the specific gravity of the oil is supposed to be 0.9, it follows that one litre of oil will need 77,000 litres of air. In other words: *Under flame conditions a spherical oil particle will need for its complete combustion a quantity of oxygen equal to that contained by a surrounding sphere of $\sqrt[3]{77,000}$ or 42.5 times its diameter.*

Suppose it were possible in some way to suspend in air a number of motionless oil particles of equal size = 1 mm. diameter, it follows from the above that the *diameter of the "oxygen supplying spheres"* would be *about 42.5 mm.* or, if taking into account the space *between* the spheres too (oil particles arranged in a tetrahedron structure), the *minimum distance allowable between two neighbouring oil particles to ensure sufficient oxygen supply to them would be about 38 mm.*, or, in general, *about 38 times the diameter of the particles.*

This makes it clear that, first of all, *sizes of 1 mm. are much too large for oil drops of pulverizing burners*, as in such cases the oxygen supply can not be established in an efficient way, whereas further it is clear why the main factors to be considered in oil burning practice regarding oxygen supply are: *Turbulency of the air* and *penetration velocity of the oil drops into the air*, in order that the drops may strip off their vapour "coats" like comets.

*Evaporation* produces, of course, *the largest surface per unit of weight obtainable*, but it is not necessary to go as far as that, it having been proved in practice that increasing the oil surface, say 100 times, by pulverization was quite sufficient to obtain good combustion.

Oil burners are often divided into two classes, *vaporizing* burners and *pulverizing* burners, but it must be borne in mind that with burners of the second class the oil is evaporated too, though not in a special vaporizing device but *in suspension by air*. This method offers the advantage of there being only a *very small quantity of oil at a time exposed to the evaporation temperature*, which reduces the time available for cracking to a minimum. At the same time it avoids carbon deposition on the "vaporizing device" (= air), which is the main trouble with vaporizing burners.

Vaporizing burners generally have special evaporation devices which are often choked with carbon after a short time of operation if suitable oils be not used. These troubles increase rapidly if the vaporizers contain more oil, because in that case the fresh oil has to wait longer before it is burnt and therefore is longer exposed to the evaporation temperature and will form more carbon residue than if less oil is present.

With regard to these carbon troubles it may be stated that as a rule the best vaporizing burners are those operating with a *minimum* quantity of oil exposed at a time to high temperatures, and this too is more or less covered by the above statement *the surface of the oil (including the oil content of the vaporizing device) per unit of weight burnt should not be too small.*

*d) Preparing the oil in order to give it a suitable temperature and fluidity.*

With *vaporizing* burners it is necessary to heat the oil at least as high as the *average dew-point* of the oil in order to evaporate it. In most *gasifying* burners the oil is heated still higher in order to crack the high-boiling hydrocarbons into lower-boiling ones. With another class of burners the oil is pulverized and then partly evaporated and cracked, forming an *oil mist* by cooling the resulting gases.

*Pulverizing burners* (operated by oil pressure or centrifugal means) for heavier oils usually need preheating too, in order to improve pulverization, because the magnitude of the drops is mainly governed by viscosity and surface tension, both properties decreasing with increasing temperature.

Moreover, it is often necessary to preheat heavy oil in order to reduce the resistance of filters or to draw it from the bunker tanks.

*e) Preparing the oil in order to give it a suitable motion.*

With *vaporizing* and some *gasifying burners* it is generally sufficient to admit the oil *by gravity* into the vaporizing or gasifying device, and it is comparatively easy to produce as a second step a thorough mixture of oil vapour or oil gas and air.

*Pulverizing burners*, however, give much more trouble with respect to this point. First of all the oil must be given such a motion that it *breaks up into fine drops*, which may be done either by means of the velocity effect of an auxiliary medium, such as steam or compressed air, or by a pressure or velocity effect of the oil itself. But this is not sufficient. It is necessary to give the oil drops a suitable *direction* and, moreover, a suitable *velocity* in order that a *mutual penetration* of oil drops and air takes place. This effect is generally obtained by producing a hollow conical spray of oil.

Thus there are three important factors regarding pulverizing burners, viz.,
1) Efficient pulverization,
2) Correct direction and
3) Sufficient penetration,
which are necessary not only for good combustion, but also to obtain a good shape of flame.

### B. *Second stage: Preparing the air for combustion*

It is necessary to prepare the air in order to:
*a)* Measure or regulate its *quantity*.

*b*) Impart suitable *motion* to it.

*c*) Give it a suitable *temperature*.

These points may be discussed briefly as follows:

*a*) *Preparing the air in order to measure or regulate its quantity.*

Attention has been drawn to the fact that it is necessary for good combustion to maintain a constant ratio of air and oil throughout the oil burning process. Whereas the oil may be fairly easily controlled or measured, the air gives much more trouble.

If the air is available at sufficient pressure, e.g. by mechanical means, compressors, blowers, etc., well-known *orifice measurement* (see fig. 93), which may be

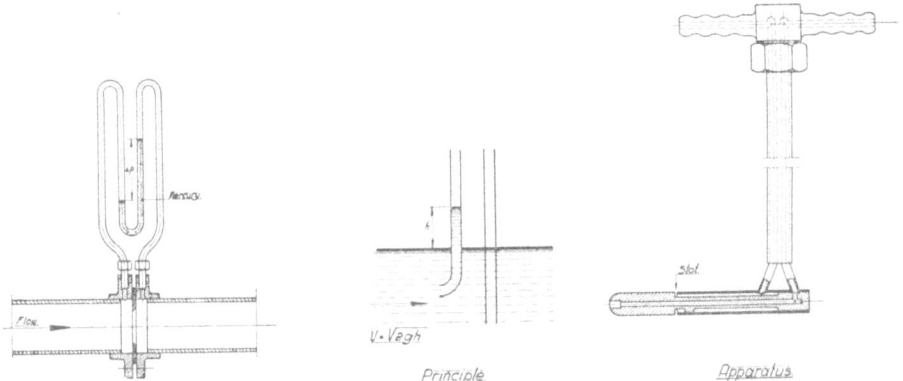

Fig. 93. The Principle of Orifice Measurement.

Fig. 94. PITOT tube Measurement.

connected with automatic adjustment, might be applied, but as an accurate measurement involves at least some 5 inches of water pressure loss, it is clear that this method is objectionable for low-pressure fan installations. Generally the delivery of air by low pressure fans is measured with PITOT tubes (see fig. 94) measuring the velocity at various points of the cross-section of the air duct from which data the total quantity of air may be calculated. This measurement, of course, causes no pressure loss, but is rather laborious, and can not be used for automatic adjustment of the quantity of air fed to the fire.

For measurements it is often preferred to accept the statements of the manufacturer printed on the shield of the fans, but it should be borne in mind that the capacities quoted generally apply to the fans blowing *freely into the atmosphere*. If, however, the fans are connected to furnaces, their capacities may be increased by the natural draught of the chimney or decreased by the friction and acceleration of the flame and flue gases through combustion chamber, boiler and chimney. In this connection it is very instructive to read the publication of SEELY and TAVANLAR in "Heating, Piping and Air conditioning" of May 1931, page 419.

It is still more difficult to measure a quantity of air sucked in by *chimney draught*, as even PITOT tubes then give readings which are too small to be reliable. As no better method has yet been found for the *direct* measurement, in this case the air is usually measured *indirectly*, e.g., from the $CO_2$ content of the flue gases.

The above shows that it is not so easy to *measure* the quantity of air consumed by an oil fire but it is still more *difficult to regulate it automatically*, which in fact forms a problem which is anything but solved. If one thinks of the complicated modern engine carburettors with several jets, holes, cones, valves, compensators etc. which

Fig. 95. Air registers of WALLSEND-HOWDEN.

are necessary for "automatic oil firing" with approximately constant fuel to air ratio in automobile engines, no further explanation is needed to show why most industrial oil burners may yet be developed in this direction.

It may be clear too that all these difficulties are aggravated in the case of weak air pressures, such as from *natural draught*, and this was one of the reasons why domestic oil burners operated by natural draught alone, although attractive because of their simplicity and cheapness, never have been able to maintain an important position on the market: *it was impossible to keep the ratio of oil and air reasonably constant under various loads and draught conditions*. Even for mechanical draught burners this was a hard job and only the introduction of the *"on-off" regulation system*, which will be discussed later, making it possible to operate the burner exclusively under full load conditions, offered a practical solution of the problem.

*b) Preparing the air in order to impart suitable motion to it.*

It is not sufficient to pass a measured quantity of air together with a measured quantity of oil into the combustion space, for the air must have a *suitable motion* too. With certain atomizing burners, for instance, it may be necessary to pulverize the oil by means of the velocity effect of the air. Various constructions for attaining this in an efficient way will be discussed in section III. A thoroughly homogeneous mixture of oil mist and air must then be produced, which can only be done efficiently by giving the air a turbulent motion. In fact *turbulency is a most important factor* in oil burning practice, not only with atomizing burners but also with most vaporizing burners.

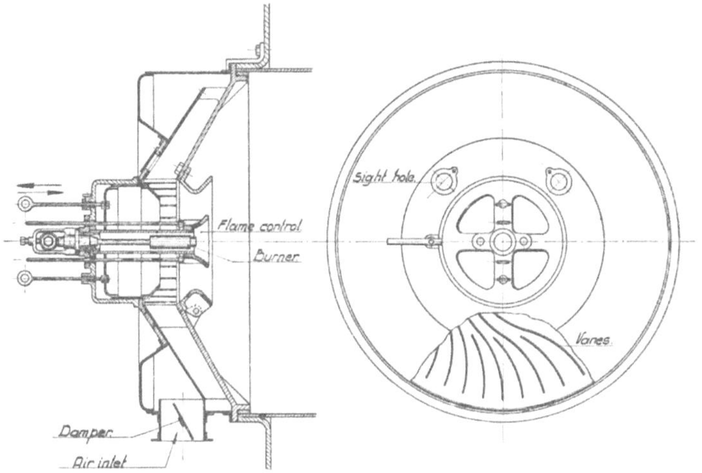


Fig. 96. The WHITE oil-front with air-duct.

If, for instance, the oil is prepared for combustion by spreading it out in a thin layer exposed to the radiant heat of the flame itself (such as is done with the PRIOR burner of fig. 61) it is necessary *to blow a strong current of air onto the oil surface in order to remove the oil gas above it* as soon as it is evolved. Very often a rotary swirling motion is given to the air in order to improve mixing, but it must be pointed out that a rotary motion of large diameter may sometimes consist of parallel layers (a circular but *laminary* flow), which do not have at all the desired effect.

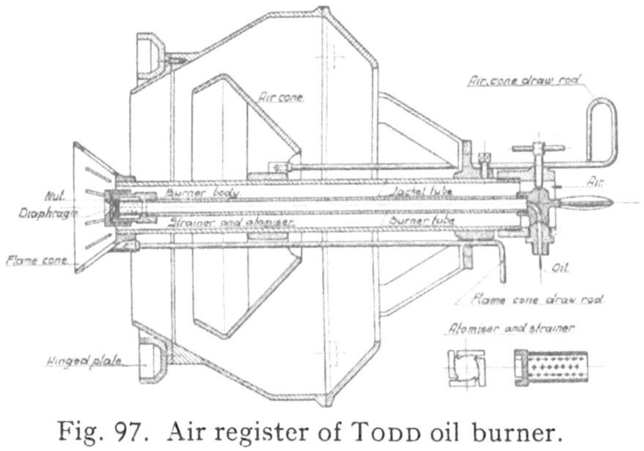


Fig. 97. Air register of TODD oil burner.

Range burners are examples of devices which ensure good combustion with laminary flow of air, which is mainly due to the soot-free, aldehydeous combustion.

Oil burning practice has shown the importance of the use of *air registers* (see figs. 95, 96, 97) and several oil burner manufacturers are aware of the necessity that air and oil should be prepared for combustion together, and they therefore sell both *the burner and the air register as one unit*, which is very good practice indeed. Figs. 96 & 99 show a common construction to give the air of mechanically atomizing domestic oil burners a rotary motion at the burner mouth by means of a number of vanes.

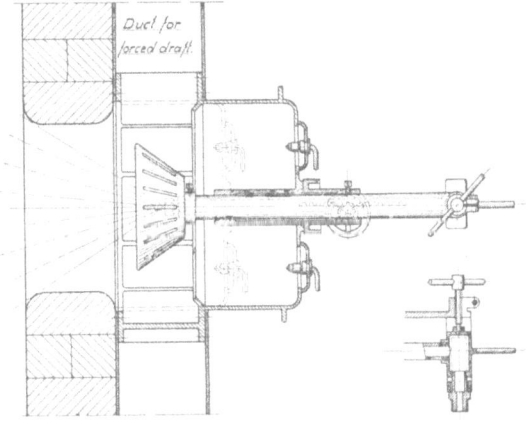

Fig. 98. Air duct of BETHLEHEM oil burner.

If the *total* quantity of air necessary for combustion of the oil is blown directly around the burner tip there will be a certain risk, especially when the air is not preheated, of the flame's being blown too far away from the tip, resulting in bad combustion. This may often lead to the occurrence of carbon monoxide and bluish-white oil smoke in the flue gases, due to too much cooling of the first part of the flame, and for this reason it has been found to be good practice to split up the air into two portions: the *primary air*, which is blown directly around the burner tip, causing a short hot zone of

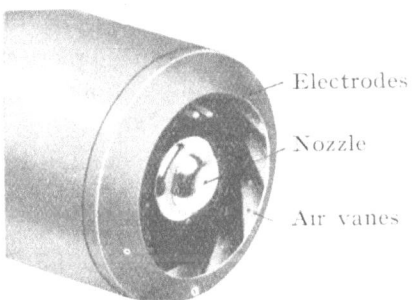

- Electrodes

- Nozzle

- Air vanes

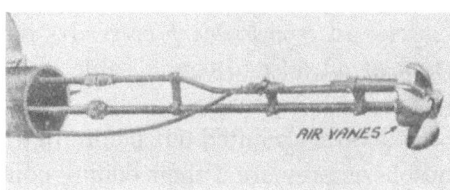

AIR VANES

Fig. 99. Air vanes of OIL-O-MATIC oil burner and of PARWINAC domestic oil burner.

incomplete combustion at the first part of the flame producing mainly intermediate combustion products (carbon monoxide, hydrogen, methane), which are burnt further, on where the remainder of the air, the *secondary air*, is admitted to the flame.

Most air registers allow for suitable adjustment of primary, secondary and even sometimes of tertiary air, permitting a perfect control of the combustion process. The same division of the air into two portions may be found with air-atomizing burners to be discussed in Chapter X of the third Section. As soon as carbon monoxide is found in the flue gases, attention should be given to the adjustment of the secondary air, which preferably should be admitted to the middle of the flame, at the same time ensuring a through turbulent mixing.

*c) Preparing the air in order to give it a suitable temperature.*

Considering that the heat capacity of the quantity of air theoretically necessary for the combustion of 1 kilogram of oil is about *seven times* the heat capacity of the oil, it follows that it is far more efficient to preheat the air by one degree than to raise the temperature of the oil that much.

Good preheating promotes good combustion by resulting high temperatures and favours the aldehydeous type of combustion, which prevents smoking. The temporary success of the so-called *mushroom type* (see fig. 53) of domestic oil burner in America was mainly due to the air being well pre-heated, which resulted in excellent ignition conditions and a good combustion, provided draught conditions were perfect.

## C. *Third stage: Vaporizing or Gasifying the Oil*

For the sake of clearness of discussion the combustion process has been divided into seven stages, but in practice it will be scarcely possible always to keep them strictly separated, as for instance vaporization or gasification may be used also as a means of preparing the oil in accordance with stage I.

Sometimes a difference is made between *evaporation*, as a purely physical *reversible* process, and *gasification*, which generally denotes in oil burning practice a combination of evaporation and a complex series of chemical reactions called "thermal decomposition" or "cracking", producing permanent gases. Gasification is, of course, an *irreversible process*. Its definition may be enlarged by including production of *oil mist* of condensable oil vapours or of partly cracked "oxygenated" or polymerized compounds.

It has been pointed out before that hydrocarbons can only be evaporated under atmospheric pressure if their boiling points are lower than the temperatures at which their thermal destruction starts. In fact it is this property which draws the line between distillates and residues. For most hydrocarbons this temperature limit lies at about 350° C.

Devices for evaporation and gasification will be dealt with in the third Lecture.

D. *Fourth stage: Mixing of oil or oil gas with air*

The same remark as was made under the discussion of the third stage may be applied here again, as very often mixing and preparation of the oil are one. If the method of pulverization is followed, the oil will be mixed with air first and gasified afterwards.

The method of *evaporation* consists in admixing the air after the oil has been evaporated, and the method of *gasification* may either gasify the oil first and mix the air afterwards or admix a *limited* quantity of air with the oil, causing an *incomplete pre-combustion* which gasifies the oil more or less completely by the heat evolved, and then mixing the rest of the air. This last method, which has found some application in *"chemical carburettors"* for gasoline engines running on kerosene or gas oil, usually produces an oil mist, because admixing secondary air cools and partly recondenses the products of incomplete pre-combustion. Devices will be discussed in Section III.

*Turbulency* is necessary for good mixing, but it causes more resistance to the gas flow and therefore *requires more applied energy* to make up for the pressure loss. It goes without saying that the weaker the source of energy is, the more difficult will it be to obtain turbulency, and thus with natural draught of the ordinary low chimneys of dwelling houses etc., if used as a motive force to operate an oil burner, many difficulties must be overcome to secure an efficient, clean combustion, especially if the burner is to be started with a cold chimney.

E. *Factors governing the Type of Combustion*

Before passing on to the discussion of the stage of ignition, it is well to remember some essentials of the theory of combustion of hydrocarbons.

We have seen that there are *two extremes:* the *aldehydeous combustion*, which produces a *blue* flame which is perfectly free from soot, and the *carbonic combustion* producing a *yellow* flame having a certain tendency to smoke. For industrial as well as domestic purposes smoky flames are detrimental, not only because of the loss of latent heat of unburnt carbon, which in fact will be comparatively small in most cases, but more so because a layer of soot on the walls of boilers etc. will reduce the heat transmission considerably and thus cause lower efficiencies by higher stack losses.

Moreover, smoking stacks are a nuisance to the neighbours and have been prohibited in most municipalities. There is little doubt that the development of oil burning has contributed considerably towards diminishing the soot content of the atmosphere above cities, at the same time probably reducing the occurrence of fog.

*Aldehydeous combustion* seems to be ideal from the point of view of absence of soot; however, it has some drawbacks: *Firstly*, its blue flames have considerably *less radiative power* and, *secondly*, up to the present the means of realizing it are rather restricted, e.g. with the "reverse flame principle" comprising flames of air in atmospheres of hydrocarbons, which method implies that *the oil should be evaporated or gasified first*. As this may be obtained with distillates only, the applications of this principle are rather narrow.

Whereas the aldehydeous combustion in its pure form consequently has little industrial value, we are able on the other hand to control the combustion conditions of oil flames in such a way that, instead of being purely carbonic and smoky, they are made "clean". This method might be called: *"promoting the aldehydeous process in a 'background' of destructive combustion"* and comprises the practical means used to reduce the tendency to soot as far as may be thought necessary. It is certain that between burner nozzle and flame a great many important reactions take place which have considerable influence on the character of the combustion.

In order to obtain *sufficient aldehydeous orientation*, it is necessary to:

1) *Ensure good pulverization* (surface).

2) *Preheat oil and air in a suitable way*.

3) *Mix oil-mist or oil-gas and air well*.

4) *Expose the mixture during not too short a time to a gradually increasing temperature* (time and temperature).

As an experienced American oil burner expert briefly expressed it: "In order to obtain clean combustion the oil must be given *time, surface* and *temperature*", which in fact are the factors necessary for "pre-oxygenation" and the aldehydeous type of combustion.

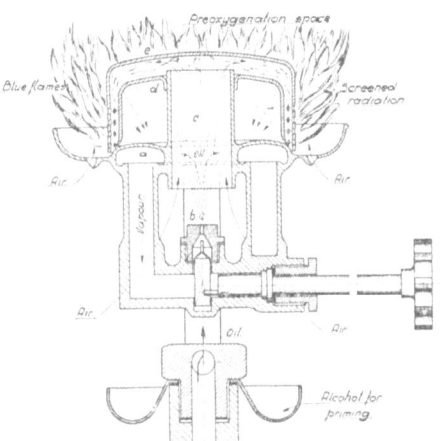

Fig. 100. Typical Modern Vapour jet Burner (SVEA).

F. *Blue flame principles*

There are three methods of burning hydrocarbons with blue flames:

*a) By exposing during not too short a time a thorough mixture of gaseous hydrocarbons and air to temperatures between 200 and 300° C.* in order to start *preliminary oxidations*, such as formation of peroxides, aldehydes or hydroxylated compounds (often called in literature: "pre-oxygenation") *without yet causing cracking*.

Examples of this method are the modern blue burner devices of SIEVERT, PRIMUS, OPTIMUS, RADIUS etc. Fig. 100 shows a SVEA burner. The oil gas generated in (*a*) issues from the orifice (*b*), mixes with air in tube (*c*) and is preheated to about 250° C. in the space between the caps (*d*) and (*e*), where previous oxidation takes place. The pre-oxygenated mixture issues from the holes (*f*) and burns according to the *alde-*

*hydeous* process with perfectly *blue* flames. The diameter of the holes in cap (*e*) has been so chosen that the velocity of flow from the holes exceeds the speed of flame propagation and consequently the flame cannot enter into the space between (*e*) and (*d*). If the cap (*e*), however, is heated too much, e.g., by an extra gas flame, a *carbonic* combustion may be started inside the space between (*d*)

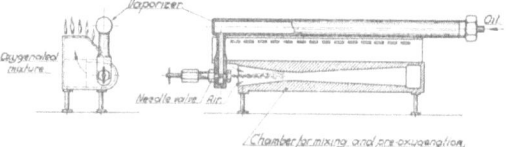

Fig. 101. Blue flame vapour jet burner (TAPP).

and (*e*), causing the apparatus to smoke badly. The same happens if the blue flames are allowed to slip backwards into the space between (*d*) and (*e*) by slightly lifting up cap (*e*).

A similar blue burner device is shown in fig. 101, which speaks for itself.

*b) By means of the "reverse flame" principle*, i.e., by burning "flames of air" in an atmosphere of hydrocarbon gas. The basic idea of this principle is that the centre of such a flame exposed to the *convergent* radiation of heat from the surrounding flame front then consists of non-decomposable air, whereas the unstable hydrocarbons outside are only exposed to a far less intensive *divergent* and dissipating radiation of heat, and are, therefore, protected against decomposition. Examples of this method are the modern *"range burner"* types of LYNN, SILENT-GLOW, STERLING, HALLER, GAS-CONE, GLARUS, etc., which devices in the last few years have enjoyed a rapid development for heating by kerosene. (fig. 84)

*c) By burning or converting the carbon particles in the oil flame almost at the very moment they are formed*, which may be done either by blowing *large quantities of excess air*, preferably preheated, into the flame or by artificially *raising the water vapour concentration* of the flame gases, thus influencing the equilibrium of the water gas reaction, according to which carbon and water vapour may be converted into carbon monoxide and hydrogen, both being gases burning with scarcely visible flames.

Although in these cases flames are almost transparent and show a blue colour, the intermediate presence of the carbon particles may be recognized by the continuous spectrum emitted. Consequently, this kind of combustion still belongs to the carbonic type and might be called a *"steam-corrected carbonic combustion"*.

### G. *Fifth stage: Ignition and Combustion*

Ignition may be started by a flame, spark or glow-plug, and if the ignition conditions of the mixture are sufficiently good, the combustion will be able to maintain itself and a continuous oil flame occurs, provided, moreover, that the form of the stream allows of establishing a stable equilibrium between combusiton speed and velocity of flow. If this last condition is not fulfilled, a discontinuous oil flame with intermittent detonations will occur, or the flame will be extinguished completely. These periodical detonations may occur either in very rapid succession, resulting in a *roaring* flame, or at much slower rates e.g. once per second, giving a so-called *"panting"* flame. To avoid this phenomenon, the mechanism of which has been explained before, page 97, it is advisable to eliminate as far as possible accelerations of flame gases by suitable design of combustion cones.

Whereas a *minimum* flash point for fuel oil was originally fixed as a safety precaution, the use of automatic spark ignition of modern domestic oil burner equipment made it necessary to prescribe *upper* limits for the flash point too, in order to ensure ready ignition.

The combustion itself has been fully discussed in previous chapters and will therefore be passed over here.

H. *Sixth stage: Transferring the heat of combustion to the materials to be heated.*

At present oil burning is used for several widely divergent purposes, such as:

A) *Steam-raising* for marine or mobile and stationary shore installations for power or heating purposes.

B) *Water-heating* for industrial (chemical industries, laundries, bleaching) or domestic use (household, central heating).

C) *Air-heating* for industrial purposes (core-drying of foundries, enamelling, artificial silk and tea-drying) or for domestic use (indirect central heating, direct room heating and heating of hot-houses).

D) *Burning* (brick and pottery industry, cement kilns and crematoria).

E) *Roasting* (coffee and cocoa roasting).

F) *Baking* (bread-baking, and also fish-frying).

G) *Melting* (glass industry, steel and other metallurgical industries, forging, tinning, welding).

H) *Special processes* (chemical industries and heat treatment).

Each application has its own peculiarities and special demands and there will, therefore, always be several exceptions to the general rules. For instance, when in the following it is stated to be of the utmost importance to keep the amount of excess air as low as possible, this does not hold good for air-heating if the hot air and flue gases are mixed together, such as is done in tea-drying. On account of limited space the following discussions will be restricted to one of the most common applications of oil burning, i.e., *heating of a steam or hot-water boiler.*

If high efficiency of a boiler equipment is desired, the *limited* dimensions of the heating surfaces of the boilers make it necessary to use the highest allowable flame temperatures, and thus in the first place to supply the smallest possible amount of excess air. For, if we had *infinitely large* heating surfaces at our disposal, the amount of excess air would be of no importance at all, as we should always be able to transfer all of the heat of the flue gases through the heating surfaces of the boiler. From this it follows that a *small amount of excess air is all the more urgent* according as the heating surfaces are more limited, or, in other words, according to the *load of the heating surface* (number of heat units to be transferred per square foot of the heating surface).

### 1. *Convection versus Radiation*

There are three different ways of transferring heat, viz., by conduction, convection and radiation.

1) *Conduction* consists of a flow of heat transferred by molecules scarcely moving from their places but vibrating around their main positions. It occurs if heat is transferred inside a solid material.

2) *Convection* is a kind of transportation of heat by means of a flow of molecules, i. e., by the contact of flowing gases or liquids with walls.

3) *Radiation* is a kind of heat transfer from the molecules of one substance to those of another without the intermediate help of the molecules between, and it takes place merely by rays. Thus it may be possible to eliminate convection by applying an absolute vacuum to the space between the double walls of a DEWAR's flask, but heat transfer by radiation will remain and can be reduced only by using mirror surfaces.

The major factors governing *heat transfer by convection* are the

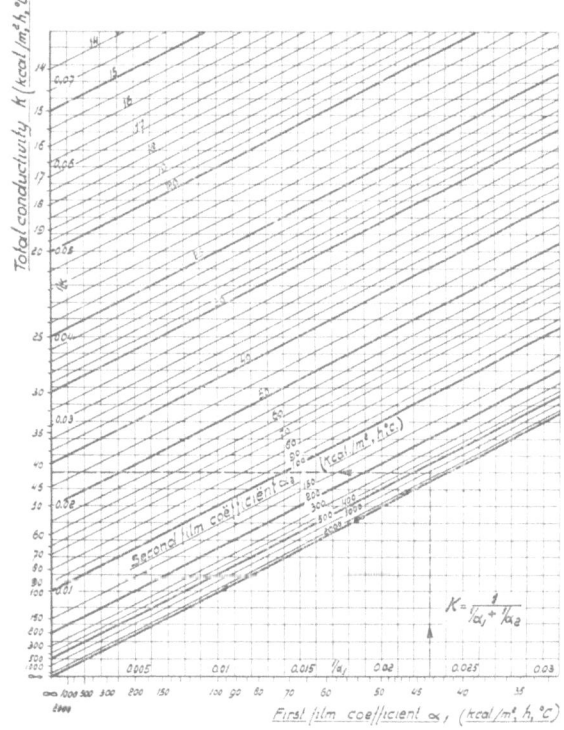

Fig. 102. Heat transfer by convection and conductoin. (After „Wärmetechnische Arbeitmappe" V. D. I. Verlag).

*temperatures* and the *velocity of the gases*, since heat transfer by convection is seriously obstructed by a film of gas adhering to the walls, and any influence tending to decrease the *thickness of this film* will also increase the heat transfer rate. Causing the hot gases to impinge on the surface to be heated is the most common method of doing this, and it is clear that *turbulency* is a most important factor not only for good combustion but also for efficient heat transfer by convection.

The major factors governing *heat transfer by radiation* are the *temperatures of the radiating and receiving bodies*, their *blackbody coefficients* and the *transparency* to heat

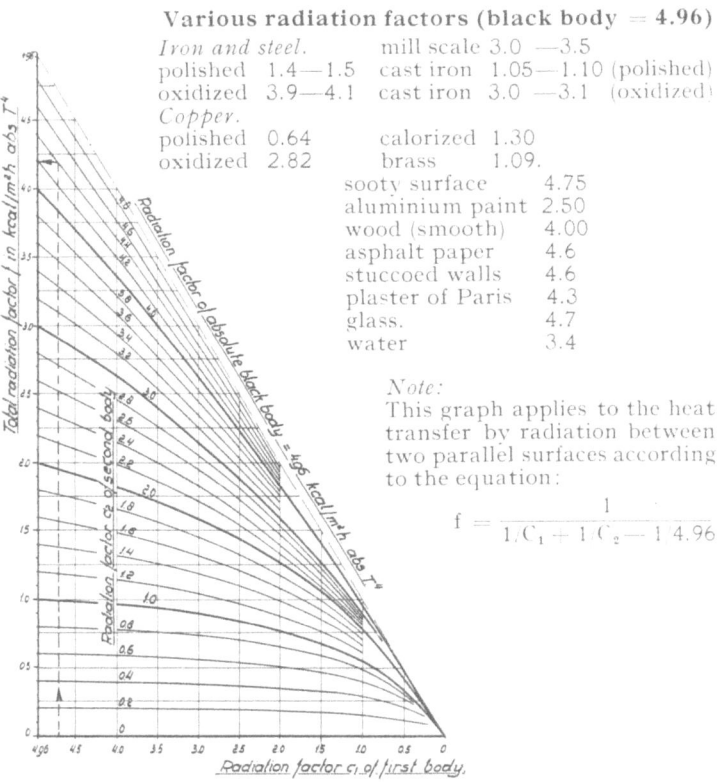

Various radiation factors (black body = 4.96)

*Iron and steel.*          mill scale 3.0 —3.5
polished 1.4—1.5   cast iron 1.05—1.10 (polished)
oxidized 3.9—4.1   cast iron 3.0 —3.1 (oxidized)
*Copper.*
polished 0.64          calorized 1.30
oxidized 2.82          brass        1.09.

sooty surface        4.75
aluminium paint 2.50
wood (smooth)     4.00
asphalt paper        4.6
stuccoed walls       4.6
plaster of Paris     4.3
glass.                       4.7
water                       3.4

*Note:*
This graph applies to the heat transfer by radiation between two parallel surfaces according to the equation:

$$f = \frac{1}{1/C_1 + 1/C_2 - 1/4.96}$$

Fig. 103. Heat transfer by radiation.
(After „Wärmetechnische Arbeitmappe" V. D. I. Verlag).

rays of the gases between. The velocity or turbulency of these gases has no influence on the heat transfer by radiation.

This kind of heat transfer has received more and more attention in the last few years, as it has gradually become recognized that tremendous heat transportation per second and per unit of surface can be obtained by it, the figures many times exceeding those obtainable by convection alone. Whereas the amount of heat transferred per second by *conduction or convection* is proportional to the *first* power of the temperature difference (fig. 102), heat transfer by *radiation* tends to be proportional to the difference of the *fourth* powers of the absolute temperatures of the bodies emitting and receiving the radiation.

The two graphs of figs. 103 and 104, which are composed from similar graphs of the "Wärme-technische Arbeitmappe'" (V.d.I.Verlag), give a good impression of the tremendous importance of radiation at high temperatures. If, for instance, the temperature of the radiating body is supposed to be 1400° C. and that of the receiving body 500° C., with a blackbody coefficient of 4.2 (oxidized boiler wall), it follows from fig. 104 (dotted lines) that the heat transfer amounts to 315,000 kg. cal. per hour per sq. metre. If for the sake of comparison this figure is divided by the temperature difference of 900° C., it results in a heat transfer of *350 kg. cal. per hour per sq. metre per °C.*, which is considerably higher than that for heat transfer by convection, which for non-condensing gases usually does not exceed *100 kg. cal./hour/sq.m./°C.* (See fig. 102.) Obviously radiation is a property which is very useful for transferring large amounts of

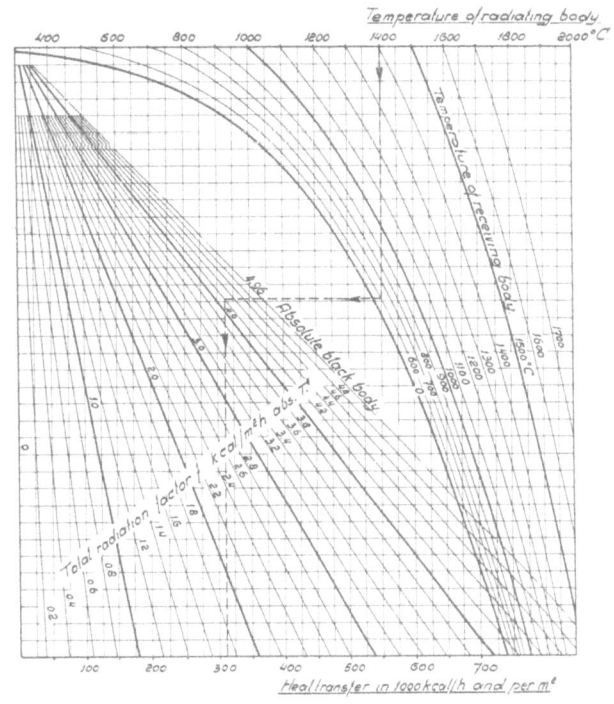

Fig. 104. Heat transfer by radiation. (After „Wärme-technische Arbeitmappe" V. D. I. Verlag).

heat units per square foot of heating surface, *but only from about 600° C. onwards.*

Radiation may, of course, take place not only between two solid surfaces but also between gases, and in oil burning practice we have to do with: 1) the oil flame "surface", 2) the boiler surface and 3) the flue gases between them. It is very difficult to give a definition of the surface of an oil flame, being a semi-transparent gaseous body, and therefore *exact* calculations of oil flame radiation are almost impossible, whilst, moreover, the available data on the reflective power of flames are meagre and mostly inaccurate. For practical purposes empirical formulae have been developed such as may, for instance, be found in "Gas Flow and Radiation" by LUBBOCK (Journ. Inst. Heating & Vent. Eng. Aug. 1933).

It is not correct to think that it is only the oil flame that radiates towards the boiler surface, since radiation takes place just as well in the opposite direction, the former sometimes being called "hot radiation" and the latter "cold radiation". In fact the amount of heat transferred by radiation is equal to hot minus cold radiation minus heat absorbed and plus heat radiated by the flue gases in between, or in other words: equal to the remainder of a *radiation equilibrium*.

Radiation is dependent not only on the temperatures but also very much on the reflective and absorptive properties of the surfaces, which makes calculations extremely difficult and unreliable, so that they will be omitted here.

As radiation follows straight lines it is necessary for an efficient heat transfer to arrange the *oil flame* in such a way that it *"faces" as much as possible of the heating surface*, and it is clear that from this point of view a refractory lining on the boiler surface may detract much from the efficiency.

## J. *Refractory Material*

In early times it was thought to be impossible to burn oil without refractory combustion chambers, which experience was founded on the bad results and damage caused by impingement of the oil flames on the unprotected boiler surfaces. If an oil flame, instead of being allowed to develop freely, is forced to touch a cold wall (in this respect a boiler wall of say 300° C. is "cold" too), an abnormal heat transfer will be established on that spot, which may cause such intensive steam production on the water side that numerous steam bubbles thus evolved will prevent the water from flowing by, and from properly cooling the metal. Thus a "dry spot" occurs, which is only cooled by steam, resulting in a local superheating of the metal.

Moreover, the combustion process may be prematurely stopped by the contact of flame gases with the "cold" wall, resulting in condensation of cracked oil products, which after being deposited, will be rapidly cracked further, leaving a coke deposit. Carbon, being a bad heat conductor, will form an insulating layer on the wall and, if the wall consists of iron, probably a part of it tends to diffuse into it, forming carbides, which are likewise bad heat-conductors and which are detrimental to the strength of the metal, making it extremely brittle. The combined action of both causes of superheating finally leads to the well-known damage.

It is obvious, however, that this evil may be avoided merely by taking precautions against oil flames touching the walls. As a rule, *"Oil flames must be allowed to have their full swing"*. If in a certain case it is technically impossible to avoid the contact of an oil flame with a wall, it is necessary to apply refractory material there in order that the temperature of this material in contact with the flame may rise sufficiently to avoid prematurely stopping the combustion process, without any danger to the strength of the structure. In this connection it must be pointed out that 200 or 300° C. must be considered as "cold", about 900 to 1000° C. being the lower limit of what may be called "sufficiently hot" for direct flame contact.

In order to avoid the contact of oil flames with cold walls some empirical figures have been found to be useful to give a rough estimate of the volume of the combustion chamber necessary to burn a definite quantity of oil. The *"combustion chamber"* is defined as that *space of a boiler which is available for free flame development*, thus excluding narrow passages such as tubes etc. For *industrial* furnaces this figure very often lies between *1 and 2 lbs. of oil burnt per hour per cub. foot of combustion chamber* (or 16 to 32 kg./cub.m./hr.), whereas *domestic* heating is often based on *20,000 B.T.U. release of heat or 1.1 lbs. of oil burnt per cub. foot per hour* (or 18 kg./cub.metre/hour).

Until about 1920 the general idea concerning the use of refractory linings for oil furnaces was that it was meant to *protect the walls*, but from that date onwards oil burner designers gradually became convinced that it served just as well to *protect the flame*, and in this they were quite right, because any glowing material causes a strong reflection of radiant heat onto the flame and has a *strong catalyzing power for combustion*, which may efface the irregularities and, to some extent, the incompleteness of the burning process. Thus a fire bridge of refractory material strongly promotes a steady, stable oil fire.

However, clean combustion is one thing, but efficient heat transfer decidedly the main point, and consequently it must be borne in mind that *refractory material must be applied as sparingly as possible* in order to leave the way free for heat transmission by radiation, as all heat not transferred by radiation will be left to the flue gases to transfer by convection, which takes far more surface to attain the same efficiency.

From the above it follows that it is not at all correct to consider refractory lining as an indispensable element of oil burning. In fact it is only a *corrective measure* which may be necessary in some cases.

The hot walls of a refractory combustion chamber should "face" as much as possible of the "naked" boiler walls in order to throw the heat onto them. It is wrong to "shut up the heat" in the combustion chamber, as this raises the temperature there unnecessarily (dissociation) and makes high demands on the quality of the refractory materials, without any advantage to speak of.

### K. *Flame temperatures*

Although, physically, it may not be quite correct to speak of *the* flame temperature, because an oil flame in fact consists of a heterogeneous conglomeration of glowing gases, which, according to the various stages of combustion, radiation equilibria and motion, may show different and fluctuating temperatures for every spot, it is technically possible to regard the flame as a gaseous body with an average temperature.

Thus the flame temperature, which is so very important for heat-transmission by radiation, is mainly governed by five influences:

1) *Degree of pre-heating.*
2) *Rate of combustion.*
3) *Degree of dissociation.*
4) *Quantity of non-reacting gases.*
5) *Equilibrium of radiation,*

which points will be considered below:

1. *Pre-heating.*

If the combustibles, oil and air, are pre-heated, say x degrees, the basis for the chemical heat-evolving process is raised x degrees too. Of course, in that case the flame temperature will be raised too, although not exactly by the same amount of x degrees, because the increased dissociation of combustion products gradually decreases the effect of pre-heating at higher temperatures, as will be seen further on under 3).

2. *Rate of combustion* may be expressed either as the speed of propagation or as the quantity of heat units set free per second in the unit of flame volume (cal./sec. per cub.m.). This last version might be called: *Combustion concentration.*

Its influence may be illustrated by comparing the combustion of *acetylene* with that of *methane*. Although the heating value of acetylene (upper: 13,830 cal./kg., lower: 11,750 cal./kg.) differs but very little from that of methane (upper value: 13,400 cal./kg., lower: 12,075 cal./kg.), the maximum flame temperatures practically obtainable by burning these gases with oxygen are about *3000° C. for acetylene* as against only *about 2000° C. for methane*. This high temperature is measured for the oxy-acetylene flame at the apex of the small central white cone, at which point the flame consists almost entirely of carbon monoxide surrounded by a jacket of hydrogen. The temperature at the apex of the flame is too high to allow the hydrogen to combine with the oxygen, and obviously the heat evolved there is caused merely by the *combustion of carbon to carbon monoxide, the hydrogen almost being a non-reacting gas*, whereas carbon dioxide is practically absent. The flame is, therefore, hot enough to melt iron and steel, and at the same time sufficiently reducing to protect the fused metal from oxidation during the welding process. This makes it clear why the oxy-acetylene blow-pipe flame, apparently not being based on the combustion of hydrogen, even shows a higher flame temperature than the oxy-hydrogen flame (about 2000° C.).

The high flame temperature of acetylene, and probably its high combustion speed too, is moreover due to an *extra storage of latent heat* which is present in the $C_2H_2$-molecule in its *threefold bond* between the carbon atoms. This "unnatural", forced situation is maintained by so much internal stress that during combustion its release liberates an appreciable quantity of heat. The heat liberated by burning the carbon and hydrogen content of one kg. of acetylene, viz., 0.923 kg. of carbon and 0.077

kg. of hydrogen, *separately* would amount only to 9670 kg. cal., whereas the *measured* heat value of 1 kg. of acetylene is 11500 kg. cal., the *difference of 1830 kg. cal. per kg. of acetylene being the extra yield from the threefold bond*. It may be observed that all heat values quoted are related to complete combustion to carbon dioxide and water vapour.

The difference between the maximum temperatures of acetylene and methane flames, which makes the latter absolutely inferior for welding purposes, is caused by the very short flame of acetylene, due to the *extremely rapid carbonic combustion*, whereas methane burns with a longer flame according to a much slower combustion process.

Moreover, methane as compared with acetylene contains four times as much hydrogen, which — as has been explained above — releases but very little heat at such high temperatures. Then, again, the heat value *per cub.m.* of methane (upper: 8940 cal./cub.m., lower: 8055 cal./cub.m.) is considerably lower than that of acetylene (upper: 15,000 cal./cub.m., lower: 12,570 cal./cub.m.) and consequently *methane will require higher velocities to produce flames of equal heat release per sec*. Thus practically the same amount of latent heat may be liberated in a *"concentrated"* form when burning *acetylene* or in a more *"diluted"* form when burning *methane*, which shows that the heat content of a fuel says very little about the flame temperatures obtainable with it.

It is very common in literature to speak of *"theoretical flame temperatures"*, which are temperatures that would be obtained if the combustion took place *instantaneously, completely and without loss of any heat to the surroundings;* that is, if all the heat of combustion were used to heat the gaseous products of combustion. The theoretical flame temperature is found by dividing the sum of the latent combustion heat and sensible heat of the combustibles (air included) by the sum of the products of weight and specific heat of the resulting combustion products (nitrogen and excess oxygen included), but it will be clear from the above that the resulting figure cannot be quite correct, one most important factor, the *time element*, having been *disregarded* in this calculation.

### 3. *Dissociation*.

A second reason why the actual flame temperatures are lower than these "theoretical temperatures" is that at high temperatures the previously formed combustion products, CO, $CO_2$ and $H_2O$, will be partly dissociated again, this being accompanied by a *consumption of heat*. This *dissociation reverses the process of combustion* and in this sense it may be said that the rise in temperature during combustion counteracts the affinity of the hydrocarbons for oxygen, and tends to reduce the temperature of combustion to below that computed on the assumption that combustion is complete. It also follows that there is a limit to the maximum attainable temperature by the combustion of a given fuel.

The dissociation products may combine again further on and thus give back again the consumed heat and, although consequently there may be no final loss of heat, this phenomenon will *flatten the flame temperature curve to a lower maximum and lengthen the flame.* This constitutes an-

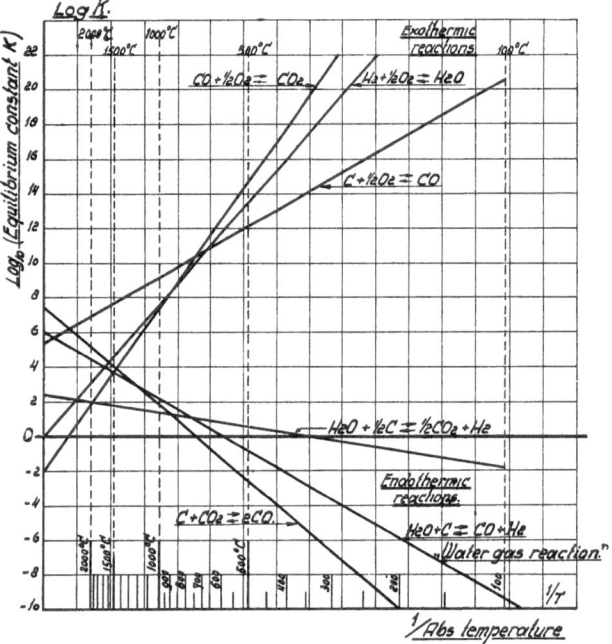

Fig. 105. Dissociation Curves.
(After HASLAM and RUSSELL)

other reason why methane containing four times more hydrogen than acetylene burns with a longer flame, water vapour being considerably more dissociated at equal temperatures than carbon monoxide.

The dissociation curves of a number of reactions, which are important for the combustion of hydrocarbons, are given in fig. 105, which graph is composed from data of HASLAM and RUSSELL.

A dissociation curve of a reaction shows how its *"equilibrium constant"* K changes with the temperature. As this relation may be roughly represented by general equations of the type:

$$\log K = C_1 \cdot {}^{1}/T + C_2 \text{ or better: } \log K = C_1 \cdot {}^{1}/T + C_2 \cdot T + C_3$$

it follows that it may be represented by straight lines in a diagram having a logarithmic scale for K and a hyperbolic scale for the absolute temperature T (inversely proportional).

For the equation:

$$CO + \tfrac{1}{2}O_2 = CO_2 \text{ the equilibrium constant } K = \frac{p_{CO_2}}{p_{CO} \cdot \sqrt{p_{O_2}}}$$

if p denotes the *partial pressure* of each component.

For the equation of the *"water gas reaction"*:

$$H_2O + C = CO + H_2 \text{ the equilibrium constant will be: } K = \frac{p_{CO} \cdot p_{H_2}}{p_{H_2O}}$$

From the corresponding line it may be seen that below about 1000° C. the formation of "water gas" (CO and $H_2$) is scarcely worth mentioning, but from 1000° C. onward there is a rapid rise in production (at 1000° C. the "equilibrium constant" = about 100, as log K = 2). The importance of this equation will be shown further in Section IV.

As the curves sloping down to the right represent *endothermic* or *heat-consuming reactions*, it follows that if they occur in an oil flame, which they undoubtedly will a part of the heat originally produced by the formation of water (combustion of hydrogen) will be consumed again by the following reaction of the water vapour with carbon, resulting in a lower temperature.

This example shows that the flame temperature-lowering influence of dissociation may be taken not only as simple decompositions of *initially* formed combustion products, according to equations as: $H_2O = H_2 + \frac{1}{2}O_2$ or $CO_2 = CO + \frac{1}{2}O_2$, but also due to more complicated *secondary* reactions.

### 4. *Quantity of non-reacting gases.*

The non-reacting or ballast gases (nitrogen) present in the combustible mixture are heated to the flame temperature and therefore consume heat without producing it, consequently lowering the flame temperature considerably. The same effect may be caused by any excess of oxygen, steam etc.

Although the above-mentioned "theoretical" flame temperatures are based on such a weak theoretical foundation as to be of little practical value, they may serve here to show roughly the influence of the quantity of excess air on oil flame temperatures, which indeed is very strong. (fig. 106).

When considering this graph it must be remembered that the actual temperatures will be considerably lower (some

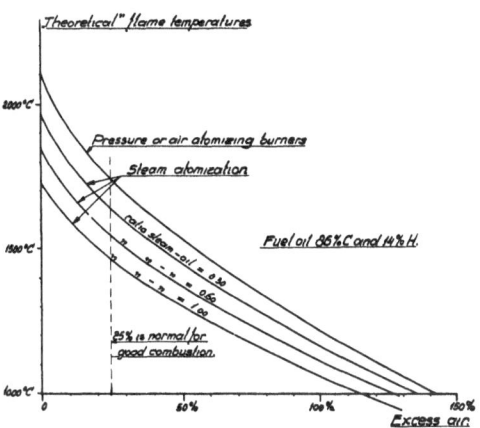

Fig. 106. Theoretical flame temperatures and excess air.

200 to 300° C.). Then again this graph only shows the decrease in flame temperature caused by excess air due to its *consumption of sensible heat*, but the flame temperature will moreover be lowered by its *flame-lengthening effect*. Anyhow the graph of fig. 106 gives a good idea of the steep decrease in flame temperatures caused by excess air and demonstrates the imperative necessity of keeping it as low as possible.

### 5. *Equilibrium of Radiation.*

It need hardly be said that if a flame is allowed to dissipate its heat freely, it will cool itself more than if it is surrounded by insulating and heat reflecting walls. An unprotected free oil flame may cool itself by radiation to such an extent that incomplete combustion takes place. On the other hand, a flame which is completely surrounded by insulating material will show a high temperature and good combustion but will not be able to transfer heat by radiation. Therefore it is the task of the oil burner designer to choose the happy medium in this respect.

## L. *Luminosity*

*Luminous oil flames radiate much more than non-luminous flames at equal temperatures*, this being caused by, for instance, the presence in the former of glowing solid particles of carbon which emit a *continuous spectrum*, as against glowing gases of the latter producing a *discontinuous spectrum of bands*.

Fig. 107 represents the intensity of energy radiated per unit of time and surface plotted against the wave length of the rays, as it would be emitted by the absolute

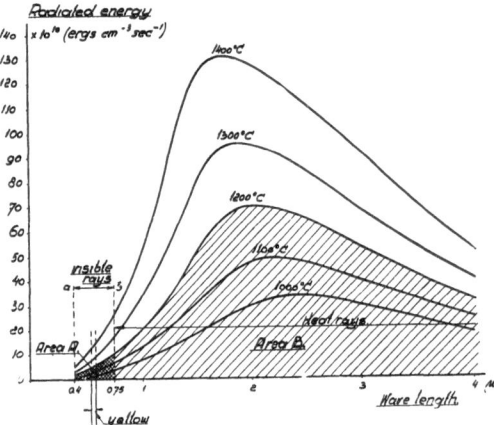

black body at different temperatures. A substance is called an *"absolute black body"* if it absorbs all radiant energy received, or, in other words, if it has *no reflective power* at all. Carbon particles may be considered as *"nearly black"* or *"grey"* bodies having only *very little reflective power*. "Grey" bodies show just the same shape of curves for the radiated energy as "black" bodies, the difference being that the former need a somewhat higher temperature to effect the same intensity of radiation than the latter.

Fig. 107. Black body radiation for various temperatures.

It will be seen from the graph that the energy emitted, for example at 1200° C., by means of visible rays (cross-hatched area between the vertical lines a and b) is very slight as compared with the total energy (hatched areas A + B), and therefore it is clear that the luminosity itself must not be considered as the *cause* of the stronger transfer of energy but only as an *indication* of it. The greater part of the radiated energy is transferred by means of invisible infra-red rays. When comparing area A with area B it follows from the graph that *carbon particles* (for a moment taken as "black bodies") of oil flames are rather inefficient radiant media for illuminating purposes but *very efficient for heating*, area A being only a small fraction of (A + B). Other substances may show entirely different radiation curves from those of fig. 107; thorium oxide, for instance, which is used for incandescent mantles of gas burners, has a very strong preference for special light rays (selective emission) and consequently has a better ratio of A to A + B than carbon.

Carbon particles may be considered as efficient means for the emission of heat rays (remember old electric lamps with carbon filament). Therefore it is needless to look for other substances or dopes to be admixed with the oil in order to improve the heat radiation of oil flames which, in fact, are capable of producing such an excellent radiant medium themselves from the very fuel merely by the adjustment of a carbonic combustion.

Another problem arises if it is desired to increase the heat radiation of a *blue* oil flame, which may under equal conditions of temperature, flame volume and surface, radiate only one-fourth of the energy radiated by a yellow oil flame. An improvement of the radiating power of blue flames may be obtained by *placing a solid material in the flame* or by blowing its dust into it. It is needless to look for any special materials, as ordinary iron, steel or the heat-resisting nickel-alloys from which the perforated concentric shells of range burners are made, especially in a slightly oxidized state, are very good substances for selective radiation of heat. This circumstance, in fact, largely contributed to the success of the range burners for heating purposes, such as cooking, in which case blue flames without any glowing metal parts show a considerably lower efficiency.

From the foregoing it follows that yellow and blue oil flames may have equal efficiencies from a *calorific* point of view, but that yellow oil flames certainly have better radiation efficiencies. *Thus blue oil flames leave more heat to be transferred by means of convection and generally will require a larger heating surface to obtain equal total efficiency.* This may be compensated for by placing a glowing medium (iron sheets) in blue flames as a substitute for carbon particles. (See figs. 71, 84 and 140).

Of course it must be remembered that with regard to radiative power there may be great differences between various *yellow* flames too, as the quantity of radiated energy depends largely on the *carbon-density* of the flame taken as a gaseous solid body (physically speaking, a body consisting of solid particles suspended by a gas is called a "smoke", or for liquid particles a "mist"), or, in other words, on the *orientation between aldehydeous and carbonic combustion.*

The radiation of yellow oil flames has been found to be especially useful for the *direct heating of steel forgings* and other applications which demand not only a sufficiently high temperature but also a quick *rate* of heating. In this connection heavy asphaltic or cracked residues, giving a strongly preferential carbonic combustion, prove to be decidedly better than waxy residues or lighter fuels.


M. *Seventh stage: Carrying away the Flue gases*


It is common practice to remove flue gases by means of *chimney draught*, which is the cheapest way of doing it. The draught of a chimney is caused by the difference in weight between the warm column of flue gases in the chimney and a column of air of ambient temperature and equal height. It is obvious that only *vertical* columns are effective, so that horizontal parts of the chimney cause friction without producing draught and therefore are possibly to be avoided.

It makes a *great difference whether the chimney draught is used only to carry away the flue gases or also to operate the oil burner at the same time.* In the latter case the draught must be controlled carefully, because the combustion process is entirely dependent on it. Such "natural draught burners" may give serious trouble when started with a cold chimney or in foggy, calm weather or in weather of unstable draught conditions (storm). Hand-operated draught regulators are useless for this purpose.

Automatic control may be established either by pinching the flow of flue gases

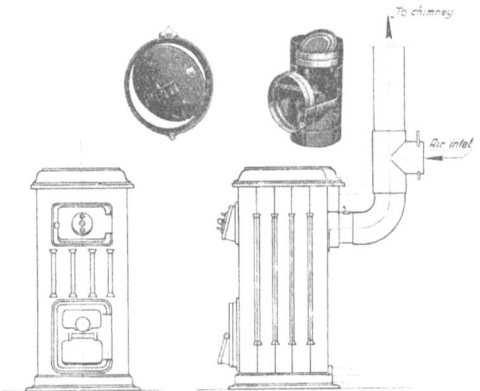

itself or by admixing a variable quantity of air from the boiler room to them, thus giving the chimney a *variable leak* to compensate for any excess of its draught. Examples of both principles are given in fig. 108, representing HOTSTREAM draught regulators, the draught being regulated by adjustment of the weight of the valves. A return valve is fitted to prevent blowing back in stormy weather.

With oil burners operated with "forced draught" the air is generally blown into the combustion chamber by a fan, so that

Fig. 108. HOTSTREAM Draught Regulators.

it would be better to speak of "forced pressure", as "forced draught" would comprise a fan at the top of the chimney. As the combustion chamber of a boiler will never be absolutely air-tight, it is advisable to keep it under a slight vacuum in order to prevent leakage of badly smelling gases into the rooms. For this reason it will be necessary to apply sufficient chimney draught to motor-operated burners too, and wrongly built or narrow chimneys will cause just as much trouble with motor burners as with the natural draught types. Variable draught may have an appreciable influence on the delivery of a fan, as it changes the head against which it is blowing, and therefore it is advisable to equip operated burners with automatic draught regulators too.

In order to guarantee good chimney draught many inventions for suitable caps have been made, but they all fail to have any effect if there is no wind and if the chimney is cold, and moreover often cause extra resistances. Some examples are the rotating chimney cap of STORM VAN LEEUWEN shown in fig. 109, which is able to produce suction even from very weak winds by use of

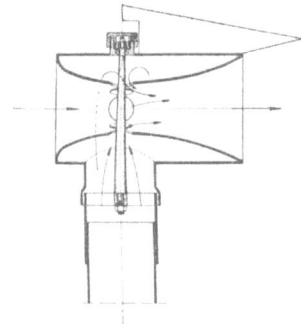

Fig. 109. Venturi cap for chimneys. (STORM VAN LEEUWEN).

the VENTURI principle, and the "CHANARD-ÉTOILE" cap (fig. 110) comprising a non-rotating star with seven points designed in such a way that suction takes place from whatever direction the wind may come.

Besides improving the natural draught, chimney caps may tend to *equalize draught variations* which are caused by wind. With regard to this important point it

must be remarked that in the first place a chimney must be high enough to be free
from the disturbance of the house. As shown by fig. 111 the wind may cause pressure

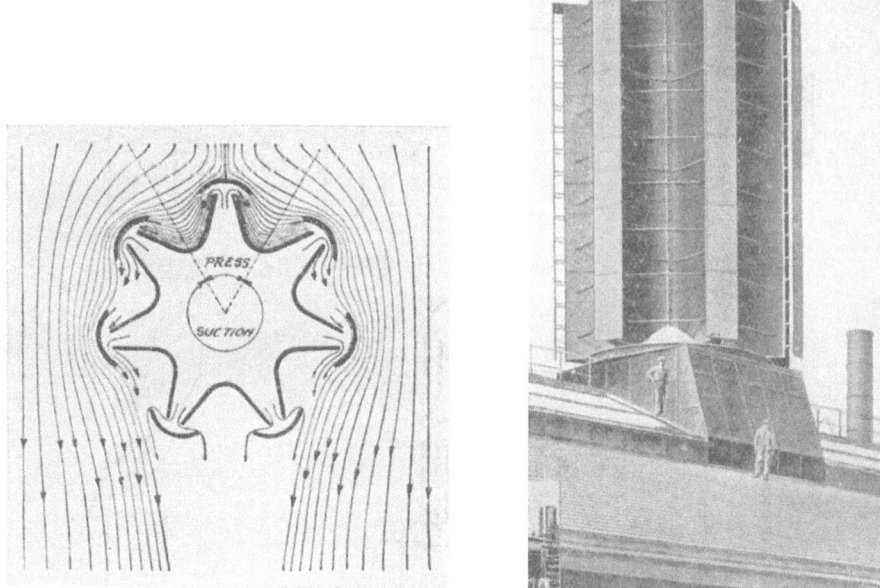

Fig. 110. Chimney cap "CHANARD-ETOILE".

to the windward and vacuum to the leeward side of the house or roof. If the chimney
ends in a pressure zone, the "temperature draught" may be seriously opposed, resulting
in what is called
"back-draught". If
the windows of a
room happen to open
towards a vacu-
um zone, the room
may be put under
sufficient vacuum to
conquer the "temper-
ature draught" of
the chimney, forcing

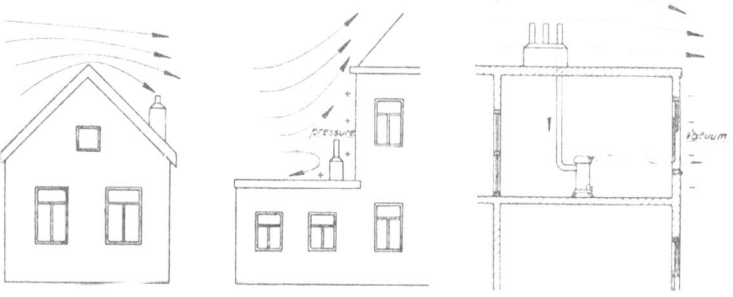

Fig. 111. Causes of "Back-draught".

combustion products back into the room. It will be clear that in such cases chimney
caps will be of little use.

Fig. 112 shows a number of chimney caps which are more or less successful in equalizing draught variations. Special attention is drawn to the HOOGENDAM cap, comprising a number of concentric perforated shells. This cap has proved to be remarkably insensitive to squalls and gusts of wind, which might be explained by the fact that the energy of the wind will be destroyed by the labyrinthine passages through the shells.

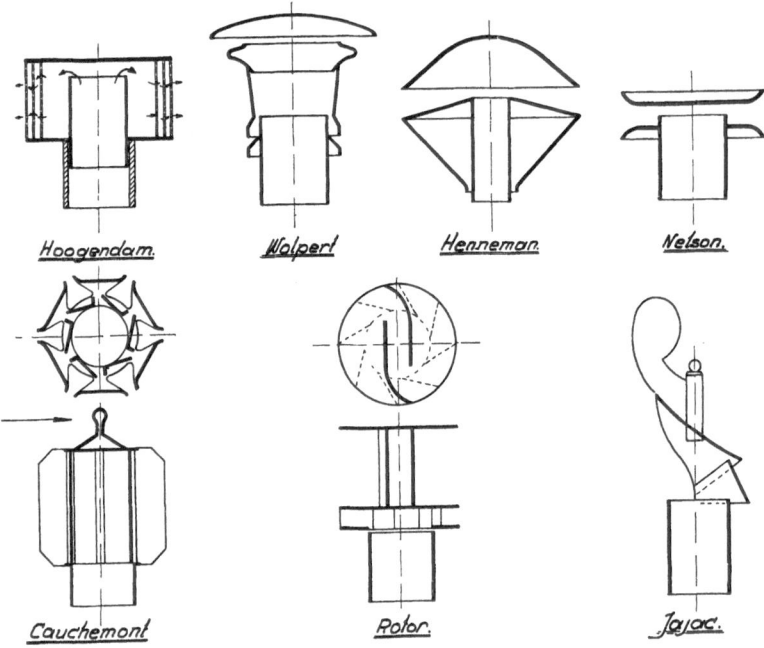

Fig. 112. Chimney caps.

If the flue gases are cooled down very low in order to obtain a good efficiency, the draught may get too small, so that it is advisable not to go too far with the cooling. Moreover if the flue gases are cooled below the dew point (for fuel oils usually not higher than 55° C. and decreasing with increasing excess air), a corrosive action by wet sulphurous and sulphuric acid may take place on iron connections etc.

# CHAPTER XI

## QUANTITATIVE CONTROLS OF THE COMBUSTION PROCESS

### A. *Quantitative Controls*

An oil burning process must be controlled not only from a point of view of efficiency but also for safety's sake, which is especially true for domestic equipments. The methods for control may be further divided into *quantitative* and *qualitative controls* and, moreover, into *automatic* and *manual controls*. In this chapter only quantitative controls necessary for the determination of the efficiency of the oil burning equipment will be considered, the regulation of the intensity of oil fires being left for the third section.

Every oil burning application has its own peculiarities, sometimes making it very difficult to calculate the *thermal* efficiency (tea-drying, brick industry, etc.), which calculation, therefore, is often replaced by a comparison of production figures previously obtained with similar installations.

For *steam-raising plants* it is comparatively easy to determine the thermal efficiency, especially if the quantity of steam produced can be measured (e.g. by measurement of the feed water supply keeping the water level in the boiler constant) and the heat content of the steam produced is known. In that case a calculation of stack losses from flue gas analysis may be used as a rough check for the radiation losses and a balance may be drawn up.

For *central heating plants* with hot water circulation, *direct* determination of the *thermal* efficiency is far more difficult, especially if the circulation is not obtained with a pump but by gravitation. In the latter case the velocity of circulation which must be known for calculation of the net heat leaving the boiler is not known and cannot be measured without great difficulty. It is also difficult to determine the heat delivered to the air by hot water radiators for, although there are many "calorie-counters" on the market for this purpose, they are not reliable enough for accurate measurements.

In that case only the following *indirect* measurement can be made: Utilized heat production per hour = (oil consumption per hour) × (heat value of the oil) — (stack and radiation losses per hour).

The stack losses may be subdivided into

a) Sensible heat loss,

b) Chemical latent loss (incomplete combustion) and

c) Physical latent loss (water vapour).

### B. *Sensible Heat loss of Flue gases*

The sensible heat loss of the flue gases per hour is equal to the sum of weight per hour times specific heat for every flue gas constituent multiplied by the difference between the temperature of the flue gases immediately after leaving the boiler and the room temperature. It follows that in order to calculate the loss it is necessary to know not only the total *quantity* of flue gases but also their *composition*.

The determination of the *composition* is generally done with an ORSAT apparatus based on the absorption of carbon dioxide and oxygen from the gas sample by suitable liquids. The *direct* measurement of the *quantity* of flue gases would entail serious difficulties (for instance measurement with PITOT tubes at the top of the chimney) and an *indirect* calculation from the quantity of oil burnt is therefore generally made instead. This, however, makes it necessary to know the *composition of the oil*.

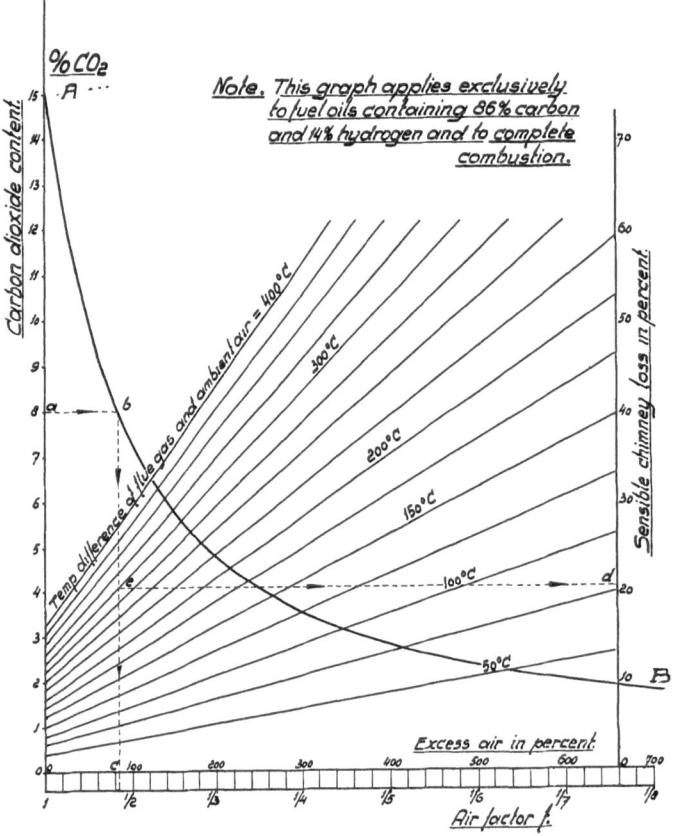

Fig. 113. Sensible chimney losses.

If the composition of the oil is known and the quantity of air used for combustion is measured directly (e.g. by means of a PITOT tube, whereby the velocities of air in a cross-section of a duct which is connected to the air registers of the burners may be determined), then the quantity of the flue gases and their composition may be calculated on the assumption, however, that the combustion is complete and that no air enters into the combustion chamber otherwise than through the air duct. These conditions are seldom fulfilled and as a rule a control by the analysis of the flue gas cannot be dispensed with.

The graph of fig. 113 has been calculated for an *oil containing 86% carbon and 14% hydrogen*, which figures might apply to a light domestic fuel oil or gas oil. If 1 kg of this oil is burnt completely with exactly the amount of air chemically required for combustion, and no excess of air whatever, it may be easily calculated that the resulting flue gases will have a carbon dioxide content of 13.2% by volume if no water vapour is condensed, or *15.2% if calculated on dry gas*. The oxygen content is zero, which case is represented by the point A.

If the above mentioned amount of flue gas is imagined to be diluted by air (= excess air), the carbon dioxide content will decrease according to the hyperbolic curve AB and thus 8% of $CO_2$ (*a*) will correspond via (*b*) to *82% of excess air* (*c*) or, if the air quantity used is expressed in the ratio of the theoretical quantity to the quantity actually used, to an *air factor f* $= 0.55$.

To calculate the heat loss it is furthermore necessary to know the flue gas temperature and specific heat of its constituents. As average values applying to the temperature range of flue gases the following figures may be taken: Nitrogen $= 0.30$, carbon dioxide $= 0.42$, water vapour $= 0.38$, oxygen $= 0.30$, and after some simplifications the following equation may be derived:

$$\text{Sensible heat loss in } \% = \left(0.8 + \frac{48}{p}\right)(T_s - T_0). \, 10^{-4}$$

in which p   = percentage of carbon dioxide, calculated on the *dry* flue gases as, for instance, determined with an ORSAT-apparatus.

$\quad$ $T_s$ = temperature in degrees Centigrade of flue gases immediately after leaving boiler, i.e. after the last contact with its heating surface.

$\quad$ $T_0$ = room temperature in degrees Centigrade.

In this connection it may be observed that the $CO_2$ determinations as obtained by flue gas analysis apparatus like ORSAT's are directly expressed in figures based on the *dry sample*, as the humidity of the sample caused by contact with the solutions and water (saturation at room temperature) magnifies all readings of volumes in a *constant ratio*: $\dfrac{p+w}{p}$ (if p = barometric pressure and w = vapour tension of water at room temperature), which consequently has no influence at all on the *ratio* of the volumes measured. This, of course, holds good only if neither the barometric pressure nor the temperature of the apparatus is changed during the determination.

It should be borne in mind that the flue gases as they flow through the chimney contain a considerable quantity of water vapour (for most oils about 1 kg of water is produced for every kg of oil burnt) which should be calculated from the composition of the oil and the quantity of oil burnt. This is especially important when for some reason it is desired to determine the *velocities of flow* through boiler channels etc., as steam has very large specific volumes, and may cause very high speeds.

The graph may be completed by drawing a series of straight lines, each corresponding to a definite stack temperature, and for the above-mentioned case it follows by intersection of (*b, c*) with the suitable line of stack temperature at (*e*) that the sensible stack loss was about 20% (*d*).

Romp, Oil burning

9

The following must be taken into account when using this graph:

1) This graph *only applies to oils consisting of 14% hydrogen and 86% carbon.* For heavier oils with higher carbon content the hyperbolic curve AB must be raised and for lighter oils lowered.

2) *This graph does not apply if incomplete combustion takes place,* as this causes a consumption of oxygen without producing the corresponding quantity of carbon dioxide or water.

3) It is not sufficient to judge the efficiency of a combustion process by the carbon dioxide content of its flue gases alone, especially if the composition of the oil is not known. The graph of fig. 113 must be considered as a first *rough control.*

Needless to say, in regard to efficiency the *carbon dioxide content of flue gases of solid fuels cannot be compared with that for oils,* as for instance an anthracite consisting of 84% carbon and 5% hydrogen and burnt completely without any excess air will show a maximum carbon dioxide content of 18% against only about 15% for oil. For coal of 86% C and 3% H this figure would be 19.5% and for coke with 88% C even 20.5%. In fact this means that a carbon dioxide content of 10% for coal will usually correspond to a considerably lower efficiency than if this percentage is obtained with oil, which difference is caused by the higher hydrogen content of the oil.

### C. *Chemical Heat loss of Flue gases*

A more accurate method of determining the quantity of excess air and, moreover, the chemical heat loss by incomplete combustion may be carried out by using the OSTWALD-diagrams, an example being given in fig. 114. For this control it is necesary to know not only the content of *carbon dioxide* but also that of *oxygen.*

This diagram is a triangle OAB, of which OA represents the carbon dioxide axis and OB the oxygen axis. The *point A* corresponds entirely to A of the graph of fig. 113, viz. a complete combustion without any excess air or air factor $f = 0$ (zero oxygen) producing the maximum possible content of carbon dioxide obtainable.

The *point B* represents pure air with 21% of oxygen and zero carbon dioxide, or in other words flue gases which are diluted with an infinitely large quantity of air.

The *line AB* represents all possible stages of dilution with air of the flue gases of A (complete combustion without excess air) and has been divided by a linear scale for the *air factor f,* which is equal to the air quantity necessary for complete combustion without excess air ($L_{theor.}$) divided by the quantity of air actually used ($L_{act.}$). By using this air factor the troublesome hyperbolic scale caused by using as a definition the "percentage of excess air" may be avoided. *Factor f* may be easily converted into the *percentage of excess air,* which is equal to $\left(\dfrac{1-f}{f} \times 100\%\right)$, by means of the additional scale drawn parallel to AB.

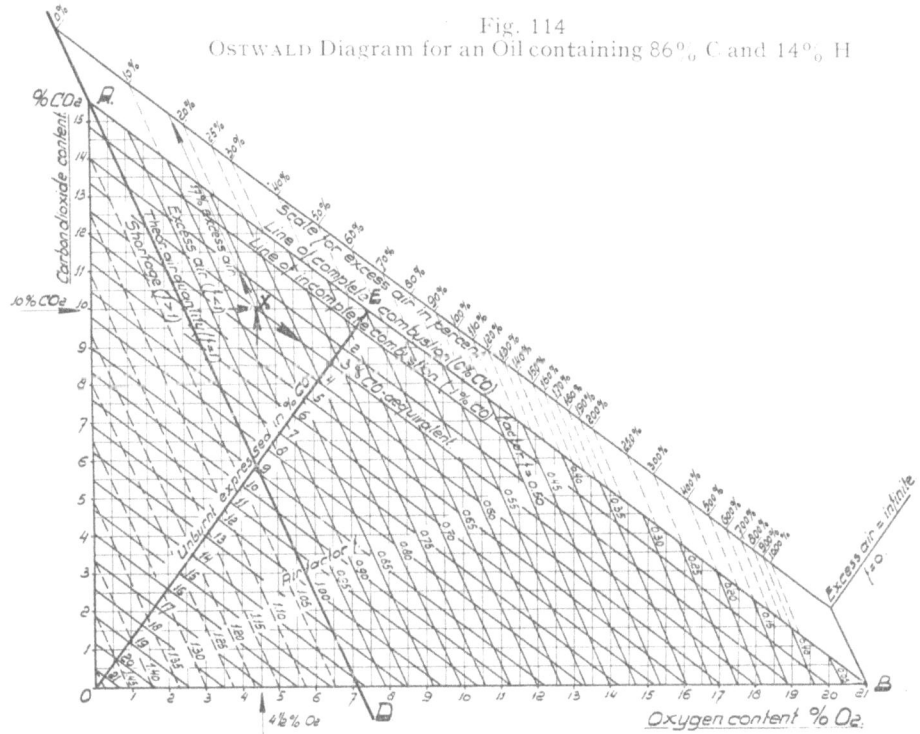

Fig. 114
OSTWALD Diagram for an Oil containing 86% C and 14% H

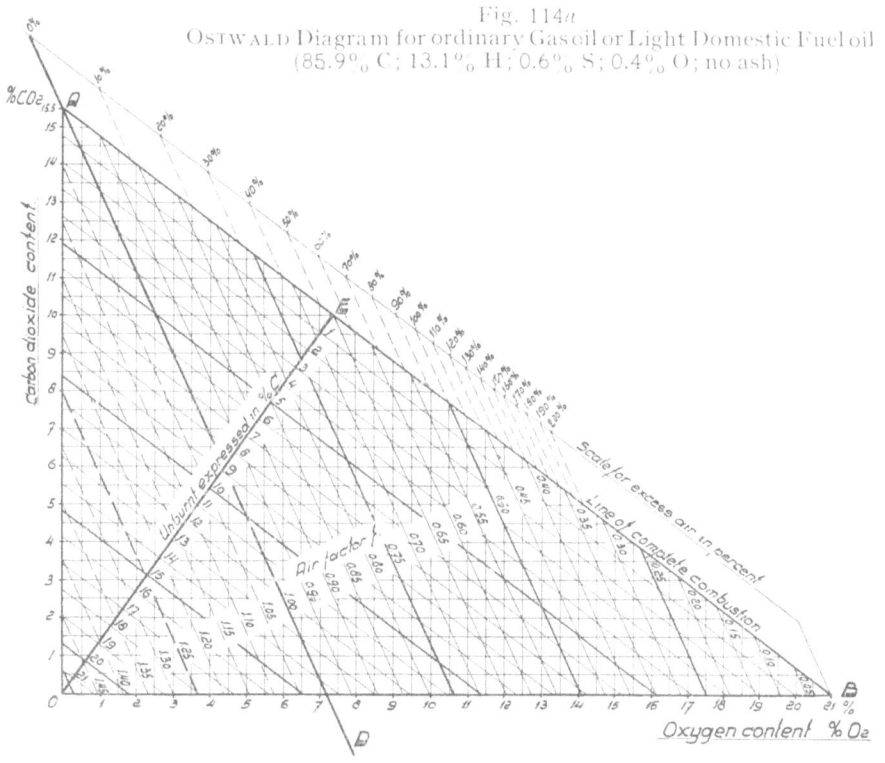

Fig. 114a
OSTWALD Diagram for ordinary Gasoil or Light Domestic Fuel oil
(85.9% C; 13.1% H; 0.6% S; 0.4% O; no ash)

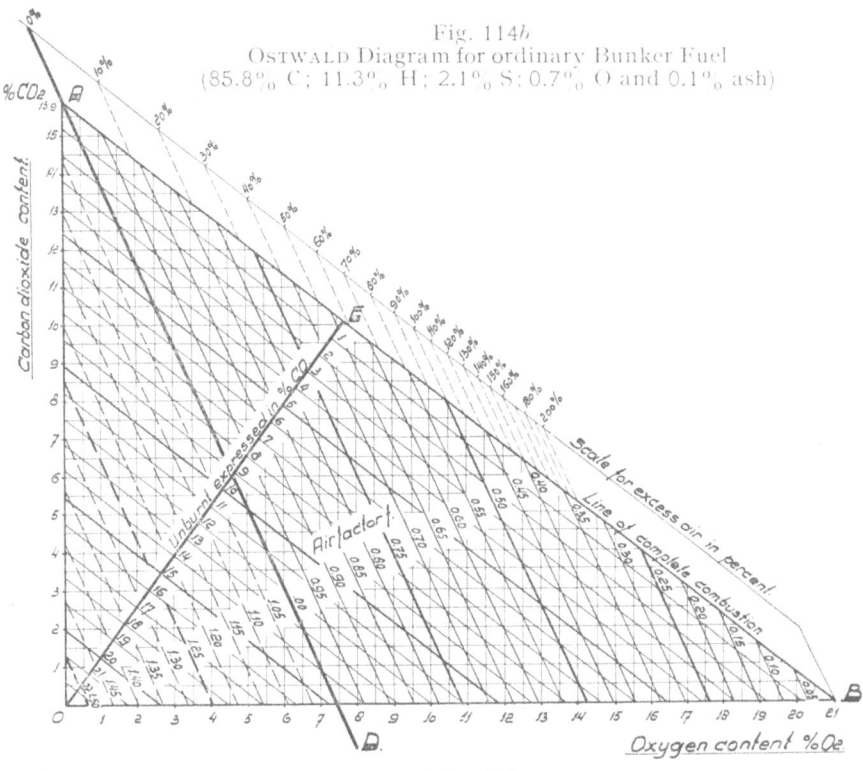

Fig. 114b
Ostwald Diagram for ordinary Bunker Fuel
(85.8% C; 11.3% H; 2.1% S; 0.7% O and 0.1% ash)

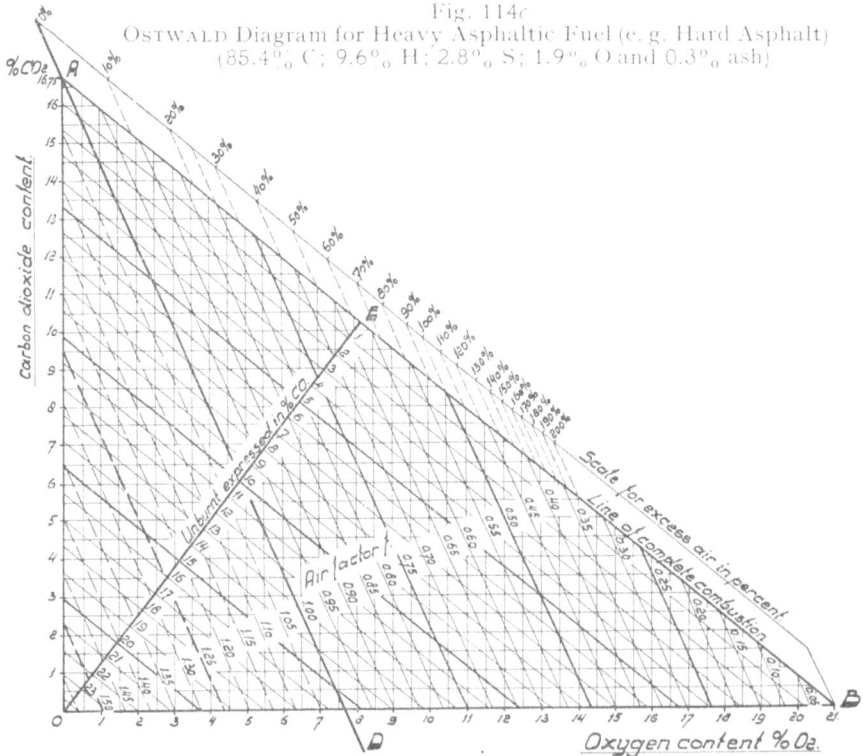

Fig. 114c
Ostwald Diagram for Heavy Asphaltic Fuel (e. g. Hard Asphalt)
(85.4% C; 9.6% H; 2.8% S; 1.9% O and 0.3% ash)

## EXPLANATORY NOTES ON OSTWALD DIAGRAMS

**GAS OIL. Analysis : 85.9 % C; 13.1 % H; 0.6 % S; 0.4 % O and no ash. (fig. 114 & 114a).**

The *oxygen* may be disregarded, as this quantity corresponds to an imaginary hydrogen consumption of only 4/32 of 0,4% = 0,05%, which is small compared with the total hydrogen content present.

The *sulphur* burns according to: $S + O_2 = SO_2$ (32 + 32 = 64) and as carbon burns according to: $C + O_2 = CO_2$ (12 + 32 = 44) and $SO_2$ is determined by the ORSAT apparatus in KOH as $CO_2$, the sulphur content must be accounted for by adding $^{12}/_{32}$ of 0.6% = 0.23% to the carbon content ,resulting in 86.1 C + 13.1 H = 99.2 or converted to 100%: *86.8% C and 13.2 H* (simplified analysis).

The *simplified formula* may be written: **$C_{7.23n}$ $H_{13.20n}$**, where n is unknown.

OB = *Oxygen axis* to be divided into 21 equal parts (air = 21% $O_2$).

OA = *Carbon dioxide axis* from which A = $CO_2$ content of gases resulting from a complete combustion to $CO_2$ and $H_2O$ with the *theoretical quantity of air* (excess air = zero).

OA = $CO_2$ maximum is found from: $CO_{2max} = \dfrac{21}{100 + 235\frac{\% H}{\% C}}$ = (in this case) = *15.5%*

AB = Hypotenuse = *Line of complete combustion* to $CO_2$ and $H_2O$, viz.:

A = same with the theoretical quantity of air, air factor *f* = 1,

B = same with infinite dilution by air, air factor f = zero.

AB = to be divided into 10 equal parts gives the *air factor division* f from 0 to 1.

f = *air factor* is defined as the *theoretical air quantity divided by the quantity actually used*, and may be converted into the often used *percentage excess air e* according to:

e = *percentage excess air* = $\dfrac{1-f}{f}$ 100% or f = $\dfrac{1}{1 + e/100\%}$

D = point on OB, which fixes the *direction of the air factor lines*. It represents an imaginary combustion with the theoretical quantity of air *to CO alone*, instead of to $CO_2$.

The equation for A reads: $C_{7.23n} H_{13.2n} + 10.53n O_2 = 7.23n CO_2 + 6.60n H_2O$

and for D: $C_{7.23n} H_{13.2n} + 10.53n O_2 = 7.23n CO + 6.6n H_2O + 3.6n O_2$

Taking into account the fact that 10.53n $O_2$ will entrain $^{79}/_{21}$ times as much nitrogen and that the water vapour formed will be condensed, one finds for the resulting combustion products of D *14.4% CO* and *7.2% $O_2$*. This fixes both D and the AD direction, after which the lines of constant air factor f can be drawn.

The lines of *constant percentage of CO-equivalents* are parallel to the hypotenuse AB and, since it is known that D corresponds to 14.4% of CO, the perpendicular distance from D to AB must be *divided into 14.4 equal parts*. For verification 21.7% CO must then be found for the origin O, as the following equation applies to this point:

$$C_{7.23n} H_{13.2n} + 6.88n O_2 = 7.23n CO + 6.6n H_2O.$$

---

**BUNKER FUEL. Analysis : 85.8 % C; 11.3 % H; 2.1 % S; 0.7 % O; 0.1 % ash. (fig.114b).**

The *oxygen content* of 0.7% would be able to consume $^4/_{32}$ of 0.7% = 0.09% of the hydrogen, giving 11.2% H.

The *sulphur content* of 2.1% corresponds to $^{12}/_{32}$ of 2.1 = 0.79 % carbon, which must be added to the original carbon content of the oil giving 86.6 C.

The *simplified analysis* will thus be: *88.4% C; 11.5% H and 0.1% ash*, whereas the simplified formula is: **$C_{7.37n}$ $H_{11.5n}$.**

Point A is found from $CO_{2max} = 15.9\%$, whereas the analysis of the combustion products of point D shows: *14.8% CO* and *7.45% $O_2$*.

The structure of the graph may be completed further along the scheme given for gas oil.

---

**HEAVY ASPHALT. Analysis : 85.4 % C; 9.6 % H; 2.8 % S; 1.9 % O; 0.3 % ash. (fig.114c).**

The *oxygen content* of 1.9% corresponds to an imaginary hydrogen consumption of $^4/_{32}$ of 1.9% = 0.24%, which must be subtracted from 9.6% H.

The *sulphur content* of 2.8% must be accounted for by adding $^{12}/_{32}$ of 2.8% = 1.05% to the carbon content.

The *simplified analysis* thus reads: *90.0% C; 9.7% H and 0.3% ash*.

The *formula* may be written: **$C_{7.50n}$ $H_{9.70n}$.**

Point A is found from $CO_{2max} = 16.75\%$, whereas the analysis of the combustion products of point D shows: *15.45% CO* and *7.70% $O_2$*.

The structure of the graph may be further completed along the scheme given for gas oil.

---

Whereas the *line AB* represents *complete combustion with various percentages of excess air*, the *area of the triangle OAB* represents all possible cases of incomplete combustion. The area outside OAB represents no really possible cases, as "*over*-complete" combustion has no meaning.

The area of OAB may be divided by two series of lines, one parallel to the hypotenuse AB according to the *percentages of incomplete combustion products* (expressed in "CO-equivalent"), and another parallel to the line AD according to the *percentages of excess air* or to the *air factor f*. Lines parallel to and on the *right* side of AD give *excess air*, whereas lines on the *left* side of AD represent a *shortage* of air (dotted lines).

If a flue gas sample obtained by burning oil of 85% C and 14% H (for which the diagram of fig. 114 has been constructed) shows 10% carbon dioxide and $4\frac{1}{2}$% oxygen, it follows from the position of point X that there happened to be *17% of excess air* (reading parallel to AB).

For a correct understanding it may be remarked that *point D* represents the case where a mixture containing the theoretical amount of air required (zero excess air) is *burnt to CO alone* without $CO_2$, thus an amount of *7.15% of oxygen* being left. Calculation shows that the resulting flue gases of that special case would then contain *14.3% of CO*.

Furthermore *point O* represents the case where a mixture with a considerable shortage of air ($f = 1.51$ or 34% shortage) is burnt as far as possible so that *no oxygen is left*, and calculation shows that the result would be a flue gas with *21.7% CO*. It will be obvious that these cases do not occur in practice and are only of theoretical interest, in fact being merely used for furnishing fixed points as bases for construction of the diagram (see the "Explanatory Notes on OSTWALD Diagrams").

Incomplete combustion products of coal burning consist mainly of carbon monoxide, and consequently several methods of analysis to determine CO have been developed, most of which, however, are troublesome and very often unreliable. Products of incomplete combustion of an oil burning process consist, moreover, of several other substances such as *aldehydes* (= *acrid vapours*), *fatty acids, methane, ethylene, hydrogen and comparatively little carbon monoxide*, and this is the reason why a CO determination is of little value for oil burning practice.

One of the best known methods for determining the percentage of incomplete combustion products of oil burning consists in *burning the gas sample for a second time* in the presence of glowing copper oxide after carbon dioxide and oxygen have been removed from it. This analysis, known as JAEGER's, is generally carried out by means of a heated quartz tube filled with oxidized copper filings, which method in fact ·forms a clever solution of the problem of how to introduce *pure oxygen having no gas volume* (see fig. 115). When the gas sample, after carbon dioxide and oxygen have been removed, is led through this quartz tube, the combustible products are burnt to carbon dioxide and water, the former being determined by absorption, whilst the latter condenses during the experiment.

If the chemical formula of the incomplete combustion products (ratio of C to H) is known, this determination will be sufficient to calculate the quantity of hydrogen present in the gas sample. As, however, the C to H ratio will not generally be known, some additional means must be found for determining the quantity of burnt hydrogen. This is often done by means of so-called *selective combustion* of hydrogen, e.g. by leading the sample, after adding a known quantity of oxygen to it, through a platinum capillary tube heated to such comparatively low temperatures that other combustible constituents are not yet oxidized. This last condition, however, is rather doubtful, especially having regard to the strong catalytic power of platinum, and moreover it is not sufficient to determine

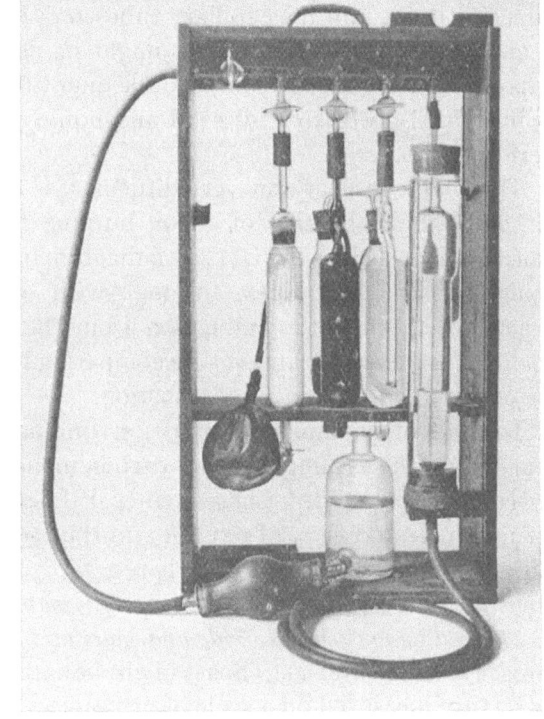

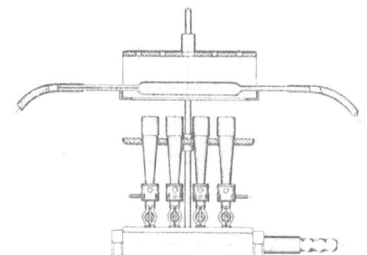

Fig. 115. JAEGER furnace and ORSAT apparatus.

the *free* hydrogen alone, for there must also be determined the hydrogen in the *combined* state as contained in incomplete combustion products and hydrocarbons.

Another kind of selective combustion has been suggested, consisting in leading the gas sample over copper dioxide at a temperature between 275 and 300° C. (BURRELL-ORSAT), and it is claimed that only carbon monoxide and hydrogen are burnt that way, while the stable saturated hydrocarbons are not oxidized and may be determined separately by slow combustion over a yellow-hot platinum coil in a combustion pipette.

An exhaustive study of several methods of selective combustion may be found in WINKLER-BRUNCK's book "Lehrbuch der technischen Gasanalysen" (Felix-Leipzig), and after reading it one might come to the conclusion that selective combustion very often works out remarkably well, because usually there is no check on the results! As a matter of fact, if applied to a *synthetic* gas sample of known composition, astonishing discrepancies may often be discovered.

Then again a *frequent product of imcomplete combustion of oil burning is the well-known bluish-white, steam-like oil smoke*. This white oil smoke consists of fine liquid particles often originating from recondensed oil vapours and will be partly kept back by the cartridge of cotton wool in the ORSAT apparatus and partly washed out by the contact with water in the capillary tubes and, therefore, will never be fully shown by the analysis. This loss by oil mist might be determined by measuring not only the composition of the flue gas but also the quantities of (1) the oil, (2) the air and (3) the dry flue gases freed from all solid and liquid particles, which test, however, is too laborious for practice.

This makes it clear how very difficult it is to determine the quantity of products of incomplete combustion of an oil burning installation, so that a comparatively simple method, such as the OSTWALD diagram in fact is, may be very useful for practical purposes. It determines, although with some approximation, the quantity of products of incomplete combustion from the *actual loss of oxygen, from whatever oxidation this loss may originate*, by comparing it with the loss of oxygen which would have occurred with *complete* combustion.

The lines of incomplete combustion running parallel to the hypotenuse have been calculated on the assumption that carbon monoxide was the only product of incomplete combustion formed. Consequently if the gas sample is found to contain 10% of carbon dioxide and $4\frac{1}{2}\%$ of oxygen and thus point X shows 3% of incomplete combustion products, this means that there are 3% *of carbon monoxide or other intermediate gaseous products (aldehydes, fatty acids etc.), which need just as much oxygen to complete their combustion to carbon dioxide and water as would be necessary to burn 3% of carbon monoxide to carbon dioxide*. Thus it might be expressed that the incomplete combustion products are measured on a scale of *percentages of "CO-equivalents"* which gives these figures a considerably broader value than real CO-figures can have.

Attention is drawn to the fact that the OSTWALD diagram only applies to oils of one fixed chemical composition, which is a drawback to its use. Moreover its construction includes some approximations, which, however, are so small that their influence disappears against the accuracy of the $CO_2$ and $O_2$ readings (see the original book by OSTWALD, "Beitrage zur graphischen Feuerungstechnik").

### D. *Physical heat loss of flue gases*

A second form of latent heat loss of flue gases consists of the *condensation heat of the water vapour* formed by combustion.

It has been explained before that it is common practice on the Continent of Europe to subtract this latent heat of water vapour beforehand from the calorific heat value as determined by the bomb method and to speak of the *lower or net heat values*. In English-speaking countries, however, people are accustomed to use the *upper or gross heat values*, and it will be clear that in this case the latent heat of water vapour must be considered as a real stack loss when calculating the efficiency, provided the water leaves the chimney completely uncondensed. If, however, the *lower heat value*, is used as a basis for the intake of heat, this latent heat of water vapour, being already subtracted from the intake, must not be considered as a loss any more.

It has been previously pointed out that, although it makes no fundamental difference, it is well to remember that in the literature of English-speaking countries on oil burning the boiler efficiencies are expressed in values which are generally 7 to 9% lower than those in European Continental literature, merely because they are based on different definitions of heat values.

### E. *General Remarks*

Practical adjustment of an oil flame is very often done by sight, as is described for instance in TAPP's "Handbook of oil burning" page 490, as follows: "At first the

Fig. 116. Carbon dioxide tester (CARBOSCOPE).

oil flame generally will be started with a considerable quantity of excess air to prevent smoking and will be almost white in colour. As the amount of excess air is cut down the white will shade down through yellow to a deep orange and then to a reddish colour, which will usually have a slight tinge of smoke. After cutting the air down to where this effect is apparent, the air shculd be increased until the smoke is cleared up and the flame has a bright reddish orange colour".

The setting should then be checked by a flue gas analysis, which, however, is only too often omitted, most gas analysis apparatus being rather clumsy and troublesome instruments. A very handy pocket apparatus suitable for rough control is shown in fig. 116 representing the "Carboscope" designed by an Austrian engineer Gross. By means of a rubber bulb aspirator the gas sample is sucked into a metal bellows of known volume. After the inlet has been shut off by a valve the sample is pumped through a cartridge filled with solid caustic soda by turning it over and back again, the cartridge itself acting as a piston. The absorption of carbon dioxide causes a reduction of the volume; the bellows, being compressed by the atmosphere accordingly, show a displacement which is measured by means of a micrometer comprising a scale for direct reading of the $CO_2$ content. To facilitate adjustment the micrometer, as soon as it touches the bellows, closes an electric circuit by which a small lamp is set glowing. This apparatus, which is sufficiently accurate for rough determinations, has the advantages of using no liquid and being a handy pocket outfit.


## F. *Sampling*


A point of great importance, which is very often disregarded but which nevertheless may be able to make any analysis absolutely worthless although made with extreme accuracy, is the method of *taking a gas sample*. It is not enough to insert a small pipe in the flue and to draw a gas sample. It is necessary to take the sample at the right place, immediately behind the furnace, in order to prevent infiltration of air through porous material of the walls, leaky dampers etc., and preferably from the centre. Moreover the sample must be drawn for a sufficiently long time and with constant suction speed (for instance with the *two-bottle arrangement* of fig. 117) in order to get reliable *time-averaged* figures.

So as to prevent the removal of carbon dioxide from the sample by contact with water, it is necessary for accurate measurements to dissolve salt in it or to use aqueous glycerine as a liquid. Instead of thus taking a large sample during a long period, it is also possible to make a great many analyses at regular intervals and to calculate the average from the results. In that case it would be sufficient to use in the ORSAT apparatus water which is completely saturated with gas, e.g. by bubbling gas through it.

## G. *Excess air*

Although for a definite fuel the carbon dioxide content of its flue gases will be mainly fixed by the amount of excess air as well as by the degree of completeness of the combustion, the former factor has decidedly the greater influence.

As a rule 10% of carbon dioxide is considered to be a reasonably good figure for the efficient operation of oil burning installations, and from the OSTWALD diagram of

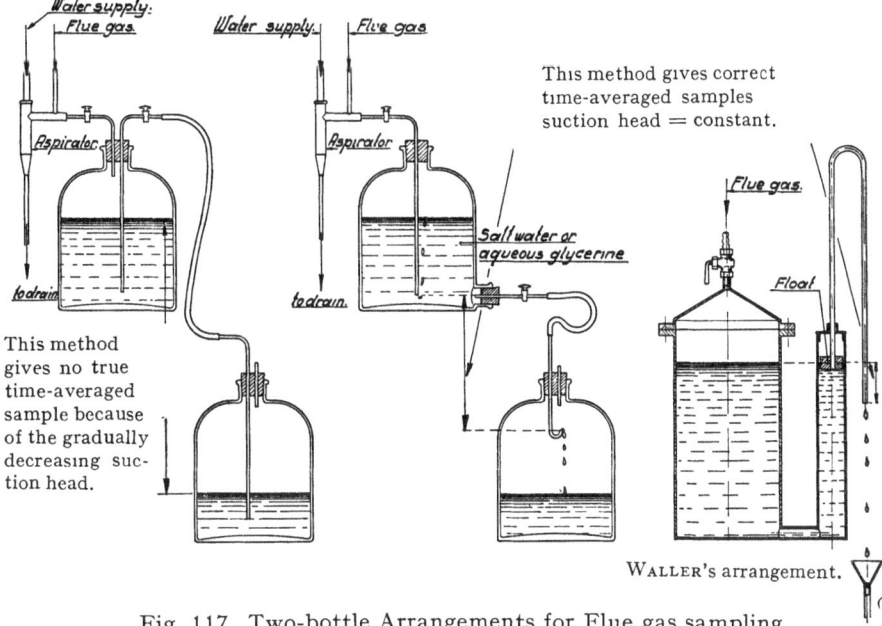

Fig. 117. Two-bottle Arrangements for Flue gas sampling.

fig. 114 it follows that for light domestic fuel oil this figure corresponds to 55% of excess air, provided the combustion be complete. With perfect adjustment it may be possible to go down as far as 25% of excess air, which corresponds to $12\frac{1}{2}\%$ of carbon dioxide, but such a fire requires very careful attention to keep the figures continuously as high as that. With regard to this point, it is of interest to note that a difference of $2\frac{1}{2}\%$ $CO_2$ between $12\frac{1}{2}$ and 10% only corresponds to 30% *difference of excess air*, whereas an equal $CO_2$ difference between 10 and $7\frac{1}{2}\%$ corresponds to 50%, and between $7\frac{1}{2}$ and 5% even to 95% difference of excess air, which clearly shows a hyperbolic increase. Consequently an increase of $2\frac{1}{2}\%$ from $7\frac{1}{2}$ tot 10% means more gain than $2\frac{1}{2}\%$ between 10 and $12\frac{1}{2}\%$.

Smokeless combustion practically without excess air, giving almost the "theoretical $CO_2$ figures" of about 15%, have been obtained only by application of the principle of *surface combustion*, which will be discussed in the last chapter of this section.

As previously pointed out, *the influence of excess air on the efficiency may be tremendous, especially if the heating surface is comparatively small for the quantity of heat to be transferred.* In such a case, in fact, it is necessary to utilize the principle of heat transfer by radiation to the utmost, which, however, requires the highest possible flame temperatures. From the graph of fig. 1C6 it may be seen that flame temperatures decrease roughly about 5C° C. for 10% increase of excess air and consequently, if the above-mentioned figures are used, a decrease from 12½ to 10% $CO_2$ correspond to 30% difference of excess air and *140° C decrease in flame temperature.* If taking the flame temperature at 12½% $CO_2$, equal to 1400° C. or 1673° abs., this means a decrease of $\left[1 - \left(\dfrac{1533}{1673}\right)^4\right] \times 100\% = 19.5\%$ *for the heat radiation.*

Moreover a decrease from 10% to 7½% $CO_2$, corresponding to 50% difference of excess air, gives a decrease in flame temperature equal to about 210° C., resulting in *44.5% decrease in heat transfer by radiation,* which again shows the same rapidly increasing hyperbolic influence.

This explains a curious fact which has been described by some authors before (SEELY & TAVANLAR in "Study of Performance Characteristics of Oil Burners and L. P. Heating Boilers", published in "Heating, Piping and Air conditioning", May 1931, page 426, and BARGEBOER: Warmtetechniek, Dec. 1931, page 144), namely, that an increasing amount of excess air initially may cause a *rise* of the flue gas temperature. This is obvious because the excess air, causing lower flame temperatures and less efficient heat transfer by radiation, makes it unavoidable that transfer by convection is more heavily loaded. Convection, however, being proportional only to the *first* power of the temperature difference, would require much more heating surface to make the flue gases leave the boiler with the same low temperature as before, and, if that extra required heating surface is not present, it follows that the flue gas temperature must be decidedly higher.

Thus the *unfavourable influence of excess air* might be called — using the words of SEELY & TAVANLAR — *"a double penalty",* as not only is this quantity of air heated from room to flue gas temperature and despatched through the chimney as useless hot air, but it also increases the flue gas temperature itself.

The following practical example may be given. Supposing an oil burning equipment for central heating for a small dwelling house consumes 3 kg. (0.8 Imp. Gln.) of oil per hour with 100% of excess air (7½% $CO_2$) and 250° C. (480° F.) flue gas temperature, a calculation may show roughly that the heat of the excess air, being heated uselessly from room temperature to flue gas temperature, would have been sufficient *to heat per hour about 1400 m³* (50,000 cub.ft.) of air from 10 to 20° C. (50 to 70° F.) *for room heating purposes.* By reducing the excess air to 25% (12½ $CO_2$), not only will two-thirds of this heat be transferred usefully to the hot water system, but also the flue gas temperature will be lowered by better radiation transfer, e.g. to 150° C. (300° F.), and consequently about three-fourths of the above-mentioned heat loss sufficient for *about 1000 m³ of room air to be heated from 50 to 70° F. per hour* will be gained by cutting down the excess air from 100 to 25%.

For natural draught burners, which, as a rule, operate with considerable quantities of excess air in order to obtain clean combustion (even 200 to 300% excess air may often be found), this may lead to excessive oil consumption.

*Regulation of the intensity of an oil fire* forms another class of quantitative control, but, as this comprises more constructional than theoretical problems, it will be left for discussion in the third Section.

---

## CHAPTER XII

### QUALITATIVE CONTROLS OF THE COMBUSTION PROCESS

Qualitative controls of oil burning may be divided into two classes, viz.:
A. Temperature Measurements and
B. Radiation Measurements,
which, as will be shown further on, are often intimately related.

### A. *Temperature Measurements*

Temperature measurements may be based on methods using as working substances: a) *gases*, b) *liquids* or c) *solids*, whilst a fourth group d) is in fact based on determinations of *radiation* converted for temperature measurement by application of some rule or other.

When considering these different methods for the *measurement of oil flame temperatures* one might easily arrive at the conclusion that but very few are suitable for this purpose. For instance:

*a) Gas thermometer* consisting of a heat-resisting reservoir exposed to the temperature to be measured and filled with gas, the pressure of which is determined and used as a basis for calculating the temperature according to the thermodynamic scale. This is a very expensive apparatus which requires the application of several complicated corrections and skilful attendance. Moreover, for oil flame temperatures up to 1600° C. a reservoir consisting of platinum-iridium filled with helium would be necessary, which, of course, affords no practical application.

*b) Liquid thermometers*, which need no further explanation, may be used up to

750° C., if filled with mercury and, above that, with nitrogen in order to prevent boiling. They are of no use for the determination of oil flame temperatures.

*c) Solid thermometers.* This kind of thermometers may be divided into 4 classes, viz.:

1) Thermometers based on the *linear dilatation by heat of a solid material* as compared with another, which principle has found extensive application in the form of a *bimetallic element* consisting of two sheets of metals of different linear dilatation welded together. The application of heat then causes the strip to bend under internal stresses between the layers, which movement may be used for the measurement of the temperature. These bimetallic thermometers cannot, however, be made for temperatures above approx. 300° C., it being necessary for reliable operation that no irreversible changes take place in the metals. In modern practice they have found a very useful application in safety controls (e.g. stack stats and flame stats) and room thermostats for domestic oil burning equipment.

2) Thermometers based on the *melting point of substances*; this principle has found extensive application in the well-known SEGER cones. Although this method may give very valuable and reliable information, especially in the case of oil flames for industrial purposes, it is felt as a serious drawback that its indication is not a continuous one.

3) Thermometers based on the *change of the electrical resistance of a substance* caused by application of heat. Gold, nickel, lead and platinum are mainly used for this purpose, whilst the electric resistance is generally measured by means of a WHEATSTONE bridge. The upper limit of application, however, is only about 1100° C. for platinum, as the relation between resistance and temperature at higher temperatures is not accurate enough to be used as a basis for measurements.

4) Thermometers based upon the *change of thermo-electrical potentiality* produced by the contact of two different metals when heat is applied. For temperatures up to about 1000° C. several suitable inexpensive combinations, so-called *"thermo-couples"*, have been found, such as *iron* and *constantan*, *chromel* and *alumel*, etc. For higher temperatures, however, more expensive thermo-couples are required, such as *platinum* and *platinum-rhodium* for temperatures up to 1600° C. or *iridium* and *iridium-rhutenium* for temperatures up to about 2000° C.

Moreover these couples are not quite reliable at high temperatures, especially when in contact with flue or flame gases, because of a mutual diffusion of both metals gradually taking place and resulting in considerable discrepancies. Therefore *thermo-couples* must be considered to be rather *unsuitable for contact measurement of oil flame temperatures*, although they may give good results when put into a small hole in materials to be heated or in a closed quartz tube surrounded by oil flames.

## B. *Measurements of Radiation*

Radiation is generally measured as a substitute for the determinations of temperatures. STEFAN-BOLTZMANN's law, assuming *the total radiation intensity of an absolute black body to be proportional to the fourth power of its absolute temperature,* allows of a conversion being made from the radiation figure to temperature. It follows, however, that this conversion only holds good if the radiating body is an absolutely black one, i.e. if it absorbs all radiant energy of any wave lengths received or, in other words, if it has no reflective power at all, a condition which in practice is never entirely fulfilled.

In fact, radiation proves to be highly dependent on the material, which is expressed by the *black body coefficient.* From the following table it may be seen that the condition of the surface has considerable influence on the values of this coefficient, which, moreover, is to some extent dependent on the temperature.

---

*Relative Radiating Power of surfaces of solids*
(Average values of black body coefficients for 500—1000° C.)

| | | | |
|---|---|---|---|
| Ideal black body . . . . . . . . | 1.00 | Copper (polished) . . . . . . . . | 0.13 |
| Cast iron (bright) . . . . . . . . | 0.22 | „   (calorized) . . . . . . . . | 0.19 |
| „   „   (oxidized) . . . . . . | 0.78 | „   (oxidized) . . . . . . | 0.57 |
| Steel (polished) . . . . . . . . | 0.29 | Nickel . . . . . . . . . . . . | 0.35 |
| „   (oxidized) . . . . . . . . | 0.79 | Platinum. . . . . . . . . . . . | 0.14 |

---

As *most optical pyrometers* (thermometers based on the measurement of radiation) *are gauged for "black body radiation"*, it is clearly seen from the above-mentioned figures that they will give enormous differences when applied to a "non-black body radiation", and this kind of thermometers would certainly not have found any practial application if there had not been a simple method of *making a black body of any material, viz. by making a hollow space in it* and eliminating all radiation from the outside. The radiation occurring in such a space is called *"black-body radiation"* and is *independent of the substance or its surface.*

Practically speaking, this means that if we want to measure the temperature of the inside of an oil furnace by means of an optical pyrometer having a scale based on black-body radiation, we are obliged to look into that furnace through a very narrow hole in order to avoid the possible entrance of daylight. Moreover it is advisable that the wall of the furnace, where it is seen through the hole, should be at an acute angle to the direction of the hole in order to prevent as far as possible reflection of daylight passing through the hole (see fig. 118). It is a well-known fact that a red-hot piece of iron looks *cold* in sunshine, which demonstrates fundamentally the same fault of observation.

If we want to measure the temperature, not only of a furnace, but also of the oil flame itself, still greater difficulty is encountered, as in that case we must avoid the

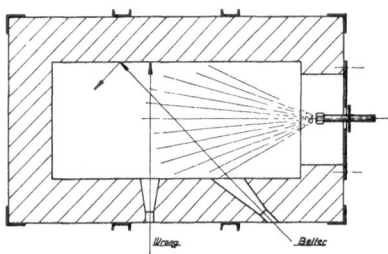

Fig. 118. Optical temperature measurement on furnaces.

reflection of the glowing furnace walls through and onto the flame, and in this connection it is found that even yellow oil flames are often far more transparant than they are thought to be. Therefore it is advisable to look at the flame under an acute angle to its direction, in order that the layer of flame gases observed may be as thick as possible (see fig. 118).

Another practical departure from STEFAN BOLTZMANN's law is that, whereas according to this law the *total* radiation intensity is proportional to the fourth power of the absolute temperature, several optical pyrometers measure or compare only a *part* of the radiated energy, namely *light rays*.

It has been shown before in fig. 107 that the energy emitted by means of *visible* rays (area between the vertical lines a and b) is very small as compared with the total energy, the greater part of it being emitted by means of *invisible infra-red rays*. Now this would not have been so serious if there were only a fixed proportionality between visible and invisible energy constant for all temperatures, which, however, is not at all true. At higher temperatures the maximum of radiation intensity moves in the direction of smaller wave lengths, as formulated by WIEN's law, which states that the wave length of maximum intensity is inversely proportional to the absolute temperature.

From the foregoing it must be clear that *optical pyrometers cannot be considered as exact instruments* for measuring absolute temperatures. On the other hand they are very valuable for *comparison* of temperatures.

Various types of optical pyrometers will be briefly discussed below:

1) *Optical pyrometers measuring total radiation.*

From a theoretical point of view these pyrometers would be better than those based on measurement or comparison of visible rays only. Pyrometers of this type all include the principle of collecting some heat radiated by the source to be measured on a black body and measuring the heating rate of that body. This may be done either by means of a small telescope placed before a peep-hole at the outside of the furnace (*external* pyrometer) or by placing the black body itself in the furnace (*internal* pyrometer).

Out of several others two will be discussed which have been developed by Dr. HASE. His *external* optical pyrometer consists of a small thermopile having a black surface which is heated by radiation of the furnace emitted through a peep-hole and concentrated by means of a quartz lens. The thermo-electric potentiality is measured by means of a milli-voltmeter directly reading in degrees Centigrade.

HASE's *internal* optical pyrometer has been described in "Archiv für Wärmewirtschaft", Dec. 1932, and consists of a *copper globe blackened with uranium oxide,* which is put into the furnace (fig. 119). *From the time necessary to heat the centre of this globe* (containing a thermo-couple) *from 300° to 400° C.* it is possible to calculate the intensity of radiation into the surface. In fact, this method, which consists, properly speaking, of a *trialheating on a small scale*, produces exactly the figure which is often asked for in several industrial

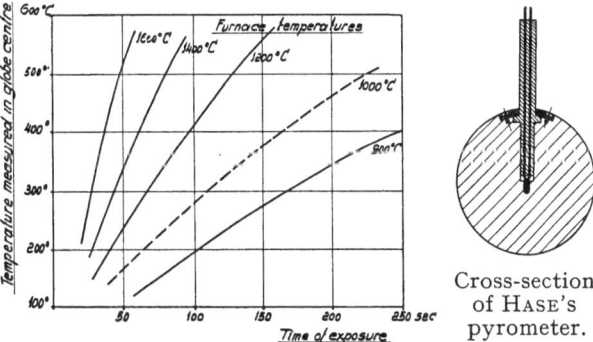

Cross-section of HASE's pyrometer.

Fig. 119. HASE's internal pyrometer.

applications of oil burning, viz., *the rate of heating by direct contact with flame gases.* Moreover, this last method avoids the discrepancies caused by neglecting the black body coefficient, although on the other hand not only radiation but also convection is measured.

If an optical pyrometer based on the measurement of total emission and having a scale which applies to radiation of an absolute black body is used for temperature measurement of a non-black body, the result will be a temperature which is too low, and an *additive correction* will be necessary. This is obvious because the total energy emitted by a black body is larger than that of a non-black body at equal temperatures (see fig. 120 A).

This decrepancy increases with decreasing black body coefficient and may even grow to tremendous proportions if flames are observed, such as has been illustrated diagrammatically by fig. 120 B. A *non-luminous flame* produces only a few bands of infra-red rays, as is shown by curve (b), and it is clear that this radiation includes only a small fraction of the total energy emitted by a black body of the same temperature. Nevertheless the pyrometer based on *total* emission erroneously calls temperatures equal if in both cases the areas below the emission curves are equal.

A *luminous flame* shows a smaller difference (curve c), but it may be taken as a rule that *this type of external optical pyrometers based on total emission should not be used for oil flames, but only for temperature measurements of glowing solid surfaces.* Of course, they may be used for oil flames too if it is only needed to compare the radiative power of two flames and not to measure the flame temperatures.

Other pyrometers of this type are those of FÉRY, FOSTER, etc.

2) *Mono-chromatic Optical Pyrometers.*

These pyrometers compare the intensity of *mono-chromatic emission* by the body of which the temperature is to be measured with that of a black body, and if both intensities are equal the temperatures are said to be equal too, which, however, in many cases may be rather doubtful.

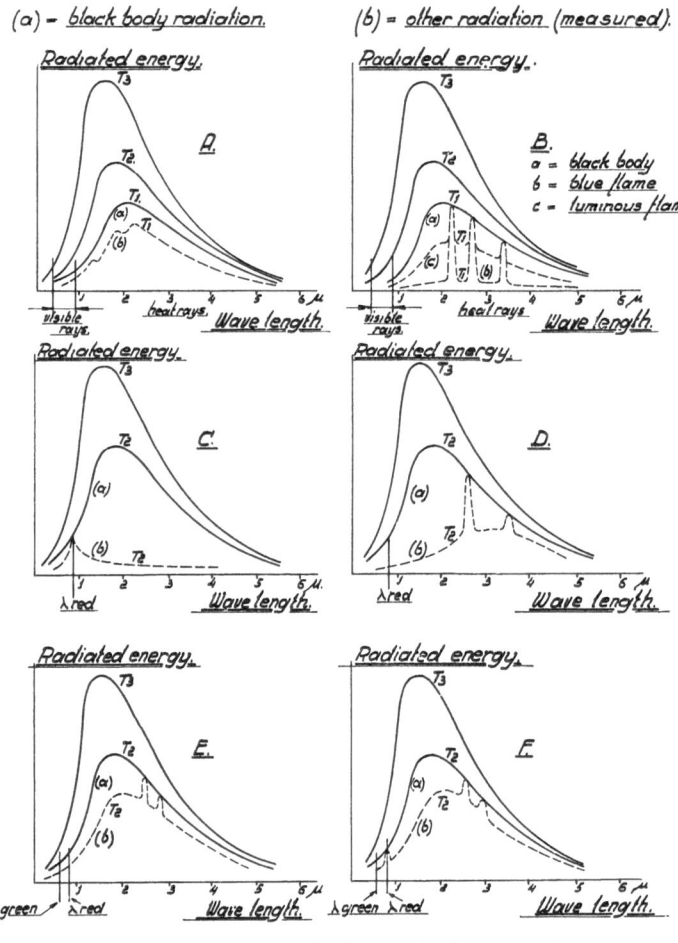

Fig. 120. Errors made by optical pyrometers.

HOLBORN and KURLBAUM have developed a pyrometer on this principle. The total emission of a furnace or flame is first made *monochromatic* by means of coloured glasses and then, by regulation of an electric current, a glowing filament of an electric lamp is adjusted in such a way that the filament disappears if it is observed against the background of the monochromatic furnace emission. An ammeter indicating the current makes direct reading in degrees Centigrade possible.

If a body shows an emission curve (b) of fig. 120 C apparently having equal intensity for red rays to that of a black body of the same temperature, then a monochromatic pyrometer which compares the intensities of *red* will give a correct reading of the temperature.

If, however, a body shows an emission curve (b) of fig. 120 D the reading will be absolutely wrong, so that it is quite clear that great mistakes are scarcely avoidable if the composition of the emitted energy is not known.

As a rule monochromatic pyrometers may give reasonable results for glowing metals (provided reflection from the outside is eliminated) *but they are useless for flames.*

3) *Bichromatic Optical Pyrometers.*

To reduce the danger of wrong applications of monochromatic pyrometers, another type has been developed which compares the *ratio of emission intensities of two colours* (e.g. red and green) of the body of which the temperature is to be measured with that same ratio of a black body, and temperatures are said to be equal if these ratios are equal, which, of course, is again open to many objections.

It goes without saying that there are some cases in which such bichromatic pyrometers may give good readings, but there are also a great many in which wrong

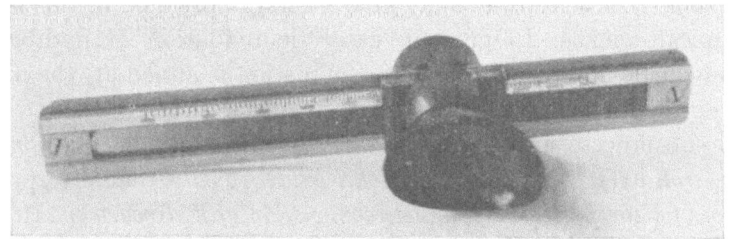

Fig. 121. Bichromatic optical pyrometer STRÖHLEIN.

results will be obtained. Fig. 120 E, which represents roughly the emission curve for oil flames as compared with that curve for a black body, shows that the ratios of the intensities of red and green are almost equal at equal temperatures. It follows that as a rule *bichromatic pyrometers may be used for oil flames*, although they may show large discrepancies if either red or green rays are produced artificially, e.g. by traces of copper salts, etc., in the flames (see fig. 120 F).

STRÖHLEIN (Düsseldorf) have developed such a bichromatic pyrometer, which is a cheap and handy apparatus. It compares the ratio of green and red and must be adjusted in such a way that a yellow tinge appears between green and red; this adjustment, however, requires some experience (see fig. 121).

# CHAPTER XIII

### CATALYSIS AND SURFACE COMBUSTION

Before closing this section a chapter may be devoted to catalytic action and surface combustion.

It was discovered as early as 1817 by DAVY that platinum was able to ignite a mixture of hydrogen and oxygen at room temperatures, obviously due to a surface action during which the platinum itself suffered no perceptible change. Although in the mechanism of this catalytic action there are still many unexplained problems, it is generally understood that hydrogen and oxygen are adsorbed by the platinum surface and thus brought into a condition which is more favourable to combustion than if they are in the free gaseous state. *Platinum might be called "a matrimonial agent"*, so to speak.

In the beginning of this century a tremendous number of catalytic agents of widely divergent nature and suitable for the most varied reactions were found. SABATIER's book "Die Katalyse" may give a good impression of this new branch of physico-chemical science. In general, catalytic oxidation of hydrocarbons finds comparatively little interest if *complete* combustion is aimed at, the main object of research being the formation of *intermediate* oxidation products.

Several substances, such as *platinum, tellurium, iridium, osmium, gold, silver, copper, iron, iron oxide, copper oxide, nickel oxide, cobalt oxide*, etc., proved to have *catalytic power for the partial or complete combustion of hydrocarbons*. However, it was often found to be necessary first to prepare the surface in a suitable way, which is often called *"activation"* and probably consists in the removal of foreign elements and the establishing of a surface of capillary porosity. *"Activated charcoal"* has in particular found widespread industrial application.

Activation may often be destroyed by the presence of certain substances, which are therefore detrimental to the catalytic action and are consequently often called *"catalyst poisons"*. Thus *sulphur* is an example of a poison for a large number of the above-mentioned catalytic agents for the combustion of hydrocarbons, and as sulphur is present in nearly all mineral oils and extremely small quantities may be sufficient to produce this poisoning effect, it goes without saying that the application of catalysts for the ordinary purposes of the combustion of fuel oil has been almost negligible.

Some patents on slow catalytic combustion, most of French origin, are the following British Patents: 253,124 (1925); 266,765 (1925); 273,668 (1926); 322,117 (1928); 364,342 (1931); 381,783 (1931); 382,472 (1932) and 392,822 (1932), which as a rule are only meant for *sulphur-free* gasolines.

Moreover, it was found that *above red heat practically all surfaces have an equal catalyzing power* and thus, although platinum, ferric oxide, etc., are good catalysts at lower temperatures, incandescent coal, glowing refractory materials, etc. are equally good at higher temperatures, and as the technical applications of oil burning usually only ask for *high* temperature combustion, there is little reason to use special catalysts to allow of the combustion of fuel oil at *low* temperatures.

The use of the *catalyzing power of glowing surfaces*, originally discovered by BONE about 1902, has found some practical application known as *"surface combustion"* (e.g. BONECOURT boilers of Surface Combustion Company, N.Y., Br. Patents: 11,957; 11,958; 23,536; 24,113; 28,477 about 1911—'13), a process comprising a combustion space filled with a porous granulated refractory material into which a gaseous combustible mixture is introduced and there burnt without flame. Thus very high temperatures may be obtained, whereas the combustion, being strongly stimulated by an intensive all-round radiation at short distances, is perfect with a minimum of excess air. This principle probably includes useful future prospects, its main drawback at present being that only *clean gases* can be used as fuel, and consequently fuel oil would have to be completely gasified first, which is not so easily done.

A *recent application of the surface combustion* principle is the frequently mentioned *range-burner*, the glowing perforated concentric shells of which may be considered as the simplest form of a porous material. Notwithstanding comparatively low combustion temperatures and the absence of turbulency, its perfect combustion is obtained by the intensive radiation of the glowing surfaces at short distances.

# SECTION III

MODERN FORMS OF CONSTRUCTION OF OIL BURNING DEVICES

———————

# CHAPTER I

## A. *Decimal System and Others*

The *Historical Development of Oil burning* having been discussed in the first Section and the *Modern Theoretical Bases of Combustion* in the second, the third Section will be devoted to the *Modern Forms of Construction of Oil burning Devices*.

Since the beginning of this century such a tremendous number of oil burners have been invented for the most varying purposes that, before any discussion of details can be begun, a suitable system has to be found for classifying the various types of burners in order to avoid hopeless confusion. An idea of the vast number of oil burner designs may be gained when it is noted that class 75 of the *"Abridgements of specifications of Patents for inventions"* as published by the British Patent Office — which class comprises *"Burners and burner fittings"* — contains no less than some 6000 patents on burners for the period 1909–1933 and, although perhaps only one half of these burners have been designed for liquid fuel, it must be remembered that these are only oil burning devices for which a British patent was granted; during the year 1933 about 45 patents on oil burners were filed in Great Britain, 150 in the United States, 30 in Germany and 15 in Holland, these together making about 240.

Comparatively few systems have been devised for the classification of oil burners, which are generally divided into: 1) steam atomizing, 2) high pressure air atomizing, 3) low pressure air atomizing, 4) mechanical atomizing and 5) vaporizing burners.

The patent offices of various countries have their own systems and index keys, which cannot always be kept up to date, it being impracticable to alter the system of filing too often. On the Continent of Europe the *Decimal Classification System* has been adopted to some extent, a specimen of which is shown in fig. 122. Every type of oil burner and even every part of an oil burning equipment is indicated by a number, which fully fixes its place in the system, figures from left to right defining still more details.

Thus 662.94 comprises all furnaces for liquid fuels and 662.95 the same for gaseous fuels. The next number added indicates a detail, and thus 662.942 comprises "furnaces for liquid fuels being pulverized by centrifugal force". Another number added indicates a division into further details, thus 662.942.2 being the indication for "oil burners used in furnaces for liquid fuels being pulverized by centrifugal force", and so on. Of course there are nine numbers available for each division into equivalent details, which offers enough possibilities for extension. Although this system has many advantages for bibliographic work, it is not so suitable for a technical survey, as indications by figures alone demand too much of the reader's attention and imagination.

CLASSIFICATION OF OIL BURNERS ACCORDING TO THE DECIMAL CODE

## 662.94 Furnaces for liquid fuels.

662.941     General problems and construction details.
    .2    Burners for liquid fuel in general.
    .3    Pipings and fitting for the transportation of the oil in general.
    .5    Devices for regulating safety in general.
    .8    Special devices for viscous or colloidal oils.

662.942     Furnaces for liquid fuel pulverized by centrifugal force.

662.943     Furnaces for liquid fuel pulverized by other mechanical means.
    .2    Burners.
    .3    Means for transportation of oil to the burners.

662.944     Furnaces for liquid fuel pulverized by means of air, steam or other gases.
    .2    Burners.
    .3    Means for transportation of oil to the burners.

662.946     Wick burners.
    .2    Burners.
    .22   Wicks.
    .3    Means for transportation of oil to the burners.

662.947     Furnaces for liquid fuel comprising an absorption surface.
    .2    Absorption surface.
    .3    Means for transportation of oil to the burners.

662.948     Furnaces for liquid fuel previously gasified. (For burners see 662.951.2).
    .3    Devices for previous gasification.

662.949     Furnaces for liquid as well as solid fuel.

## 662.95 Furnaces for gaseous fuels.

Fig. 122. Example of the Decimal Classification

In America another method called *"mechanical analysis"* has been found to be useful. This system was originally composed by the *"Fuel Oil Journal"* especially for

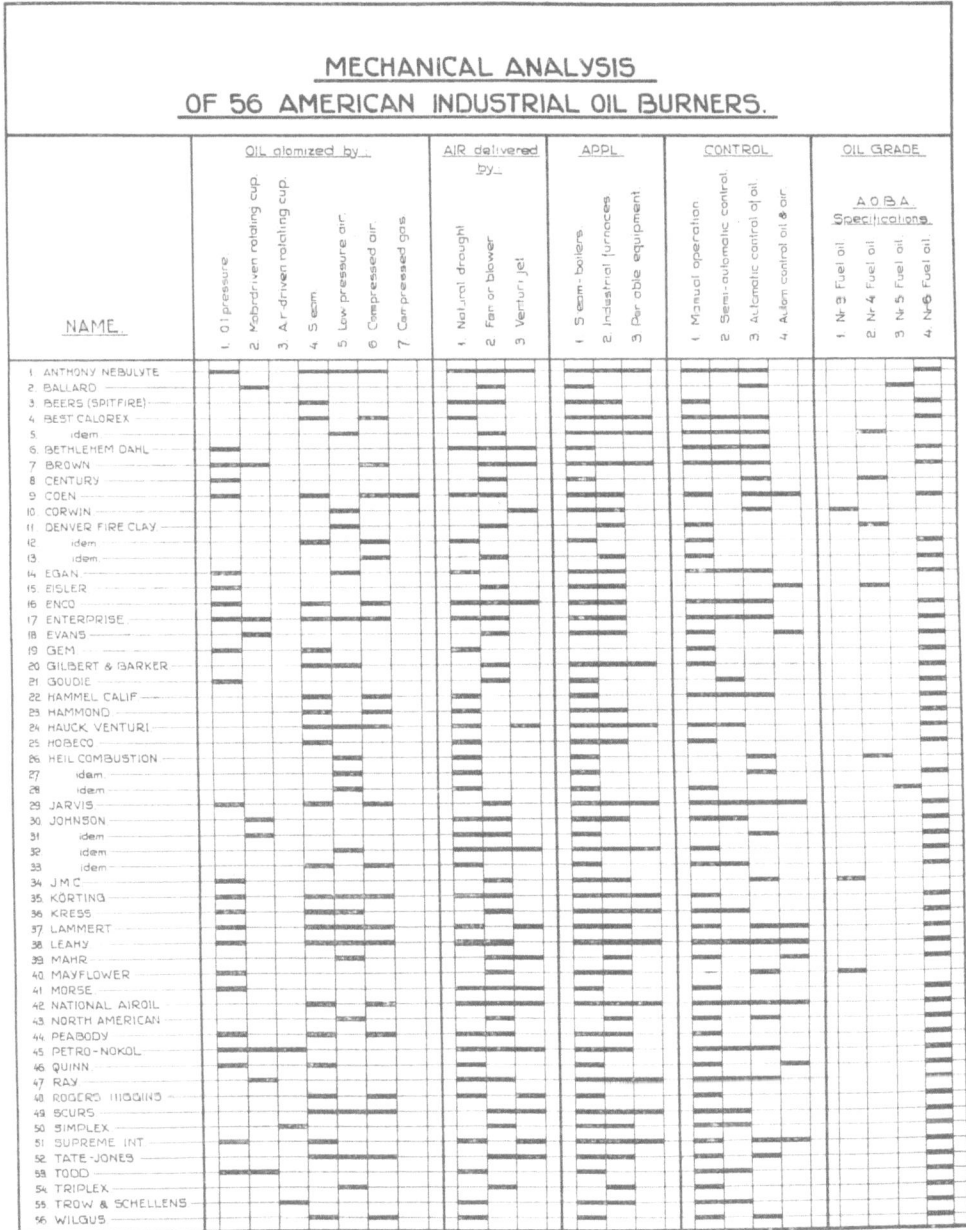

Fig. 123.

Mechanical Analysis of 56 American Industrial Oil burners.

domestic oil burners. An oil burner is described by means of a combination of letters and numbers, the former indicating groups, the latter details thereof, which of course, being fundamentally a simplified decimal system, has the same disadvantage of tiring the reader too much. To meet this objection a graphical indication has been worked out in Table I based on the figures collected by the "Fuel Oil Journal" for *179 American domestic oil burners* which were on the market in 1932. This table affords a quick survey and is especially instructive as regards the various methods of operation most commonly used. Fig. 123 represents a similar survey for *industrial oil burners*.

Although such mechanical analyses may be very useful for commercial purposes, they are too superficial for deeper technical studies, and this is the reason why the author looked for another system.

Systems may be composed on several basic principles, for instance:

1) *The method of preparing the oil for combustion.*

This basis has been found to be most fertile, because from it there spring forth many essential factors.

2) *The method of preparing the air for combustion.*

This qualification offers no great advantages, as the air conditions show less variations in the combustion process than those of the oil. The possibilities of indications such as: Preheating or not, using compressors, low pressure blowers, fans or natural draught, are soon exhausted.

3) *The method of combustion.*

Although this is a very useful indication, the most important factor, viz. the preparation of the oil, is antecedent to the combustion. Indications such as high or low temperature combustion, free or surface combustion, yellow or blue flame combustion, flame form, etc., are mainly governed by the method of preparing the oil, and therefore such systems would not be able to show the basic principle of various combustion processes in a clear way.

4) *The source of energy used.*

This offers little variation.

5) *The application.*

Although every application has its own peculiar demands, this indication does not allow of enough fundamental differences being shown.

6) *The method of regulation.*

This forms a comparatively small detail, which is only of fundamental importance in some special cases and, therefore, is not suitable for use as a basis for general classification.

### B. *Classification according to the method of preparing the oil*

A classification based on the *method of preparing the oil for combustion* obviously offers the best general possibilities, and this is the reason why that indication has been chosen for further elaboration. However, it must be pointed out that the other five indications, although of minor importance, may often give valuable information too, and consequently an ideal system of classification should really comprise all six sub-systems used simultaneously. For the sake of simplicity, however, these will be omitted and only one system based on the first indication will be drawn up, with some remarks belonging to the other systems added as explanatory notes now and then. Thus the Table III, called CLASSIFICATION OF OIL BURNERS ACCORDING TO THE METHOD OF PREPARING THE OIL FOR COMBUSTION, also contains here and there information regarding the preparation of the air, methods for regulation, etc. Moreover, attention is drawn to the fact that this table is only given as a suggestion, which is anything but complete and certainly open to further improvements.

Of course it is impracticable to discuss every existing oil burner and, anyhow, this would certainly be confusing rather than instructive for the reader; consequently the idea of the author in composing the Table III has been to give just enough to show the fundamental principles of each typical application and to treat the subject on broad lines, not entering into a minute discussion of various details.

In accordance with the above-mentioned table, the various types of oil burners will be classified into 9 classes, as follows:

| Division | Group | Class | Heading |
|---|---|---|---|
| A | I | | Preparing the oil without auxiliary agents or mechanical means (see Table IV). |
| | | 1 | By means of evaporating the oil |
| | | 2 | ,,    ,,    ,, gasifying the oil |
| A | II | | By mechanical means (see Table V) viz., |
| | | 3 | By means of dynamic pulverization |
| | | 4 | ,,    ,,    ,, kinetic        ,, |
| B | I | | Preparing the oil with auxiliary agents by using steam (see Table VI). |
| | | 5 | Steam produced by burner itself |
| | | 6 | Steam produced by foreign source |
| B | II | | By using air (see Table VII) viz., |
| | | 7 | ,,    ,,    high or medium pressure air |
| | | 8 | ,,    ,,    low pressure air |
| B | III | 9 | By using flue gases. |

# CHAPTER II

## VAPORIZING BURNERS COMPRISING OPEN SYSTEMS UNDER ATMOSPHERIC PRESSURE

The class A-I-1: *Preparing the oil by means of evaporation without using auxiliary agents or mechanical means,* may be subdivided according to whether evaporation takes place in *open devices* of atmospheric pressure or in *closed vessels* under higher pressures. In the first case the gas or vapour evolved is not able to develop any appreciable kinetic energy, whereas in the second case part of the heat applied for vaporizing the oil may be transformed into kinetic energy, e.g., by causing the vapour to issue from a small aperture, so that burners of the latter so-called "vapour jet type" may dispose of enough energy to draw in sufficient air for combustion without any foreign source of energy (fig. 124).

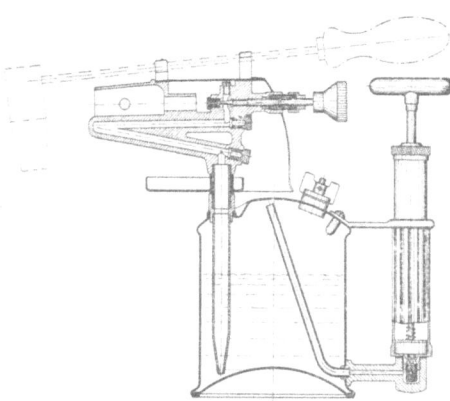

Fig. 124. SIEVERT soldering lamp.

### A. *Open systems under atmospheric pressure*

It has previously been pointed out that evaporation occurs if the oil is heated to temperatures equal to, or higher than the average dew point of the oil used. *Most hydrocarbons with boiling points below 350° C. may be evaporated without any carbon being formed, provided their vapours are not heated above this limit of 350° C. either. Higher-boiling hydrocarbons,* however, such as are contained in residues, *are cracked before being evaporated* (at atmospheric pressure) and, as this cracking process is usually accompanied by the formation of coke, these fuels tend to cause *carbonization* if used in burners based on the principle of pre-evaporation of the oil.

In this connection it has been explained before that such *"vaporizing burners"* giving the least carbon troubles are those in which the *smallest quantity of oil is exposed at a time to the evaporation temperature,* because in that case the oil, after being admitted into the vaporizer, has not to wait so long before being evaporated and burnt and consequently there is less opportunity for carbonization, as the amount of carbon formed increases with the time of exposure to heat.

Therefore it is a logical consequence to make another subdivision according to whether *the quantity of oil exposed at a time to the vaporizer temperature* is relatively:

1) *large*, 2) *medium* or 3) *small*, the last group being further subdivided according to three other principles of practical application, viz., where the oil is present a) *in drops*, b) *in an oil film* or c) *in a porous material*.

Another important characteristic of this type of burners is the *method of supplying heat to the vaporizer*, which may be done

1) By *direct radiation*.

2) By *indirect* or so-called *screened radiation*.

3) By *heat conduction*, or

4) Under control of a *thermostatic device*,

each of which is divided into three subdivisions, according to whether the heat is transferred:

*a*) By radiation, convection or conduction of the flame heat in backward direction *without the flame being turned itself*.

*b*) By radiaton, convection or conduction of the flame heat where *the flame is turned backwards (vortex or swirl)*.

*c*) By radiation, convection or conduction of a *pilot flame* using the same fuel.

Sometimes pilot flames of city gas are used, but strictly speaking this includes the use of an *auxiliary* agent, which departure from the system will be disregarded here for the sake of simplicity.

When combining the above-mentioned five subdivisions in a vertical sense (columns) with the other twelve groups in the horizontal direction (rows), we have no less than *60 different types of oil burners, all based on the principle of pre-evaporation* of the oil. It would have been possible to draw a diagrammatic scheme for every type, but for the sake of simplicity this has been omitted, only some patents having been quoted as examples for every column (see Table IV).

We have seen that *superheating* of oil or its vapours may cause carbonization, resulting in a carbon deposit in the vaporizer which often interferes with the regular operation of the burner. To avoid this evil it was explained that the best way was to keep the quantities of oil present at a time in the vaporizer as small as possible, and consequently with respect to carbonisation troubles *the three columns to the right will as a rule include better possibilities* than the other columns to the left.

Another point of special importance which has only too often been seriously neglected by designers of this type of oil burners is the *maintenance of the vaporizer temperature at or slightly above the average dew point of the oil*. This might be called the *"golden rule" for vaporizing burners*, as it is entirely on this that reliable operation depends, and consequently again with respect to carbonization troubles the fourth sub-class: control by a thermostatic device, will as a rule include better principles than the other rows above.

Instead of discussing every representative of the above-mentioned 60 different types of vaporizing burners, working under atmospheric pressure, only some of the most interesting will be considered here.

### B. *Directly radiated kerosene wick burners*

The ordinary kerosene wick lamp belongs to the class 1 of evaporation at atmospheric pressure, the quantity of oil exposed at a time to evaporation being small and absorbed in a porous material (upper wick surface) (column 3c), whereas the heat required for evaporation is obtained by direct radiation of the flame, which is not returned (row 1a).

As a rule direct radiation is not favourable to close control of the vaporizer temperature, because the heat radiated, being proportional to the difference between the *fourth* powers of the temperatures of flame and oil, will increase considerably, even with a slightly increasing flame temperature. Moreover the radiation of heat depends in a marked degree upon the flame surface and its carbon density (luminosity), which also tends to make a close regulation of the vaporizer temperature extremely difficult.

We shall see further on that kerosene wick lamps provide another very simple means of automatic control of the vaporizer temperature, viz., by the *rate of evaporation*.

Fig. 125 shows some stages of development of the kerosene wick lamp. The first lamps using *round massive wicks* were probably inspired by the wax candles. Their capacity for soot-free combustion was rather limited because larger diameters of the wick very soon caused a core of heavy cracking in the flame resulting in a high tendency to soot. An improvement was obtained by adopting *flat* wicks.

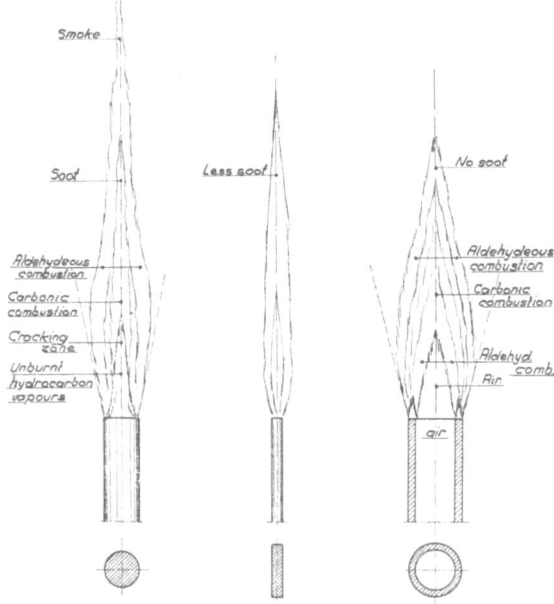

Fig. 125. Some stages of the development of oil lamps.

The next stage using *annular* wicks with a hollow centre for air admittance reduced this sooting tendency considerably, obviously by admitting non-decomposable air to the core of the flame, thus reducing cracking on the one hand and promoting soot-free aldehydeous combustion on the other. A lower tendency to soot meant greater capacities, larger flame surfaces and therefore more light per unit. The distribution of the air being of the utmost importance, this was guaranteed by making use of the natural draught of a glass chimney of suitable length and diameter.

A further stage of development comprises the *deflector blue burners*, which sprang to life when kerosene was gradually superseded for lighting by gas and was increasingly used for heating alone. With this type of burners the air, admitted in the centre of an annular wick, was moreover *blown into the flame* by means of a metal disc, called *deflector*. The air, being entrained by the natural draught of a round sheet-iron portable stove, usually without chimney connection, was blown with such force into the flame that an almost blue combustion of the *aldehydeous type* occurred. The circumstance that this blue flame could be obtained only by using *excessive amounts of excess air*, did no harm to the application for room heating, as the combustion gases were allowed to flow freely in the rooms, an efficiency of 100% being thus guaranteed whatever excess air used, provided combustion were complete. The aldehydeous combustion was very advantageous because of the absence of soot, which would have been inconvenient when heating rooms in this way (see fig. 126).

Yet another stage of development is represented by the *reverse flame blue burners* ("range burners") which have been discussed before in Section II. The *"reverse flame principle"*, according to which a *"flame of air"* is burnt in an atmosphere of hydrocarbon gases or vapours, allows of establishing a *perfect aldehydeous soot-free combustion without large quantities of excess air*. Consequently this is more economical and, moreover, the heat transmission onto the surfaces to be heated is greatly increased by the presence of glowing shells of steel, which serve as efficient "radiators" (fig. 71).

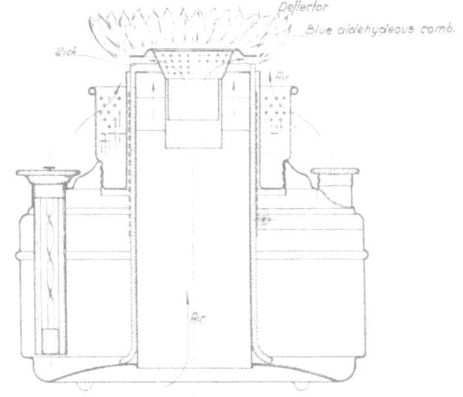

Fig. 126. Blue burner of the deflector type.

All these kerosene burners have a common means of automatic regulation of the *vaporizer temperature*, viz., by the *rate of evaporation*. If a little more heat should be applied to the vaporizer, this will be immediately answered by increased vaporisation which, reduces the temperature again, and vice versa, resulting in an automatic regulation of the temperature to just the *average dew point of the oil*.

For this method of regulation, however, it is necessary that:

1) The *fuel should be comparatively light* and should vaporize readily at atmospheric temperature without appreciable chemical changes, which leave residues behind.

2) The *fuel supply* to the upper wick surface must be practically *unlimited* to be able to respond to any increase in temperature by a corresponding evaporation.

These were the reasons why it was found necessary to fix the following specifications for kerosene, viz.:

*a*) It may not contain any high boiling hydrocarbons which will not readily evaporate, as otherwise they are apt to accumulate on the wick surface and thus spoil the automatic temperature control.

*b*) It must have a rather low viscosity, ensuring an unlimited oil supply through the wick.

It has often been thought to be an improvement to use *mineral wicks* (asbestos) instead of cotton wicks, but it must be pointed out that certainly more than 90% of the carbon of a dirty wick originates from the oil and not from cotton, and that as a consequence it is far more important to protect the oil than the cotton from carbonization, so that the use of mineral wicks does not in the least solve carbonization troubles. Moreover cotton wicks, as a rule, have much better fuel transporting qualities than mineral wicks, which is a very important factor for the control of the wick surface temperatures.

Other constructions are known using a *quartz sand layer* as an absorptive medium, especially for heavier oils (gas oils) the viscosity of which is too high for ordinary wicks. (See Am. Pat. No. 1,624,943 of GIBBS and GUILLERMIC, "Le chauffage par les Combustibles Liquides" page 42.)

The range burner type uses no absorptive substance at all, the *free oil surface* controlled by a liquid level device being exposed directly to the radiation of the red hot shells. The asbestos wicks which are present in the oil grooves are only used for priming.

## C. *Indirectly radiated kerosene wick burners*

One representative of this type of burner was discussed in the preceding paragraph, viz., the *deflector blue burner* (see fig. 126). The object of putting a *screen*

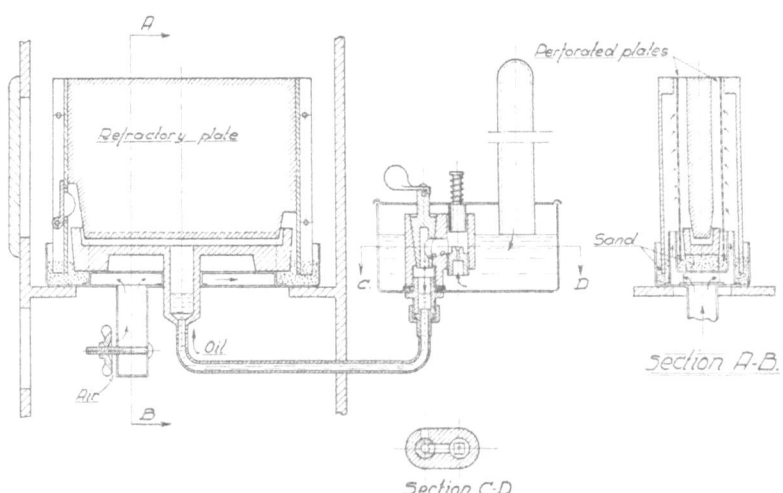

Fig. 127. Blue burner of ERICHSEN-FALKENTORP.

between flame and vaporizing oil is *to temper the heat transfer* in order to reduce the risk of carbonization of the oil. For this type of burner, however, the screen had another purpose, viz., that of obtaining blue flame combustion.

The yellow ring wick flame is transformed into a blue flame merely by placing a so-called *flame deflector* at some distance above the wick surface. Although this deflection was originally intended to blow air into the flame from the inside, some additional advantages were obtained with regard to the control of the wick surface temperature, viz.,

1) The surface of the flame, now being spread horizontally, no longer had as much influence upon the temperature of the wick surface as it had with a vertical cylindrical flame.

2) The radiative power of the blue flame is considerably less than that of the yellow flame, which also reduces the direct influence of the flame upon the wick surface temperature.

3) The wick surface was then mainly heated by the radiation of the metal deflector, the temperature of which is governed by an equilibrium between the radiation of heat from the flame on the one hand and cooling by air on the other. Thus the quantity of air used was given some levelling influence upon the wick surface temperature, which was particularly favourable for getting higher temperatures at low load and lower temperatures at high load.

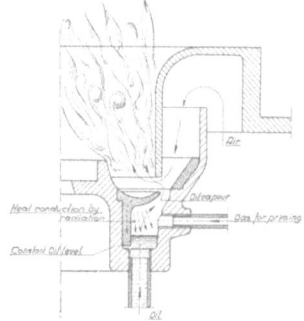

Fig. 128. Example of screened radiation.

Other examples of such *"screened" radiation* are given by fig. 127 comprising the ERICHSEN-FALKENTORP burner (Dutch Patent no. 18,293) and fig. 128.

### D. *Burners using heat conduction for vaporizing the oil*

The object of following this method is to reduce the influence of the flame

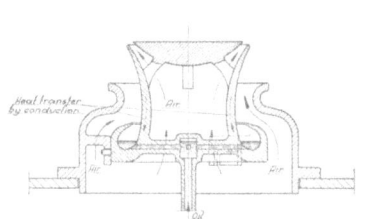

Fig. 129. Heat transfer to the oil by conduction.

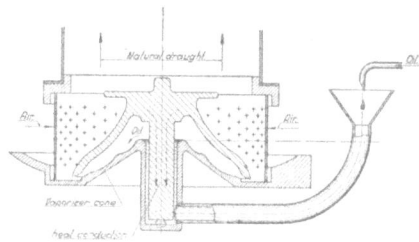

Fig. 130. French vaporizing burner "Flamme bleue".

temperature upon the temperature of the vaporizer, because *heat conduction* is only proportional to the *first* power of the temperature difference, whereas *radiation* is proportional to the difference of the *fourth* powers of the temperatures. Thus conduction has a levelling influence upon the temperature fluctuations of the vaporizer.

Examples of this sub-class are represented by figs. 129 and 130 (*"Flamme bleue"*), the last being a combination of conduction and screened radiation.

Considerable sums of money have been spent on research work in attempts to fix suitable dimensions, proportions and materials in order to obtain the right vaporizer temperatures under normal conditions. However, for other than normal fuels with different average dew points, other temperatures and other dimensions will be required, so that this type of burner is only suitable for the particular fuel for which it is designed, *if no means are provided for adjusting vaporizer conditions accordingly in a simple way.*

### F. *Burners with thermostatic control of the vaporizer temperature*

We have seen that kerosene wick burners in fact belong to this sub-class comprising automatic control of the vaporizer temperature by the automatic adjustment of the rate of vaporization. This method, however, only applies to fuels which are easily completely vaporized, e.g. gasolines and kerosenes. For gas oils carbonization troubles arise, which make operation unreliable; other means of control, such as thermostats, are necessary.

Although this principle comprises methods for burning even the heaviest distillates without carbon troubles, merely by adjusting the thermostat at or slightly above the average dew point of the distillate fuel used, there are not, as far as the author is aware, any burners of this type to be found on the market or in patent literature.

As representatives of this principle, which may hold promising prospects, two devices may be mentioned here: One burner, designed by an Austrian engineer RÖDL, with *electric* temperature control, and another suggestion by the author for the use of a *boiling bath* of diphenyl amin, a substance which boils at approx. 300° C. Since these devices are still in the experimental stage, discussion of them will be left to the fourth Section.

### F. *Burners the flames of which are turned backwards in a swirl or vortex*

There are a number of burners known, usually comprising a hot plate or some other surface on which the oil is spread out in a film, the flame of which heats that surface not only by radiation but also by *contact with the flame gases.* In order to effect this it is necessary to return the flame towards the vaporizer, which may be done by *flame deflectors* or by *imparting a swirling or vortexing motion to the flame gases.*

Swirls may be cylindrical, spiral, screwing or "onion-like" and moreover single, double or multiple. Such swirling or vortexing flames have often been used as a means of heating and evaporating the oil and at the same time burning the vapours by clean combustion. Of the various applications the following may be quoted: Swiss patent No. 85,701 (1918) of GENERAL ENGINEERING Co. (double swirl); D.R.P. 492,520

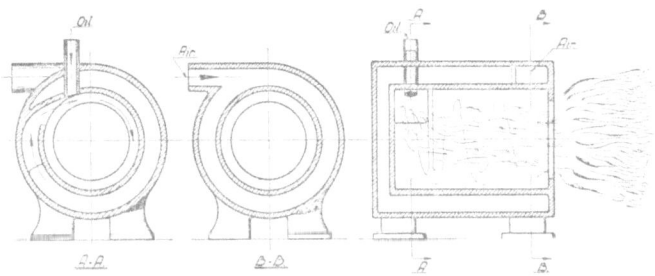

Fig. 131. STRACK oil burner.

(1930) (simple swirl); Br. Patent No. 224,511, FULLER FUEL Co. (1923) (simple swirl); Br. Patent No. 204,951, BRETELL (1922) (simple swirl); French patent No. 690,455 (1930), STRACK (simple vortex) fig. 131; Am. Patent No. 1,782,050 (1930), POWERS (double swirl); Am. Patent No. 1,624,943 (1927), GIBBS (double vortex) (fig. 132).

Of course this method may be used for all three categories, direct radiation,

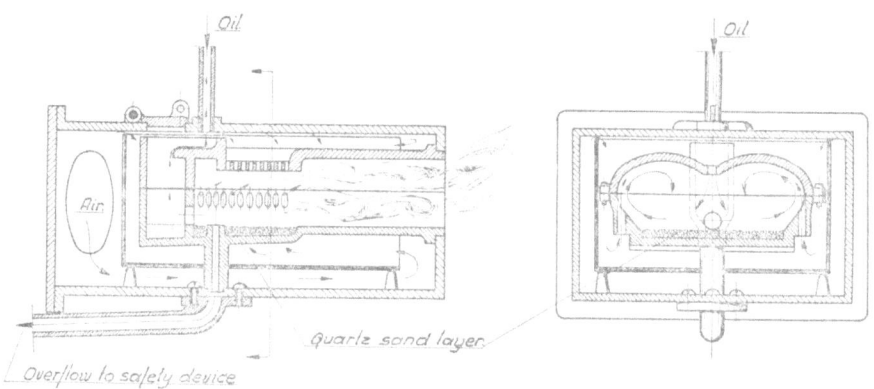

Fig. 132. GIBBS' oil burner with double vortex.

screened radiation and convection, but the swirl or vortex principle has certainly more advantages in regard to improvement of combustion than it has for automatically adjusting the vaporizer temperature to the dew point of the oil, which regulation is, in fact, usually rather bad. In this connection it may be remembered that so-called "flame deflectors" were successfully applied to blue-flame wick burners because they reduced the influence of flame radiation upon the vaporizer (wick surface) temperature, by keeping the flame *away* from it, whereas the object of a swirl or vortex is just the opposite and consequently cannot be favourable to efficient temperature control.

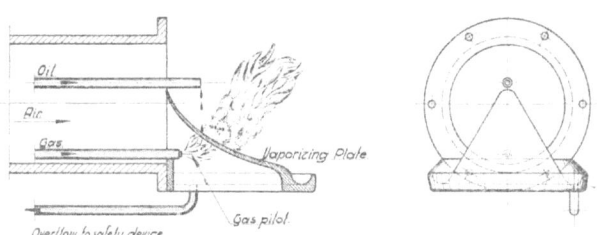

Fig. 133. Burner vaporizing by heat of pilot flame.

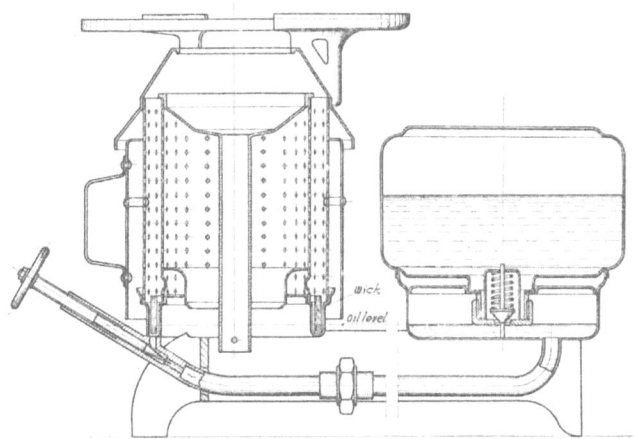

Fig. 134. A typical range burner.

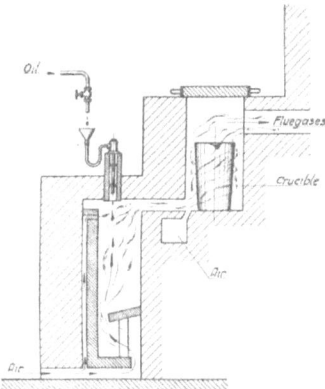

Fig. 136. Metallurgical drip
furnace (FRANCE).

Fig. 135. BRADLEY & HUBBARD range burner
with refractory covers.

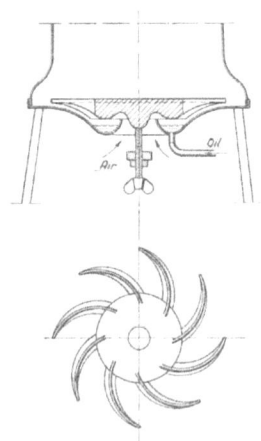

Fig. 137. CAUVET-
LAMBERT burner.

G. *Burners with vaporizers heated by pilot flames*

A "pilot flame" means a small auxiliary flame, which assists the main burner in some way, e.g. for preheating or vaporizing the oil or for ignition. Very often such pilot flames are fed by auxiliary fuels, e.g. by *city gas*, an example being shown by fig. 133 reproduced from TAPP's "Handbook of Oil Burning". On the one hand this makes the burner dependent on a foreign source, but on the other hand it offers an opportunity of regulating the vaporizer temperature fairly well, as thermostats governing a gas flame are readily available on the market and cost very little. Yet another modification consists in using a small pilot flame of a wick lamp or range burner with *kerosene* as an auxiliary fuel to heat the vaporizer for a heavier main fuel, but this again necessitates the use of two fuels, without the advantage of easy automatic control. RÖDL's burner, which will be discussed in the fourth section, comprises *pilot-heating by electricity* assisted by a small *pilot range burner*.

It being impossible to discuss within the scope of this section every representative of these 60 different burner types, this discussion will be closed by giving a number of pictures representing some typical burners of this class, viz., BECKER (fig. 45), MAYER (fig. 47), KNUPFFER (fig. 48), KERMODE (fig. 49), STRACK (fig. 131), GIBBS (fig. 132), Mushroom-type (fig. 53), PRIOR (fig. 61), FLAMME BLEUE (fig. 130), Range burner (fig. 134), BRADLEY & HUBBARD range burner (fig. 135), Metallurgical Drip furnace (fig. 136), CAUVET-LAMBERT (French Pat. 666,123) (fig. 137).

H. *Carbonization troubles*

Whilst most manufacturers of vaporizing burners claim that there are no carbonization troubles whatever with their particular burners, there are some who honestly admit carbon formation and either provide means for removing it or make the vaporizer easily accessible for cleaning, which, of course, is a better policy than ignoring the carbon and leaving the troubles to the buyer.

Anyhow it is a good principle when designing oil burners of this type *to provide a large open space where carbon may be collected* if it should be formed at all occasionally under particularly unfavourable conditions. The total quantity of carbon formed is often less important than the *place where it is formed*, which point will be considered further in section IV.

We will close the discussion of this class of atmospheric burners by drawing attention to some *carbon cleaning devices*, in the first place that invented by STENFORS according to the British Patent No. 347,350 (1929), the object of which is to remove the carbon by burning it (see fig. 138). The oil admitted at (*b*) is vaporized in the chamber ($a_1$), and the oil vapours evolved are mixed with air at (*c*), which air is preheated to some extent when travelling through the passages (*d*) before entering into the combustion chamber (e). If the vaporizing chamber ($a_1$) becomes fouled (which will probably occur, as no means are provided to control its temperature properly), the combustion chamber is turned from the x to the y position and thus a clean

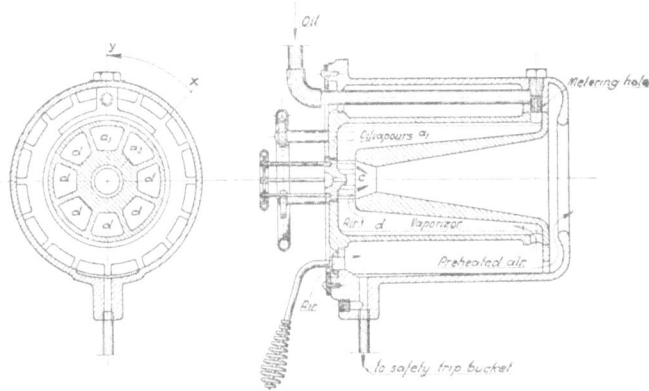

vaporizing chamber $a_2$ is put into operation, whilst the *fouled one is heated in a stream of air which burns the carbon away.*

Fig. 138. STENFORS vaporizing oil burner.

A similar idea has been followed by the engineers of the DELFT Testing Station of the BATAAFSCHE PETROLEUM MAATSCHAPPIJ resulting in a design which has been baptised *"Vortex burner"* because, as is shown diagrammatically in fig. 139, the flame follows a screw-like path through a cylindrical combustion chamber (Dutch Patent No. 34,379).

The oil drips and vaporizes on a *hot plate* (a) heated by the radiation and convection of the flame itself. The air for combustion enters tangentially into the combustion space, skimming the surface of the hot plate, and the mixture, after being ignited, by its

kinetic energy travels along a spiral path (vortex) before leaving the cylindrical combustion chamber in an axial direction.

Although this construction was not a great success — owing to 1) excessive con-

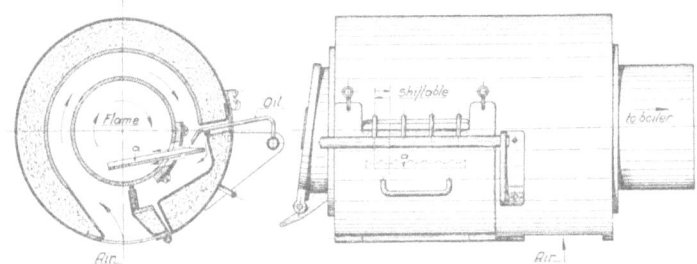

Fig. 139. Old design of the DELFT VORTEX burner.

sumption caused by large quantities of excess air, 2) too small range of regulation and 3) the greater part of the flame radiation being "shut-up" in the cylindrical combustion tube and causing the flame to "face" too little of the heating surface of the boiler, which resulted in low thermal efficiencies — it may nevertheless be mentioned here because of its good *carbon-removing device* (Dutch Patent No. 36,241).

The sloping hot plate was divided into a number of parallel gutters and the oil allowed to drip only into the *even*-numbered gutters, the *odd* ones being left dry. When, after a certain time, the wet *even* gutters were found to be covered with oil coke (which, of course, could scarcely be avoided, as the temperature control of the vaporizer—hot plate—was rather primitive and incapable of keeping the temperature of the oil practically at its dew point), these gutters could be *cleaned merely by shifting the oil-dripping device*, so that the oil trickled onto the *odd* gutters, the carbon on the even gutters at the same time being burnt away in a hot stream of air.

There is no doubt that the main point regarding the prospects of atmospheric vaporizing burners lies in solving the problem of how to get rid of carbonization in some efficient and simple way, and oil burner designers should realize that it is a bad policy to "solve" this problem by ignoring the occurrence of oil coke. On the contrary, they should either avoid its formation entirely by means of a suitable temperature control of the vaporizer or, if this method should involve too many complications, design a simple means of removing the carbon in one manipulation without the necessity of taking the burner to pieces or dirtying the fingers. The STENFORS burner and the DELFT vortex burner may be taken as suggestions in that direction.

---

# CHAPTER III

### VAPORIZING BURNERS COMPRISING CLOSED SYSTEMS UNDER PRESSURES ABOVE ATMOSPHERIC PRESSURE

### A. *Principle and Examples of Pressure-jet Burners*

If the oil is evaporated under pressure, a part of the heat which is applied to it will be available as kinetic energy if, for instance, the vapour is forced to issue from a small aperture, and consequently this type of burner will be able to entrain air for combustion without the assistance of natural or forced draught. Moreover, if suitable sets of cones are used, a thorough mixing and perfect, though somewhat noisy combustion may easily be obtained with this kind of burner.

Although on the face of it this seems to be very favourable, this burner type has found comparatively little application, first of all because *only distillates* could be burnt with it, which is *too expensive for industrial use*, and, secondly, because of the difficulties and dangers connected with the necessity of having *oil reservoirs under pressure*, which is admissible only for small quantities of oil for portable domestic or commercial emergency heating devices, such as for camping, on board of ships, blistering paint, soldering, etc. This difficulty might be avoided by using an oil pump, which, however, requires a foreign source of energy.

The oil reservoirs are put under pressure either by closing the container and utilizing the vapour pressure of the moderately heated fuel — which method is only suitable for gasolines — or by forcing air into the reservoir by means of a small hand pump, which method is generally adopted for the heavier fuels.

Some constructions of this type, which is a typically Swedish development, are the PRIMUS, OPTIMUS, RADIUS, WELLS, SIEVERT, KITSON and other burners. The original design, which is shown in fig. 42, consisting of a *vaporizer directly heated* by the flame, was apt to get rapidly choked with carbon, but there is a later improvement, now generally adopted by all manufacturers of this burner type, as shown in fig. 100 (SVEA).

This construction not only provides for a *blue flame combustion* by arranging a suitable *pre-oxygenation of the oil vapours* followed by *aldehydeous combustion*, but it also avoids excessive direct heating of the vaporizer by using *screened radiation*, reducing the frequency of carbonization troubles considerably.

Nevertheless *kerosenes* with practically no carbon residue content (CONRADSON

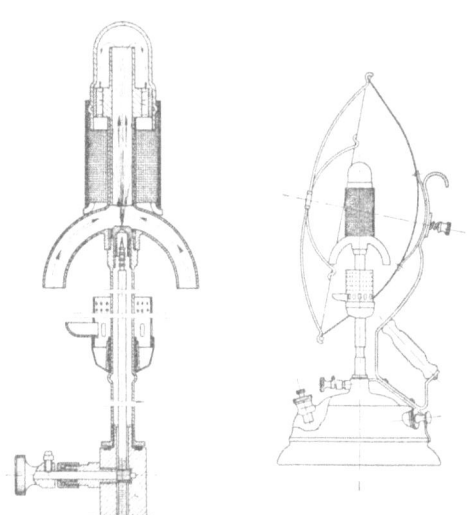

Fig. 140. PRIMUS radiator.

test = approx. zero) have been found to be the heaviest fuels with which reliable operation of this type of burner could be guaranteed under all conditions, so that apparently the vaporizer temperature is still too high for *gas oils*, i.e. too far above their average dew points and lower cracking limits. With the present constructions this unfavourable circumstance, however, could not be so easily avoided, it being necessary to transfer a considerable quantity of heat for vaporization of the fuel through a comparatively small surface of the vaporizer in order to allow of a sufficient burner capacity. This limited heated area causes a high load with respect to heat transfer, necessitating a high temperature gradient through the metal of the vaporizer and high temperatures for the flame side of it.

Fig 140 shows a PRIMUS heat-radiator which is different from the conventional form of vapour-jet burners. In order to increase the rather poor radiation of heat of the blue flame a mantle of metal gauze is placed in it, which is heated to incandescence. In this connection attention is drawn to the remarks on page 123.

An arrangement which is used in several burners of this class is shown in fig. 101 taken from TAPP's "Handbook of oil burning", a book to which the reader is referred for much useful information on the subject of domestic oil burning. The oil is vaporized under pressure in a tube heated by flame radiation; this vapour, then issuing from an orifice, entrains primary air through a number of adjustable holes and the mixture, after being what is often called: *"pre-oxygenated"* by exposure to moderate heat in the presence of oxygen, is burnt with a number of small blue gas flames.

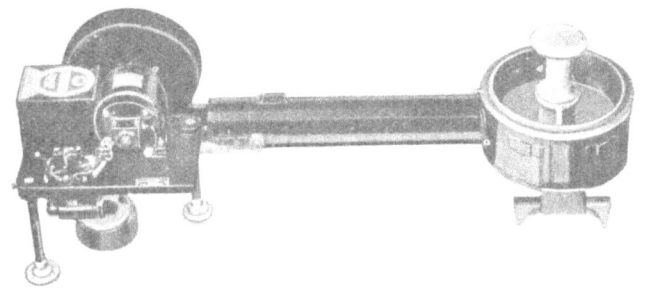 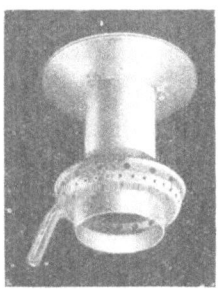

Fig. 141. MARR pre-oxygenation burner.          Pre-oxygenation Cap.

The blue-combustion effect of pre-mixing air with oil vapours is due not only to *preliminary oxidations*, but also to the *dilution* of the rich oil gas to a lean gas, which favours a slower *aldehydeous* combustion, as has been previously explained in the second lecture.

Fig. 141 shows the MARR domestic oil burner, also based on the principle of pre-

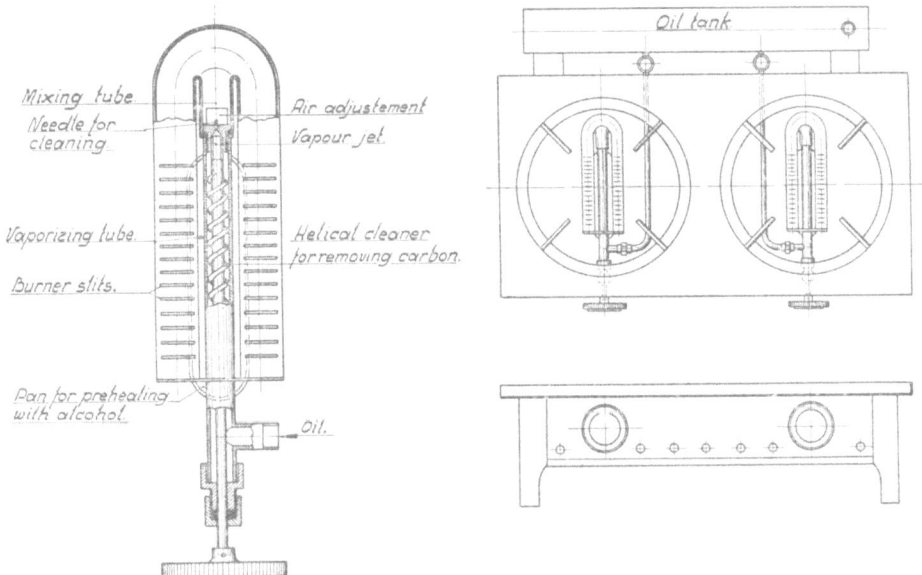

Fig. 142. GOOLD blue-burner for ranges.

oxygenation, the air being delivered by a fan. The oil is fed by gravity to the vaporizer placed in the centre of a fire pot of heat-resisting material and refractory in-

sulation. The oil vapours evolved are mixed with the air and are exposed to preheating inside a "pre-oxygenation cap" placed in the centre of the fire pot. The preheated mixture issues from a number of holes into the fire pot, where it is burnt with
perfectly blue flames.

Fig. 142 shows such a burner, covered by the British patents of GOOLD 170,938
and 170,939 (1920), known under various trade names (OKA burner, PRESTO burner
etc.), to be used as a substitute for gas ranges for cooking. This device, which speaks
for itself, moreover comprises a helical screw for removing the carbon from the
vaporizing chamber and a needle for clearing the vapour hole, and consequently it
belongs to the class of burners, the designers of which admit the occurrence of carbon
troubles and take measures to remedy them.

### B. *Pressure-jet Burners having no Reservoirs under Pressure.*
SCOTT-SNELL's *thermo-pumps*

It has been pointed out before that the main objection to the use of pressure-jet
burners is the necessity of keeping oil reservoirs under pressure. In this there is
certainly some danger, for, although the reservoirs may quite well be made strong
enough to stand the necessary working pressures, usually not higher than about one
atmosphere, there still remains the *danger of leakage* in the event of the flame's being
accidently extinguished and the oil supply not shut off. Consequently this method is
considered too dangerous for larger installations, especially if they are not continuously watched. An *automatic trip-bucket*, such as shown in fig. 62, which, in the event of
leakage, collects the leaking oil and, by the weight of it, either shuts off the oil supply
to the burner or releases the air pressure on the oil tank, might be successfully applied
in this case.

Another means of avoiding this disadvantage consists in using an *oil pump
which stops as soon as the flame is extinguished*. An electrically driven pump would
require not only a foreign source of energy but also some device, a so-called "flame-
stat" or "protectostat", which stops the motor as soon as the flame happens to be
extinguished.

In the SCOTT-SNELL patents (British Patents 283, 681 (1926) and 327, 971 (1929))
a clear description may be found of the principle of so-called *thermo-pumps*, i.e.
pumps operated by means of a *continuous heat supply*. It is obvious that if such a
pump is used for pressure jet burners and the flame itself is used to activate the
pump, the oil supply will be stopped automatically as soon as the flame is extinguished, which provides just the safety measure desired.

The design of the SCOTT-SNELL thermo-pumps is based on the *lability of a liquid
column in a U-tube*, and its principle may be explained with reference to fig. 143,
which is a glass model for water, and to fig. 144, showing a kerosene pump as it is

used in the PETROMAX Stoves developed by the International Oil Lamp and Stove Company (London) and manufactured by EHRICH and GRAETZ A. G. (Berlin). One of its advantages is that, except for a couple of ball-valves, it has no moving parts.

The model of fig. 143 consists of a U-tube (abcd) of special shape. The tube is first supposed to be completely filled with water. A suction ball-valve (e) is placed at the bottom and a pressure valve (f) at the top. The space in the U-tube between the two valves may be compared to the working chamber of a plunger pump. If the end (a) is heated by means of a flame (i), steam will be developed, which although partially condensing on the cold water surface, will be able to displace the water from (b) through (c) and force it through the valve (f) ("pressure stroke"). The intention is that, as soon as (b) is empty, fresh water shall be sucked in through valve (e) ("suction stroke"), and this switching over from pressure to suction stroke must be established without removing the flame, which is the object of the patents referred to.

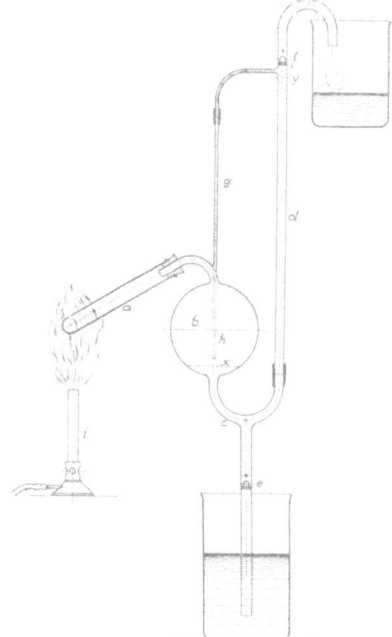

If flame (i) could be removed and thus the heat supply stopped, it would be easy enough to bring about this suction stroke, but this *intermittent heat supply* would require some mechanical operation, which complication it is desired to avoid. Then, again, it would be possible to effect a suction period even from a continuous heat supply if it were possible to boil tube (a) completely dry, in which case the steam production would be automatically stop-

ped, and consequently the system would come to a standstill, cool down and a vacuum would be created inside. But this method would involve

Fig. 143. Glass model of thermo-pump.

considerable *superheating* of the dry tube (a), which would tend to cause cracking and the formation of oil coke if the pump were used for oil.

The SCOTT-SNELL patents overcome the difficulties in the following way: By connecting (a) with (d) by means of a small tube (g), steam can be admitted under valve (f), which, however, will take some time, as tube (g) is only a capillary offering to the steam a large surface and consequently causing considerable loss of heat per unit of steam volume passing through. Thus much steam has to be condensed in (g) before any steam can get through it to (d). In the meantime the water level will be forced down, e.g. to a level (x). As soon as steam is admitted under valve (f) via the capillary tube (g) the columns of water (x) and (y) cannot be in equilibrium any longer and consequently the water from (y) rushes into the bulb (b), creating a vacuum by initial condensation, fresh water being sucked in through valve (e) with such force that the whole system, tube (a) included, is refilled with water. After that, the pressure period starts again, and so on. The duration of the pressure period depends on the heat supply and, moreover, may be increased by lengthening tube (g) as shown by the dotted line (h), this causing a delay of the moment at which the first steam bubble is admitted into the capillary tube (g). The glass model of fig. 143 comprising a 200 cc. bulb, if heated by an ordinary Bunsen gas flame, is able to pump about 25 litres of water per hour against a head of 3 metres. As, however, the specific vapour volumes of hydrocarbons are considerably smaller than that of water, the capacity of such pumps, if used for oils, is considerably smaller too, and, of course, for kerosenes or gas oils higher temperatures are required.

Fig. 144. SCOTT-SNELL thermo-pump.

The small pump of fig. 144 is used in the PETROMAX stoves (see fig. 145), which, though based on the same pressure jet principle as shown in figs. 140, 141 and 142, have the advantage of *requiring no oil reservoir under pressure*, a special safety feature being that the oil supply is stopped as soon as the flame is extinguished. It seems certain, however, that reliable operation only can be guaranteed for comparatively light fuels such as kerosene, as heavier fuels requiring higher temperatures are likely to cause sticky valves through gum formation.


### C. *Liquid head Pressure-jet burners*

Vapour-jet burners are usually operated either by oil-pump pressure or by putting pressure on the fuel tank, or again by using the pressure caused by the height difference of jet and oil level. In this last case it is generally understood to be essential that the oil level should be higher than the jet, but a recent French development comprises just the reverse, *the jet being higher than the oil level*, which of course means a valuable safety feature over the usual arrangements.

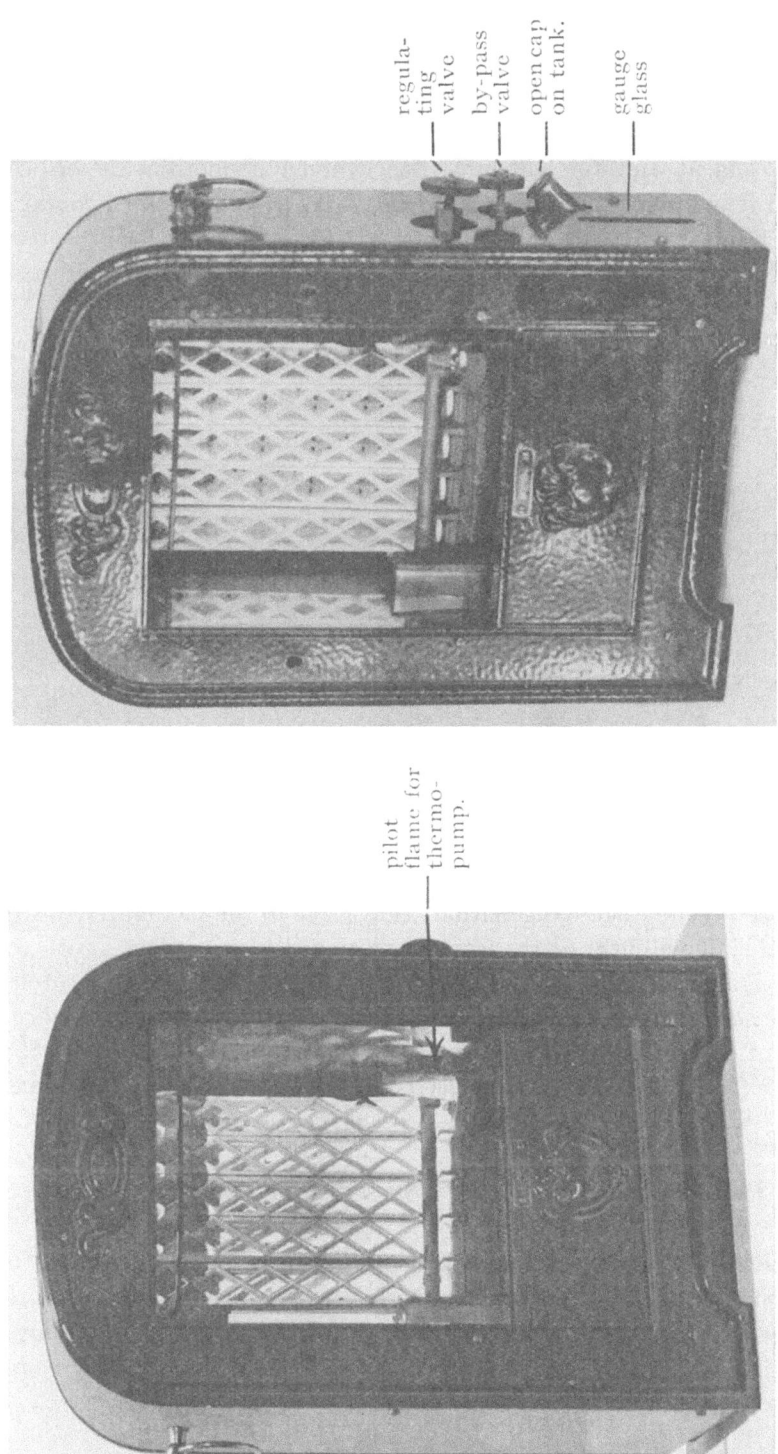

Fig. 145. Petromax vapour jet stove equipped with Scott-Snell thermo-pump.

This principle is described in the French patents of GAMARD (Nos. 38,548; 675,924; 747,674) and is embodied in the PÉTRÉLEC burner (Système ROZIÈRES-GAMARD, Bourges-France). Fig. 146 gives a scheme of operation. The fuel (kerosene) present in a tube (c) in open connection with the tank (a) is heated, for instance, by an electric element and the vapours thus evolved issue from the orifice (d), the resistance of flow forcing the oil level down until an equilibrium is obtained, thus building up a certain pressure in the vaporizer tube.

It is clear that the pressure thus available at the jet will be equal to the liquid

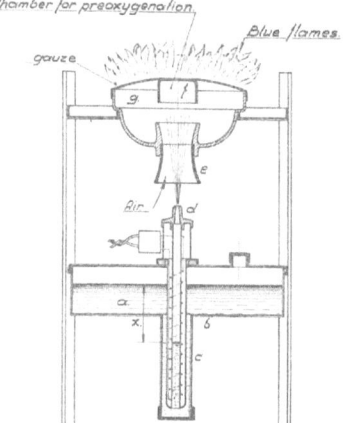

Fig. 146. Scheme of GAMARD's          Fig. 147. PÉTRÉLEC burner of ROZIÈRES-GAMARD.
          principle.

head x, which pressure may be just enough to ensure good combustion. The rather weak injecting action of the jet is assisted by the air being drawn in by the natural draught of the tube (e). Pre-oxygenation takes place in the moderately heated space (f) and the blue flame burns at the upper surface of the gauze (g).

Fig. 147 is a photograph of this burner. It requires a constant supply of electricity, although the same principle may be worked out for evaporation by *heat-conduction* of the flame itself (see French Pat. 675,924 of 1929 and British Patents 355,683 and 355,684 of 1930), in which case electricity is only used for starting the burner.

### D. *Pressure-jet Burners with Thermostatic Control of the Vaporizer.*

In Section IV some experimental burners will be discussed based on the previously mentioned idea of *maintaining the vaporizer temperature at about 300° C. by means of a boiling bath of diphenylamin* and it will be shown there that it is not only possible to burn gas oils without carbonization troubles, but even *residue-containing fuels* with such devices.

# CHAPTER IV

PREPARING THE OIL BY GASIFICATION WITHOUT USING AUXILIARY MEANS

## A. *Vaporization and Gasification*

"Vaporization" and "gasification" are often used synonymously, but in this section *"vaporization"* will be used to indicate the *reversible physical process*, whilst *"gasification"* is to signify the *irreversible chemical process* known as gas-cracking or a combination of both processes.

It has been explained before that for good combustion it is necessary to give the oil a certain surface per unit of weight, but attention was drawn to the fact that it is not necessary to carry this principle so far as to *vaporize* the oil completely, which means increasing the surface about $10^9$ times the ("molecular surface") as a surface increase of say 100 times as obtained with *pulverization* has been found to be quite sufficient to obtain good combustion.

Now it will be clear that from the point of view of good combustion alone there is little reason *to increase the surface per unit of weight to more than the "molecular surface" by gasifying the oil*, which in fact even goes beyond evaporation; therefore, if gasification is required, there must be other reasons besides merely a demand for good combustion. Such a reason may exist, for instance, where the oil gas is not burnt immediately after it is produced and where it is meant as a *substitute for city gas*, such as for heating tinning baths of tin-packing factories, bakers' ovens etc. In such cases the gas produced will be cooled down considerably before it reaches the burner and it will, therefore, be necessary to produce a mixture of *true* or so-called *"permanent" oil gases* which are not condensed at ordinary temperatures under atmospheric pressure.

## B. *Rich Oil Gas Processes*

Oil gases may be produced by cracking mineral oils in retorts at atmospheric pressure and at temperatures between 600 and 1000° C. The gases thus obtained consist, except for a small percentage, of pure hydrocarbons and are called *rich oil gases* as distinguished from *lean* oil gases consisting only up to $\frac{1}{3}$ or $\frac{1}{4}$ of hydrocarbons.

The oil gas process was originally developed by YOUNG, who, moreover, subjected the resulting gases to washing by the inflowing oil and thus caused the condensable vapours to be removed, allowing only permanent gases to pass on to the gasometer. The oil which has absorbed the condensable vapours was then run into the retort and decomposed, so that there was no loss of material. In order to restrict the amount of carbon left in the retort, only such oils were used for this purpose as have a *low carbon residue content* (CONRADSON figure), and these distillates were called *"gas oils"*.

Other rich oil-gas processes are those of PINTSCH and GRAETZ, which gases are mainly used for lighting railway carriages and buoys. The "Blau gas" process differs from these only in that, being carried out at lower temperatures (550–600° C.), it mainly produces saturated and unsaturated hydrocarbons of the *propane and butane type*, which are condensed under pressure in steel cylinders and gasified again as soon as the pressure is released. The principal advantage of the use of "Blau gas" is the possibility it affords of storing a large weight of gas in steel containers.

As a recent development which runs almost parallel to "Blau gas" there may be mentioned *Butagas* and *Propagas*, usually consisting of butane or propane fractions obtained from rectifying columns producing natural gasoline.

## C. *Combined Oil Gases*

If the oil gases are to be used for *"enriching"* other gases, e.g. coal gas or water gas, to improve their luminosity, raise their heating values or to give them a bad smell as a precaution against suffocation, they are not necessarily first produced in a pure state. Such a combined process for so-called *"carburetted water gas"* has been developed by LOWE, who decomposes the oil in the presence of a current of hot water gas. In the first stage a charge of solid fuel is blown to incandescence by an air blast and the producer gas then formed is burnt in the carburettor and superheater, which are heated by the combustion to the necessary temperature for decomposing the oil. As a second stage the air blast is shut off and steam is blown through the red-hot generator, the water gas thus produced passing to the top of the carburettor, where it meets the oil, both then being carried down through the carburettor, where the oil is thoroughly cracked.

Another *combined rich oil gas* is produced by the JONES process, in which a mixture of oil and steam is passed through generators filled with checker brick which has been previously heated by oil flames.

## D. *Lean Oil Gases*

Lean oil gas processes produce gases which are *diluted by incombustible ballast gases* to such an extent that the calorific value is comparatively low, viz. one-third or one-fourth of that of rich oil gas. This dilution is usually due to the fact that cracking of the oil is obtained by *incomplete precombustion* of the fuel itself instead of by applying a foreign source of heat.

According to this principle the oil, after being suitably prepared (pulverized or vaporized), is burnt with a *shortage of air* in such a way that mainly *incomplete com-*

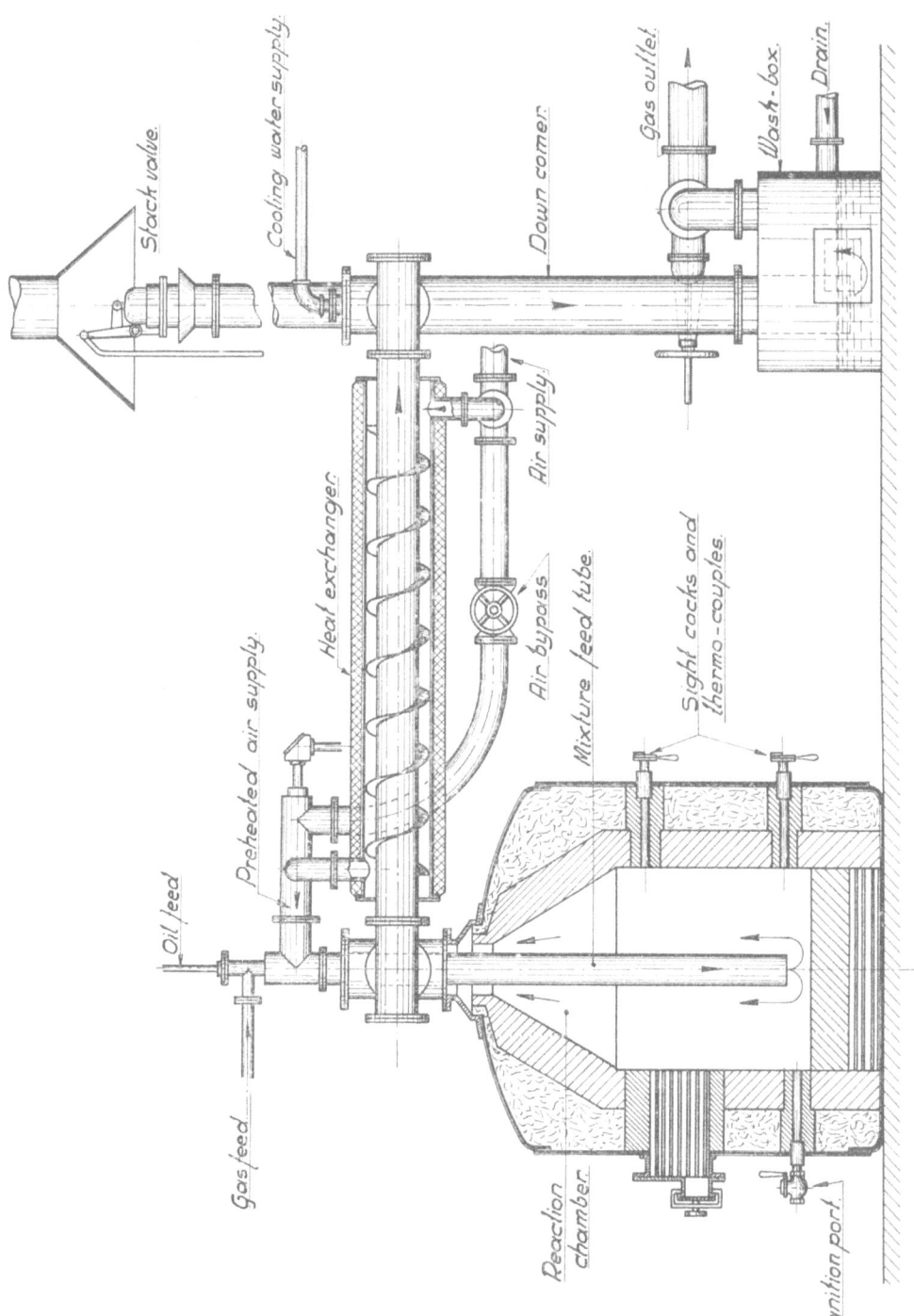

Fig. 148. Scheme of a DAYTON-FABER gas generator unit.

*bustion products* are formed. Such lean oil gases are as a rule characterised by a nitrogen content of approx. 60%, the remainder consisting of methane, hydrogen, ethylene and carbon monoxide, which are gaseous combustibles, and further a small percentage of carbon dioxide and oxygen.

a) *Industrial Lean Gas Generators.*

These processes have been developed on a large industrial scale for the production of city gas, especially in oil districts, and in fig. 148 a diagrammatic scheme is given of a DAYTON FABER plant, which speaks for itself.

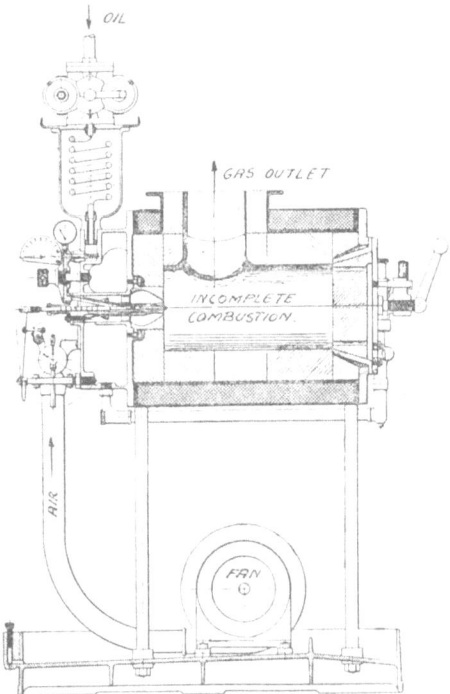

Fig. 149. The HAKOL-ZWICKY oil-gas generator.

An attempt to use this principle of oil-gas production by partial precombustion on a smaller commercial scale is shown in fig. 149, which is a section through a HAKOL-ZWICKY oil-gas generator comprising an oil burner, fan and combustion chamber of refractory material in which incomplete combustion takes place.

Although in this principle there may be attractive prospects, such as "Have your own gas factory at home", some objections attach to it when applied to small units, firstly because it is found difficult to keep the gas composition reasonably constant, and secondly because it is difficult to free the gases from minute particles of tarry constituents. For units of larger capacity it is possible to meet these objections by installing scrubbers for gas cleaning and gasometers for equalizing the quality.

b) *Chemical Carburettors.*

Other applications of lean oil-gas production by means of incomplete precombustion on a still smaller scale are the so-called *chemical carburettors*, a representative of which, the RECTOR gasifier (WAUKESHA, Wis.), is shown in fig. 150.

From time to time inventors have succumbed to the temptation of *using kerosene or gas oil as fuel for gasoline engines*, and a study of patent literature will show that several devices have been invented which are generally based on previous gasification of the oil by precombustion, the lean oil gas mixture thus produced being used as a gaseous fuel for the gasoline engine.

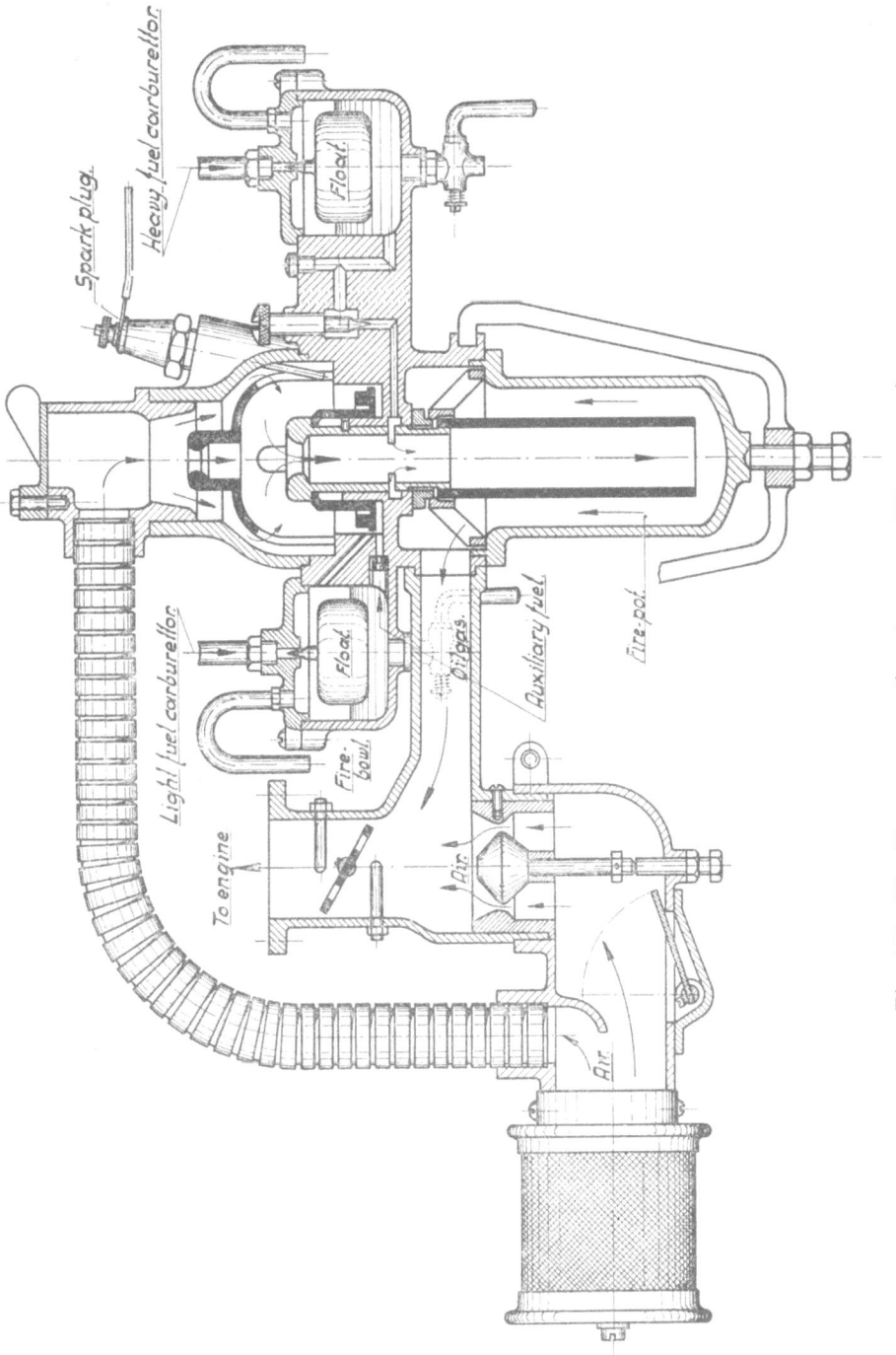

Fig. 150. The RECTOR gas oil carburettor (WAUKESHA).

Such a contrivance for "gasifying" oil by partly chemical reactions is often called a *"chemical carburettor"*, as distinguished from an ordinary gasoline carburettor, which embodies the *physical* process of evaporation only, although attention is drawn to the

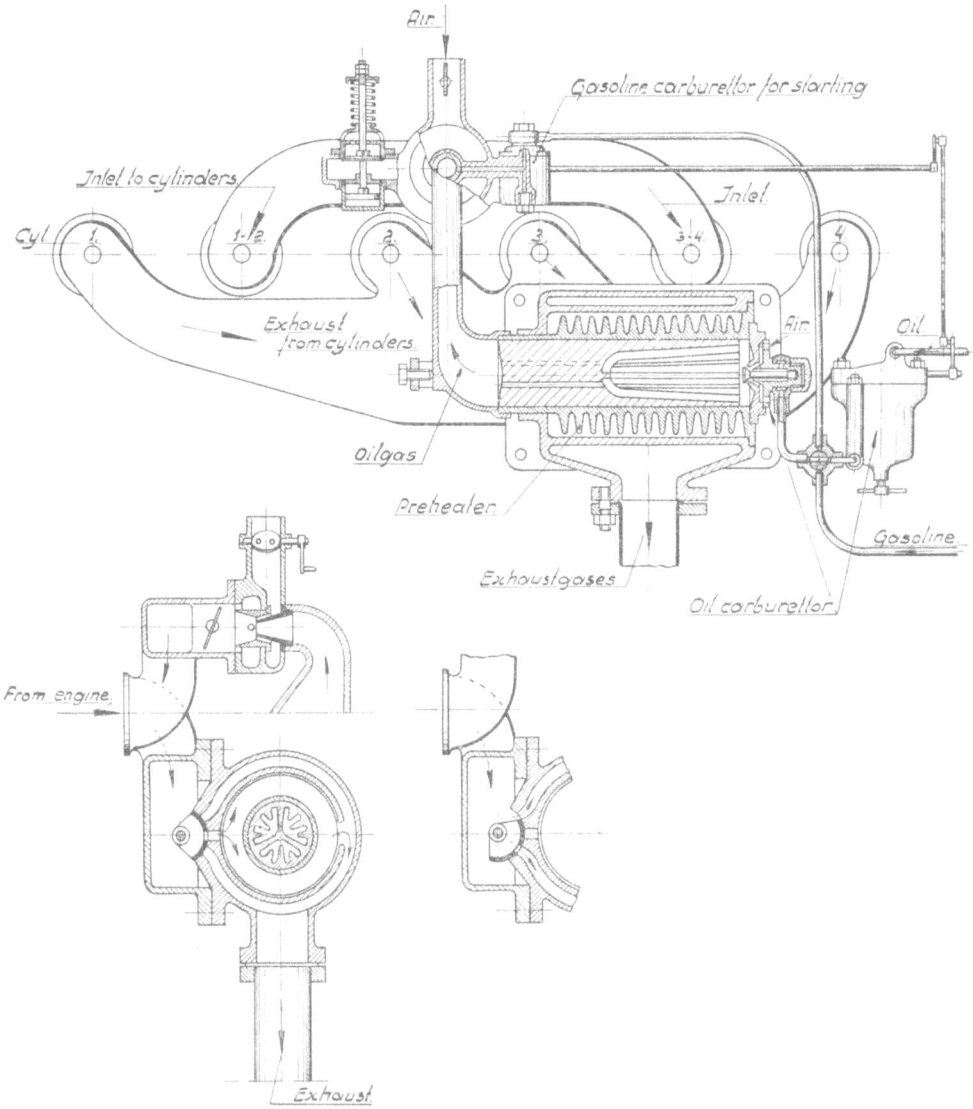

Fig. 151. CATALEX oil-gas carburettor.

fact that these chemical carburettors generally produce anything but permanent gases, the greater part of the "gas" produced in fact consisting of *oil mist*. This offers no serious objections as long as the gas is to be burnt immediately after being produced, but such "oil gases" cannot be stored in gasometers nor conducted through long gas lines. In this respect even the short time between gas production and its combustion has given rise to objections to the use of chemical carburettors for engines, the relatively low temperatures of manifolds, valves, pistons and cylinder walls causing condensation of the heavier constituents of the gases, resulting in sticking valves and piston rings, dilution of the lubricating oil and many other troubles, which detract from the reliability of the system.

The operation of the RECTOR gasifier will be clear from fig. 150. The oil, which is pulverized by a jet in a current of primary air, is ignited by means of a continuously sparking plug causing precombustion in a fire pot. The products of incomplete combustion are mixed with secondary air and drawn into the cylinders of the engine. The engine is started on gasoline. The MASSON carburettor is a similar design.

Another chemical carburettor is the CATALEX, shown in fig. 151, which uses the heat of the exhaust gases for vaporization and cracking of the oil, the rich mixture thus obtained being mixed with air resulting in the *formation of an oil mist*. It is claimed that some *catalytic actions of copper* are utilized, which explains the name. Fig. 152, taken from a report of Trial No. 739, carried out by the Royal Automobile Club, may give an impression of the

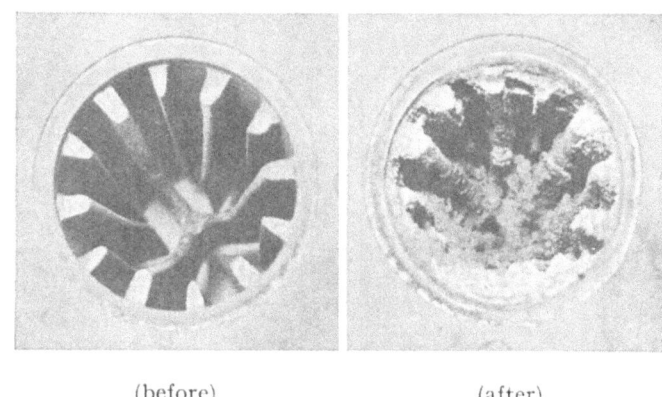

(before)                    (after)

Fig. 152. Carbon formation in CATALEX gasifier after trial No. 739 Royal Aut. Club.

kind of dangers which may accompany the use of gas oil in this device: the vaporizing chamber is badly carbonized at the end of the test. Indeed it is very difficult to maintain exactly all the conditions required for clean cracking. As a matter of fact, it was generally the inventors of these chemical carburettors themselves who, by their exceptional experience and devotion, got the best results; these, however, could scarcely be even reproduced by less experienced people under less favourable circumstances.

It is *rather difficult* to ensure continuously all stable conditions necessary for clean *combustion* when burning ordinary fuel oils; it is *still more difficult* to maintain all the necessary conditions in such a way that a clean and complete *evaporation* of the fuel oil is ensured in a vaporizing burner; but *excessive difficulties* are encountered in trying to obtain a clean *gasification* of fuel oil, and experience has shown that in order to overcome these difficulties it is necessary to make as a rule ample allowances with regard to the quality of the fuel used, i.e. by using fuels with *very low* CONRADSON *figures*, such as kerosenes.

c) *Gasifying oil burners.*

An example of a gasifying oil burner for domestic use is the CRYSTAL BLUE FLAME burner of fig. 153, which comprises a special cracking chamber heated by the flame and which explains itself.

There are still inventors fascinated by the idea that wonderful prospects are latent in the principle of previously gasifying the fuel oil but, although from a theoretical standpoint they are quite right, the practical realization has miscarried

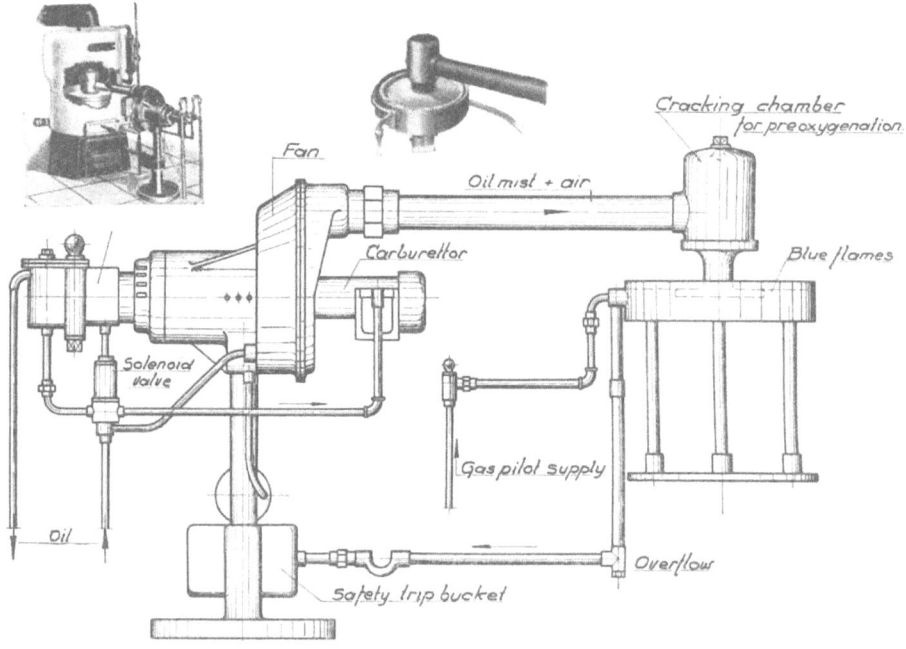

Fig. 153. CRYSTAL oil-gas burner for domestic use.

up to the present day on account of carbonization troubles and fouling of the apparatus by tarry residues. Consequently, their efforts should be directed to preventing the formation of carbon or tarry products or, if this is not practicable, to designing suitable apparatus by which carbon and tar formed may be accumulated so that their presence does not hamper operations.

On the other hand, it should not be forgotten that an ordinary oil flame in fact comprises an excellent solution of the problem, viz. by keeping the carbon and tarry products *in suspension*, thus avoiding their deposition on the surfaces of the apparatus. This method, moreover, provides means for the removal of these troublesome products, viz. *by burning* them, and consequently the *ordinary flame of pulverized oil* really must be considered as an advanced development of the gasification method. Its drawback, viz., that the gases produced are immediately burnt, might perhaps be overcome in some way, e.g. by entraining intermediate combustion products from the centre of the flame and cooling these gases rapidly down to lower temperatures.

## CHAPTER V

### MECHANICAL PULVERIZATION ON THE DYNAMIC PRINCIPLE
(Pressure-atomizing burners)

### A. *Principles*

In this chapter one of the most important forms of preparing the oil for combustion will be dealt with, and before discussing any details of construction it is advisable to give a survey of the basic principles of this kind of pulverization.

The name *"dynamic"* pulverization has been given to it because *pressure* is the primary agent causing pulverization, as contrasted with *"kinetic"* pulverization which includes *velocity* as a primary agent, although as a matter of fact pulverization is basically obtained by a velocity effect in both cases (see Table V).

In the section on Historical Development it was explained that oil burning on board vessels did not enjoy general application until in 1902 KÖRTING succeeded in pulverizing fuel oil merely by the application of pressure, using a helical screw which

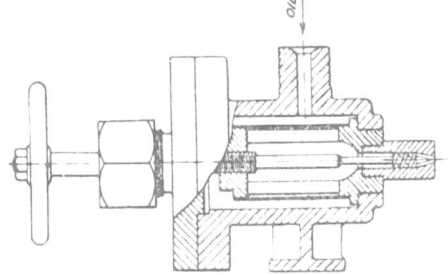

Fig. 154. KÖRTING atomizing burner.

imparted an intense rotary motion to the oil before leaving the orifice (see fig. 154). About the same time, SWENSSON tried to solve the problem in another way, pulverizing a stream of oil by causing it to strike against a sharp metal cutter (fig. 34), which method, however, did not prove to be as successful as the other. Attention is drawn to the fact that these two burners are based on entirely different principles, the former using *internal* forces to pulverize the oil, as contrasted with the latter using *external* forces.

Encouraged by the success of the KÖRTING pressure atomizing tip, in course of time inventors developed a large number of similar devices, all based on the idea of imparting an intense rotary or turbulent motion to the oil before forcing it through

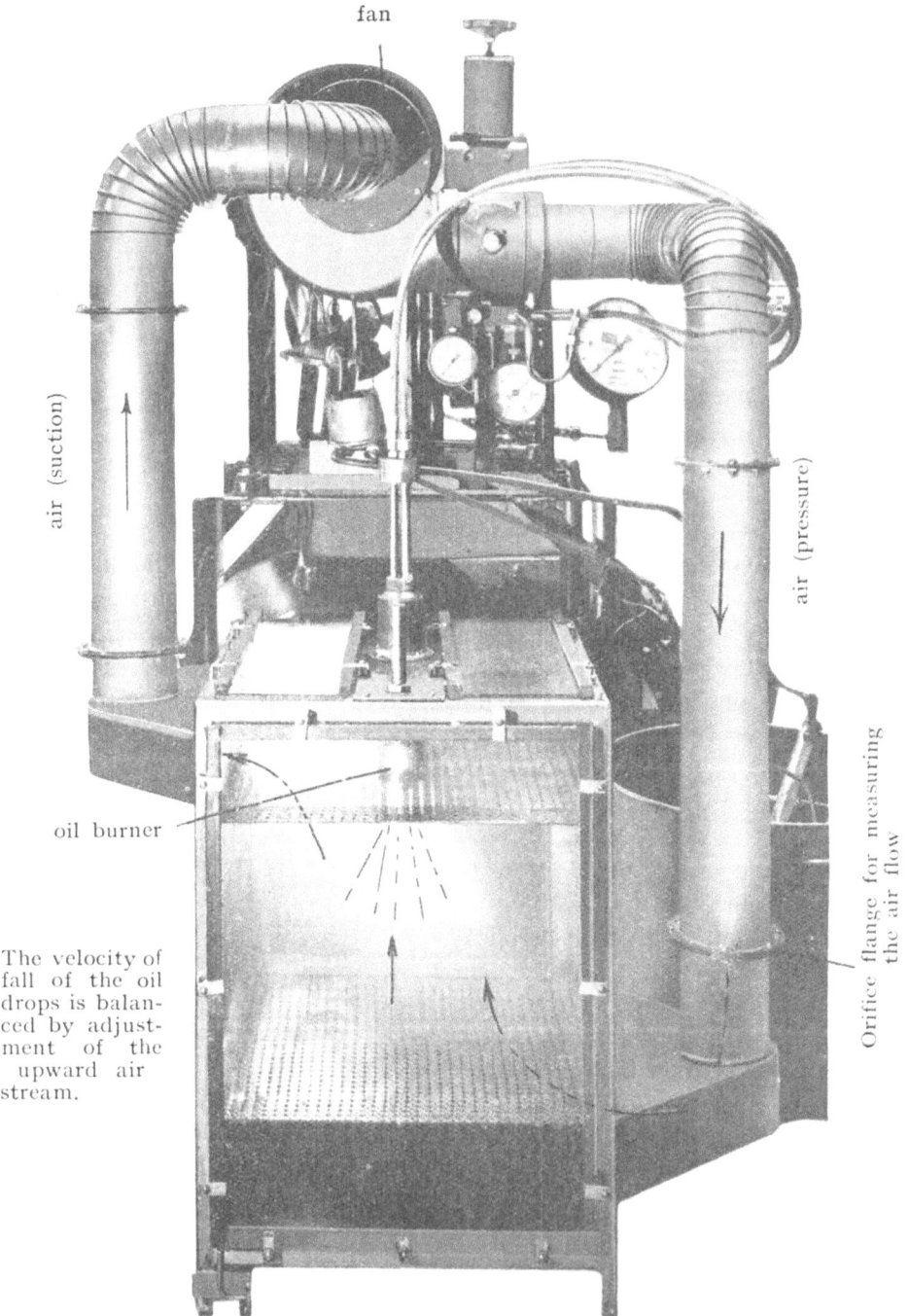

Fig. 155. BARGEBOER's atomization tester for oil burners.

an orifice. Nevertheless some 20 more years elapsed before the fundamental principles were studied more intensively, and even then this progress was mainly due to
research work done for the application of pressure atomization or so-called "solid
injection" for Diesel engines.

## B. *Pulverizing Effect*

There are three essentials for good dynamic atomization, viz.:
1) Efficient *pulverization* (magnitude of drops).
2) Suitable *direction* of the oil drops (e.g. sufficiently wide conical spray).
3) Suitable *penetration* of the oil drops through the combustion air.
The last point is more important for Diesel engines than it is for free oil burning
as in the latter case bad penetration may easily be corrected by giving the air suitable
motion, a method which, moreover, has recently found considerable application for
engines too.

*a) Magnitude of Oil Drops.*
The determination of the magnitude of oil drops is no easy matter. It has been

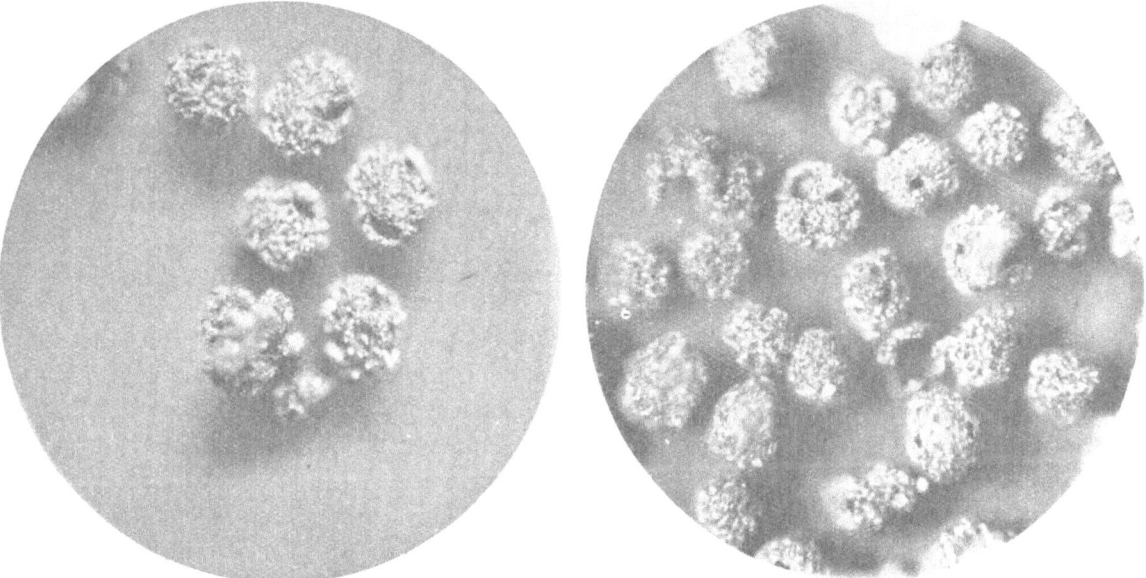

(Magnitude 200—300 μ)                              (Magnitude 150—200 μ)
Fig. 156. Microscopic pictures of ash from residue containing fuel.

proposed to calculate it from the *velocity of fall* in atmospheric air, but this method
involves a great many suppositions and approximations. It forms, however, a very

simple useful method for detecting coarse pulverization, and the standard equipment of an oil burner testing station generally consists of an *oil pond* above the surface of which the *"raining effect"* of oil burners (not ignited) may be observed. Fig. 155 represents BARGEBOER's atomization tester, which is based on the above-mentioned method of compensating the velocity of fall.

For residue containing fuels a *microscopic study of the ash*, separated from the flue gases by means of a filter or a cyclone, may give useful information regarding the pulverization of the burners. Fig. 156 shows some ash samples magnified about 50 times from which it may be seen that the particles are hollow spheres of coke and ash, each of which particle probably originates from an oil drop.

*b*) KÜHN's *Experiments.*

An apparatus by which the average magnitude of oil drops can be determined

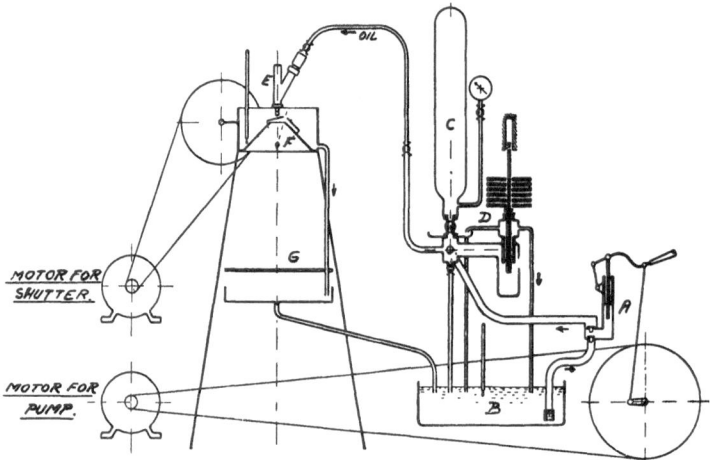

Fig. 157. KÜHN's apparatus for measurement magnitude oil drops.

with considerable accuracy is described in KÜHN's doctor-thesis entitled "Über die Zerstäubung flüssiger Brennstoffe" (Technical College Danzig University). Fig. 157 shows the arrangement of the apparatus in which A is an oil pump putting the oil from the reservoir B under a pressure which is equalized by an air chamber C and adjusted by a pressure regulator D. The oil burner E is fixed in a vertical position above a shutter mechanism, which is more clearly shown by fig. 158. The shutter, which consists of an adjustable slit, is moved from right to left at a certain known speed, thus giving the oil stream an opportunity of slipping through the slit during a very short interval, which might be compared to the *time of exposure* in photographic work. The oil drops which have been able to slip through the slit fall onto a glass plate G of known weight covered with a thin layer of soot in order to fix the oil drops. Immediately after exposure the glass plate is weighed again and the quantity of oil on it determined by subtraction. The number of oil spots is then counted, which is of course a most tiresome job. In order to keep the number of drops within a reasonable limit (say about 10,000 on a $3\frac{1}{2}'' \times 4\frac{3}{4}''$ photographic plate) and to prevent a confluence of neighbouring oil spots, it is necessary to reduce the time of exposure to about 0.0005 second. This time may be calculated by comparing the weight of oil measured on the plate with the weight of oil leaving the burner per minute. The average weight per oil drop and its diameter may then be easily calculated, assuming that the drops are spheres. It is necessary to carry out a great many determinations in order to get average values.

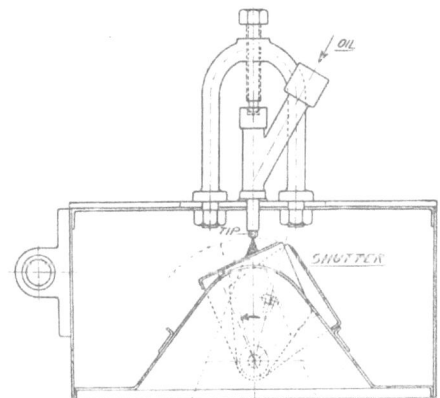

Fig. 158. Detail of shutter mechanism.

#### c) *Influence of Viscosity and Rotation on the Magnitude of Oil Drops.*

Some of Dr. KÜHN's results have been plotted in the graph given in fig. 159, from which the following facts may be deduced.

The determinations represented by *circles* ( ○ ) were carried out on gas oil pulverized by a pressure-atomizing oil burner comprising a rotation element as shown in the sketch of fig. 160, and it may be seen that an increase in the pressure gradually reduces the average magnitude of the oil drops, which behaviour, of course, is quite plausible. The determinations represented by *black dots* ( · ) were made on the same gas oil with the same oil burner, but with the rotation plug left out, and it is curious to observe from the distribution of circles and dots that in this particular case the presence of the rotation element has apparently had practically no influence on the average magnitude of the oil drops.

On the other hand considerable influence on the average magnitude is shown by the determinations represented by *circles enclosing dots* ( ☉ ), which were made on *kerosene* pulverized by the original complete oil burner. When comparing the properties of the gas oil and kerosene used, it will be seen that this difference is obviously due to a *lower viscosity*:

| Property | Gas oil | Kerosene | Units |
|---|---|---|---|
| Specific gravity/15° C. . . . . . . . . . | 0.877 | 0.837 | |
| Kinematic viscosity/19° C. . . . . . . . | 0.074 | 0.023 | cm²/sec. |
| Viscosity constant/19° C. . . . . . . . . | $0.645.10^{-4}$ | $0.100.10^{-4}$ | gr.sec./cm² |
| Kinematic capillarity . . . . . . . . . . | 32.8 | 32.6 | cm³/sec.² |
| Capillary constant . . . . . . . . . . . | 2.85 | 2.70 | mgr./mm. |

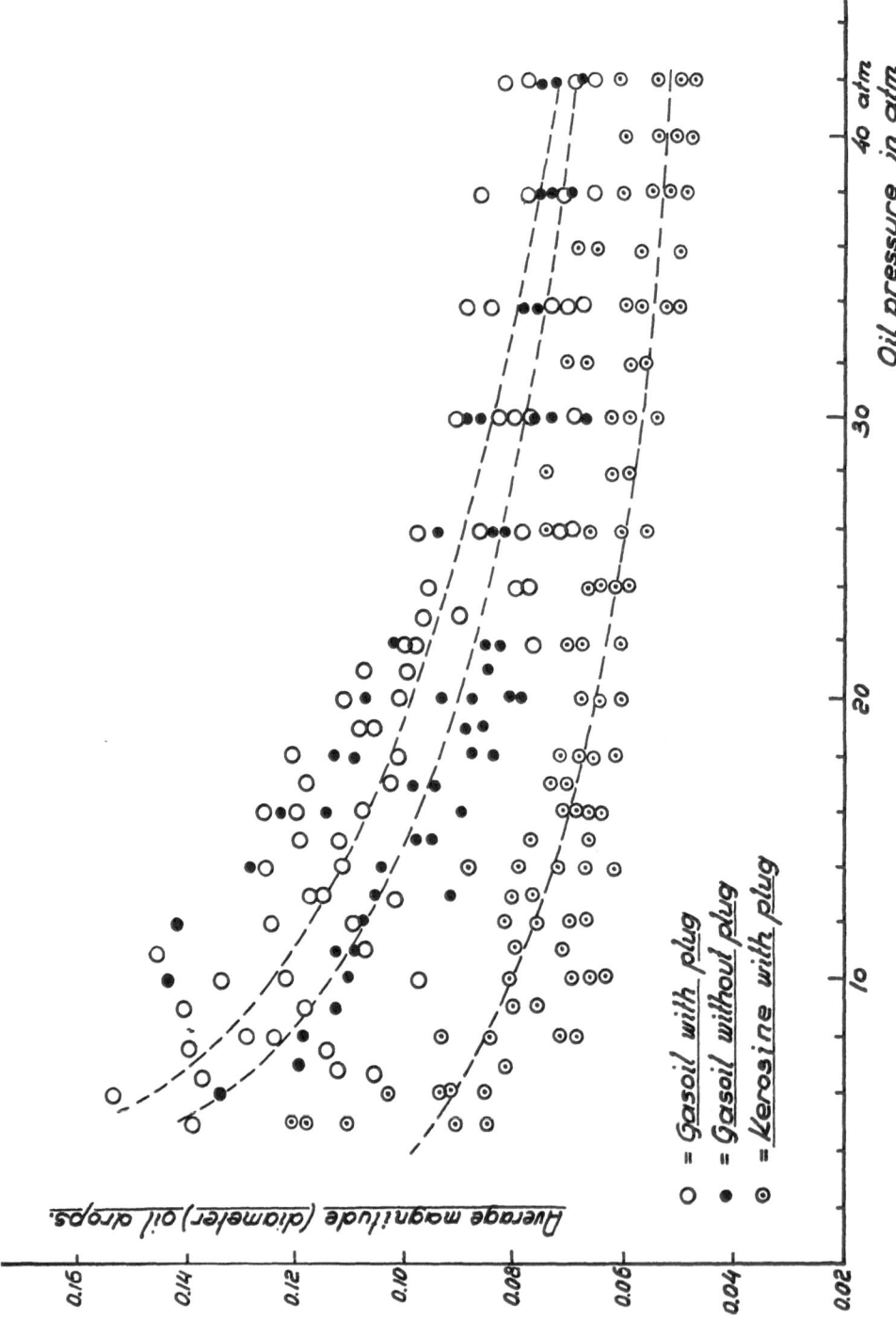

Fig. 159. Some of Kühn's experiments on the magnitude of oil drops.

If it is really true that the rotary motion of the oil when leaving the orifide has so little influence on the magnitude of the oil drops, it may be asked why, nevertheless, practical experience has shown rotation elements to be extremely useful parts of mechanical oil burners. This may be explained by the graph of fig. 160 representing the *angle of spray* plot-

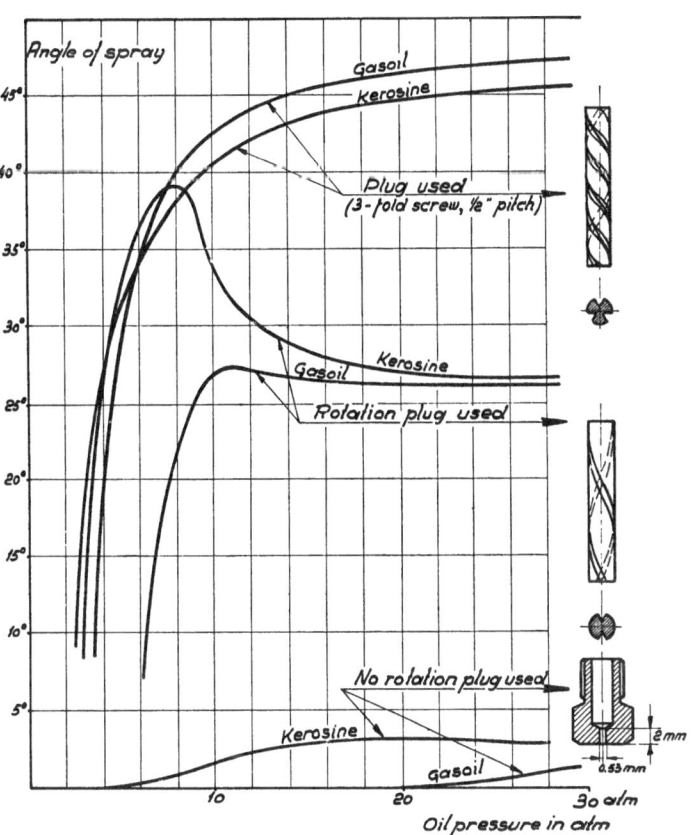

ted against the oil press-ure, from which it may be seen that, whereas at 20 atm. pressure gas oil and kerosene both give spray angles of approx. 45° when the burner is fitted with a rotation element consisting of three-fold screw thread with $\frac{1}{2}''$ pitch, both fuels show practically no spray angle at all when that rotation device is left out; a double screw thread with $1''$ pitch gives a spray angle of about 27°.

From this it may be concluded that *for this particular oil burner the magnitude of the oil drops was mainly govern-ed by the pressure and the viscosity, whereas the spray angle, and con-*

Fig. 160. Angle of spray depending on pressure and rotation plug. (after KÜHN).

*sequently the direction of the oil drops, was mainly governed by the rotary motion.* However, this may certainly not be considered as a rule without exceptions for all pressure-atomizing oil burners. In fact, the pressure-atomizing oil burners must be divided into *two sub-classes*, which, being based on different principles, also show different performance with respect to the above-mentioned points, viz.:

1) Oil burners in which pulverization is caused by a *mainly irregular rotation or turbulent motion* of the oil, and

2) Those in which this is obtained by a more or less *regular rotary motion.*

It is to the first type, or so-called *turbulent tip*, that the above-mentioned observ-ations may be applied, but not to the same extent to the second type, the *rotary tip* or *centrifugal nozzle*. Both types will be discussed in succession.

C. *Oil burners based on Irregular Rotation or Turbulent Motion of the Oil* ("*turbulent tips*")

a. *Various stages of flow.*

Let us first make an analysis of what happens to a free stream of oil flowing vertically down from a cylindrical orifice without previously passing through a rotary device, the pressure being gradually increased. The different stages of this hydro-dynamic phenomenon are shown in fig. 161, which has been taken from the above-mentioned thesis by Dr. KÜHN.

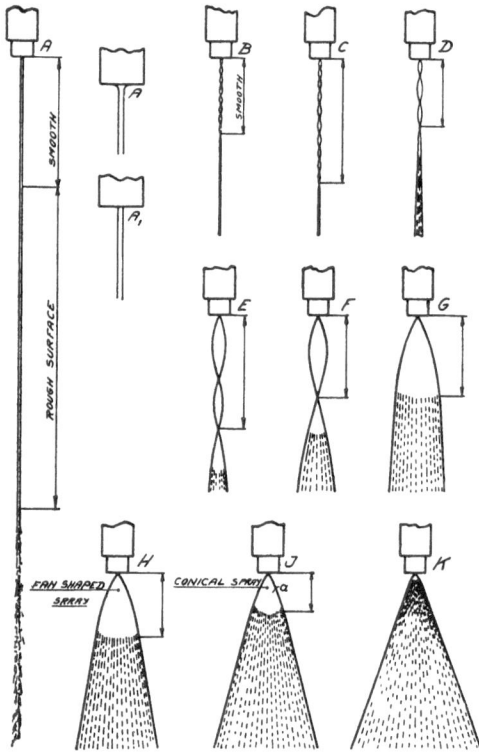

Fig. 161. Various stages of flow (KÜHN).

At very small pressures the oil is not able to form a continuous flow, and drips from the orifice. The number of drops per second and their magnitude depend mainly on the *surface tension* (TRAUBE's "Stalagmometer"), which is the molecular force tending to make the surface of a liquid per unit of weight as small as possible. Such dripping devices are often used as a means for regulating the quantity of oil for vaporizing burners, and one form is shown in fig. 162 (SAUER).

If the pressure is somewhat increased, a *continuous flow* of oil may be obtained which, after leaving the orifice, is slightly contracted but shows a smooth and perfectly transparent surface (A). On the pressure being further increased, the contraction is reduced until the flow of oil has obtained a diameter approximately equal to that of the orifice (A₁). Up to this point the flow has been what is called "laminary", all oil particles having approximately parallel velocities.

A still further increase in pressure causes a kind of *"hesitation" between two forms of flow, the laminary and the turbulent flow*, a phenomenon which becomes evident on the oil surface's turning from transparent and smooth to untransparent and rough, caused by the first traces of turbulency.

A slightly increased pressure causes the formation of a *twisted ribbon* of oil (B), which may be explained as follows. As long as the velocities of all the oil particles were parallel, there was no reason for any particle to leave the flow of oil, but as soon as turbulency occurs, comprising velocities of all directions, an oil particle may be able to leave the stream of oil as soon as its velocity is large enough to overcome the attraction of its neighbouring particles (surface tension). As a preliminary stage these attempts to leave the stream will cause *a complex of internal stresses resulting in a deformation of the stream* from a cylindrical to an oval or flat cross section (ribbon), whilst either a left or a right-hand twist may be caused by some accidental influences. The next stage will be that oil particles actually succeed in breaking through the oil surface and leave the stream.

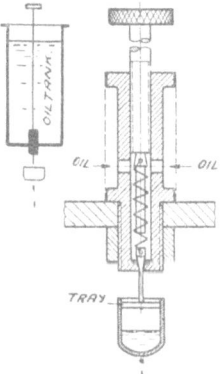

It is interesting to note that if all irregularities of the flow inside the reservoir towards the orifice could be eliminated, the initially rotation-free flow would change into a *counter-clockwise rotation* (seen in the direction of flow) if the experiment were carried out in the *Northern* hemisphere against a *clockwise twist* occurring in the *Southern* one. This phenomenon, which is caused by the rotation of the Earth and expressed by the law of BUYS BALLOT, is similar to the metereological phenomenon of cyclones.

Fig. 162. SAUER's automatic drip device (Archiv für Wärmewirtschaft Nov. 1933).

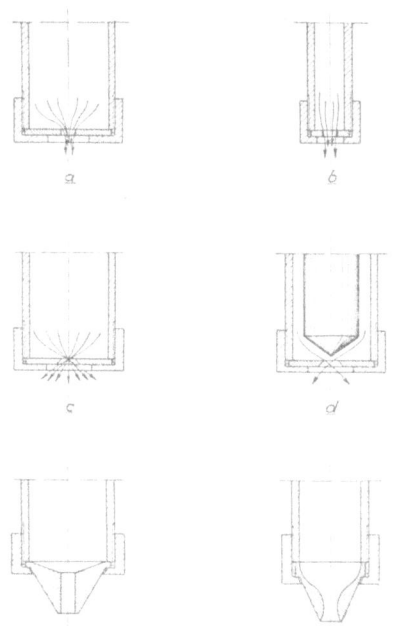

Increase in pressure gradually *reduces the number of turnings* of the twisted ribbon (C, D, E, F), at the same time making them broader and thinner, until finally only one flat blade is left (G), which is so thin that it tears itself apart and diffuses into oil drops. Further increase in pressure transforms the fan-shaped stream of oil drops (H) first into a conical spray (J), after which the little cone of non-pulverized oil (a) disappears and a perfect conical spray of oil drops is obtained, as shown by (K). This description applies to turbulent pulverization by means of a cylindrical orifice.

The same successive stage of development may be obtained at lower pressures if suitable orifices of other shape are used, such as are represented for instance in fig. 163. As pulverization is obviously caused by *oil particles having velocities which are not parallel to the main velocity of flow, every circumstance causing deviations from the main velocity will assist pulverization.* A wide tube shut off by a flat disc having a cylindrical

Fig. 163. Various shapes of orifices for atomization.

hole (a) will obviously stimulate pulverization more than a narrow tube (b). Furthermore a cylindrical hole (e), tending to force the velocities back again into the direction of the main velocity of flow, will be less effective than a divergent "knife-edged" hole (c). A stream-lined hole (f) is a very ineffective form.

### b. *Energy of Pulverization.*

It will be known to the reader that the velocity of a liquid flowing through an orifice may be theoretically calculated from: $V = \sqrt{2\,g\,h}$, where h denotes the pressure expressed in head of the same liquid. The theoretical quantity of liquid flowing through in a unit of time follows from $Q_{theor.} = F\sqrt{2\,g\,h}$ (where F denotes the area of the orifice), which liquid possesses a *kinetic energy*: $E_{theor.} = F\,d\,h\sqrt{2\,g\,h}$, where d denotes the density. The quantity of liquid *actually* flowing out, however, is often much smaller, due to what is called *"contraction of flow"*, which is expressed by putting in a contraction coefficient $\mu$, thus $Q_{pract.} = \mu\,F\sqrt{2\,g\,h}$, and consequently the kinetic energy of the stream will also be less, viz.: $E_{pract.} = \mu\,F\,d\,h\sqrt{2\,g\,h}$.

As, however, the energy originally applied to the liquid is still equal to $E_{theor.}$

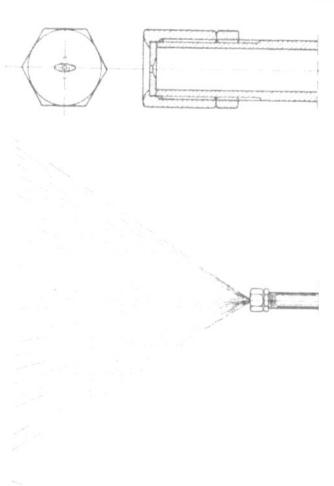

the difference $(E_{theor.} - E_{pract.})$ *must be produced in some other form, viz. as the energy of an irregular rotation or turbulent motion. It is this amount of energy* which is *available for pulverization*, and consequently the orifices which show the *largest contraction*, viz. what are called "knife-edged" orifices, are *best suited for pulverization* and least suited for flowing purposes.

A remarkable effect, which needs no further explanation, may be obtained if the knife-edged orifice

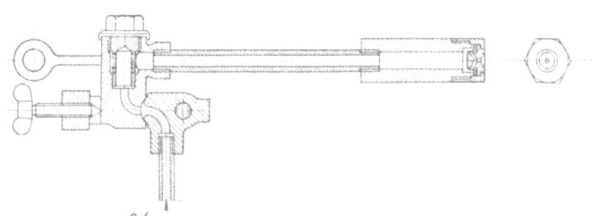

Fig. 164. Flat fan-tailed spray by oval orifices. (after TRINKS).

Fig. 165. Turbulency-tip of TATE-JONES.

is made *oval* instead of circular, as has been shown in fig. 164 taken from TRINK's "Industrieöfen".

It is, of course, possible to improve the above-mentioned forms of "natural turbulency" by using special turbulency-plugs before the orifices, an example of which is shown in fig. 163 (d), consisting of a cone which prevents the oil from leaving the orifice in the main direction of flow. Other forms comprise crossing channels or slots, such as in the design of TATE-JONES (see fig. 165).

c. *Pressure-atomizing tips for Diesel Engines.*

There is a considerable difference between the essentials for pressure-atomizing tips for *continuous* oil burning and those for *discontinuous* operation in Diesel engines, as for the latter only a *short time* is available to set up the pulverizing motion. For this reason most pressure-atomizing *tips for Diesel engines*, especially those for high speed, *belong to the turbulency type* operated by means of a high pressure pump (usually 100 to 300 atm.), as in the first place the *time available for injection is too short* to set up a perfect regular rotary motion and, secondly, because Diesel engines *require a certain penetration of the oil injection into highly compressed air*, which is obtained more effectively by means of straight shot atomizers of the turbulency type than by the "softer" spraying of the regular rotation types.

Another great difference between pressure atomizing tips for Diesel engines and those which are to be used in the free atmosphere is that the atomization into compressed air of the former type is considerably assisted *by another pulverizing effect,*

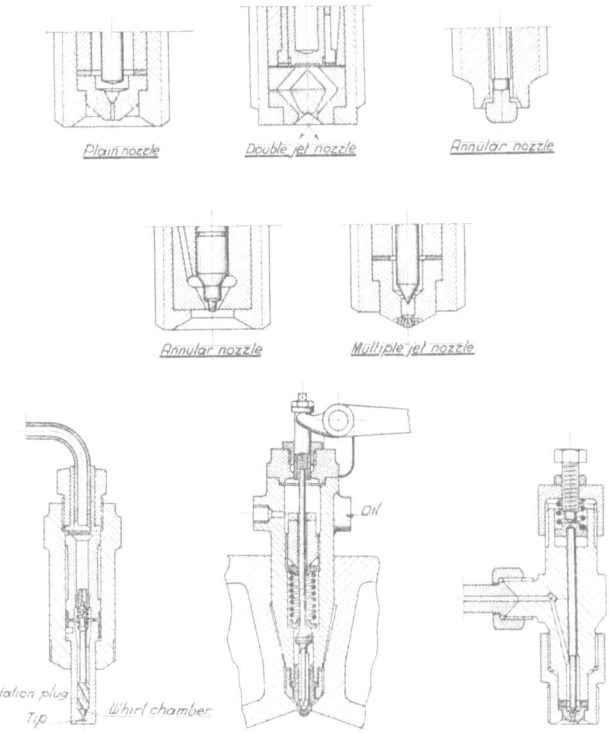

Fig. 166. Various DIESEL tips.

viz. by the *collision of oil drops with the air,* and consequently somewhat less attention may be given to the other pulverizing factors of Diesel tips in order to improve the penetration.

So as to ensure the proper distribution of oil drops in the cylinder it is now common practice to use *tips with more holes* or special conical stream deflectors in a central hole and to give the *air,* moreover, a *swirling motion* (fig. 88). Some examples of modern Diesel engine tips are represented in fig. 166, taken from Report No. 520A of the National Advisory Committee for Aeronautics (by LEE, Washington DC) in which numerous interesting experiments on the pulverizing effect of various types may be found.

d. *Preheating.*

The influence of viscosity and surface tension on pulverization may be readily understood from the following: A definite amount of mechanical energy being imparted to the oil by the pump, the viscosity not only determines the pressure drop from pump to burner tip but also how much of the velocity imparted to the oil when entering into the whirl chamber will be absorbed by friction.

This is especially of great importance for the design of the whirl chamber, which may be illustrated by fig. 167. With burner tip A, comprising a flat cylindrical box serving as a whirl chamber, the upper viscosity limit of soot-free combustion with a heavy bunker fuel was found to be 2° Engler. By changing the design of the whirl chamber as in fig. 167 B this limit was raised to 7° Engler for the same fuel and at the

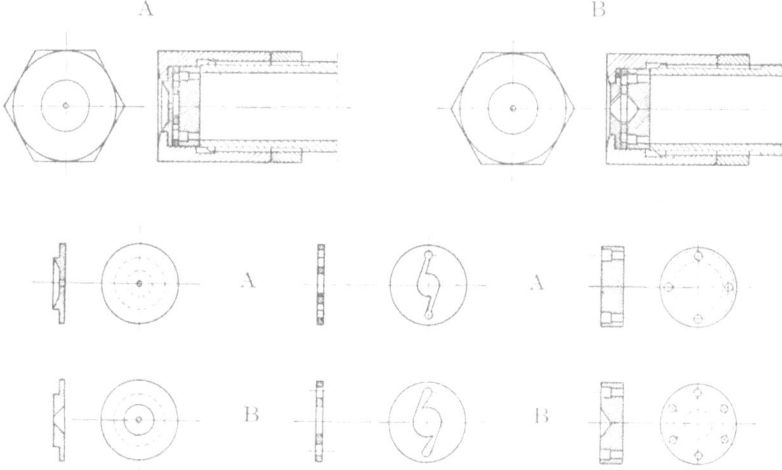

Fig. 167. Whirl chamber design and wall friction.

same oil pressure, which effect was due to reduction of friction caused by the walls in the centre of the whirl chamber, where the velocities of the oil are extremely high. This difference in viscosity in this special case meant lower preheating by about 70° C. Fig. 181*d* shows a whirl chamber design which is particularly unfavourable in this respect.

The kinetic energy applied to the oil will be consumed in four ways:

1) In *overcoming the surface tension*, which tries to keep the oil surface as small as possible, whereas pulverization tends to enlarge the original oil surface a considerable number of times.

2) By *collision* of oil drops and air molecules.

3) By *internal friction* of the oil inside its drops, whilst

4) A certain amount is left as *penetration velocity*.

The internal friction losses being entirely governed by the viscosity, it is clear that viscosity is the most important factor for mechanical pulverization, not only with respect to what takes place *before* and *in* the burner tip, but also in regard to what happens *after* it. In practical terms this means that *preheating of the oil, which lowers the viscosity and surface tension, has a highly favourable influence on mechanical atomization,* though it may lower the maximum capacity (see page 288).

However, there seems to be an *upper* limit for preheating too, which is first of all fixed by the temperature at which the first vapour bubbles are evolved from the fuel at atmospheric pressure. Such vapour bubbles may cause irregular operation of the whirl chamber, resulting in an unsteady pulverization. Moreover, extremely fine pulverization obtainable by strongly preheating the oil may lead to trouble because of the oil mist thus formed not having enough "body" or "own velocity", which means that the oil vapours as a whole are too easily carried away by the air without sufficient mutual penetration and mixing. Such fine pulverization may result in *unsteady, fluttering, long gaseous flames,* which are only kept free of smoke with the utmost difficulty.

### D. *Oil burners based on a Regular Rotary Motion of the Oil* ("rotation tips")

a. *Increasing the energy of pulverization.*

It has been shown that the *energy which is available for pulverization with pressure-atomizing burners is equal to the difference between what may be called the "theoretical" and the "practical" amounts of regular kinetic energy of the oil when leaving the orifice, which energy is directly proportional to the contraction coefficient* $\mu$.

As a matter of fact, this difference is the *definition of* $\mu$, because this factor must be seen as "an empirical correction of an incomplete theory", which only holds good for perfectly parallel velocities for every particle of the liquid flowing through the orifice. In practice, however, the oil particles will also receive velocities which are *not parallel* to the main direction of flow, causing contraction and turbulency, and it follows that a certain quantity of kinetic energy represented by these non-parallel velocities will be lost for the maintenance of the outflow, which consequently will be less than what would follow from the "theory".

The *lowest limit for* $\mu$ obtainable being about 0.60, it follows that only about 40% of the total energy initially imparted to the oil can be utilized for pulverization, the remaining 60% being represented by the main velocity of flow (projection). This means that atomizing tips based on the principle of *"natural turbulency"* (velocities non-parallel to the main direction of flow originate only from turbulency, friction etc., no special rotary devices being used) will produce a *straight shot* of oil drops (60% available for projection) having a comparatively small angle of spray. As explained before, this type of nozzle is especially favourable and therefore is generally adopted for Diesel engines, because it facilitates a better penetration into heavily compressed air, but it is considered to be less suitable for oil furnaces, because the resulting long flames would require long combustion chambers to avoid impingement upon the walls and, moreover, because with such flames it is difficult to ensure thorough mixing with the air.

A method enabling the amount of pulverizing energy to be increased far higher than the above-mentioned "natural" limit of 40% consists in *giving the oil a regular rotary motion*, as was originally done by KÖRTING. Such a regular rotation imparted to the oil before it leaves the orifice represents an *artificial method of reducing the contraction coefficient μ far below the lowest "natural" value of about 0.60.*

It is evident that the quantity of liquid flowing through an orifice at the bottom of a reservoir may be decreased to any extent wanted by rotating the oil above the orifice and the flow may even be completely stopped if the rotation imparted is so strong that a hollow whirlpool is formed above the orifice, which keeps the liquid away from its edges. In this case the contraction coefficient μ would have the value *zero*.

Rotation burners contain *three essential elements* of design, viz.:

1) The *"plug"*, which imparts a more or less regular rotation to the oil.

2) The *"tip"*, comprising the orifice through which the oil leaves the burner, and

3) The *"whirl chamber"*, this being the space between plug and tip.

The same considerations concerning the influence on the pulverizing effect of various constructive forms of the tips as have been discussed for the turbulency tips may be applied to rotary tips, and consequently a sharp "knife-edged" orifice will be better than a cylindrical or stream-lined one, as the former causes more "natural" contraction of the stream than the latter. Moreover, *a cylindrical part of an orifice will have considerable brake-effect on the rotation of the oil passing through it, without "inducing velocity"*. For a clear understanding of this argument it will be necessary to give first a kinematical analysis of the movement of the oil in the whirl chamber, in order to explain the peculiar function of this device.

### b. *Kinematical analysis*

Fig. 168 represents a pressure atomizing oil burner in its simplest form, viz. a cylindrical orifice and a cylindrical whirl chamber in which the oil enters through a tangential slot. The oil particles, which are first supposed to undergo neither external nor internal friction forces, are subjected to two simultaneous movements, the first being a *spiral motion* from the outer wall of the whirl chamber to its centre, which takes place in planes at right angles to the burner axis, whilst the second movement is a *translatory motion* of the oil parallel to the burner axis, which forces the oil to travel through the orifice. Moreover the oil particles may possess individual rotations, thus spinning around their centres.

The first movement may be expressed for a frictionless liquid by means of BERNOUILLI's theorem, and the pressure p and velocity v of any particles of the oil inside the whirl chamber may thus be calculated from the equation:

$$p + \frac{v^2 \cdot d}{2g} = p_1 + \frac{v_1^2 \cdot d}{2g},$$ where d = density of the oil, g = gravitation con-

stant. As $p_1$ and $v_1$, being the pressure and velocity of the oil when entering the whirl chamber, may be regarded as constants, it follows that the relation may be written:

$$p + \frac{v^2 \cdot d}{2g} = \text{constant},$$ which means that *if at any point within the whirl chamber*

*the pressure decreases the velocity will increase, and vice versa.* Consequently the velocity will be the maximum where the pressure is minimum (atmospheric), that is to say in the centre if the whirl chamber is completely filled with liquid, or, if there

happens to be a hollow core or whirlpool, at its inner surface, having a radius $r_1$ (see fig. 168).

Thus we see that the rotation velocity rapidly increases towards the centre and it may easily be infer-

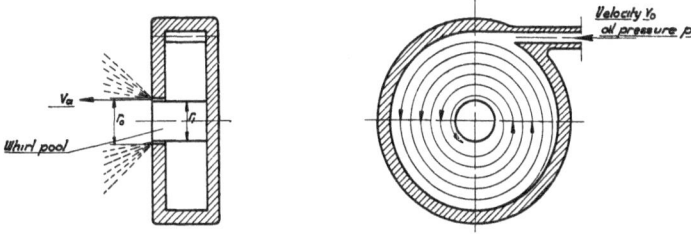

Fig. 168. Simple scheme of a whirl chamber.

red from BERNOUILLI's theorem that this increase must obey the following rule: *The products of rotation velocity and radius must be equal for all oil particles in the whirl chamber,* or *v.r = constant,* which means that the *velocity increases hyperbolically towards the centre.*

In fact this is KEPLER's *astronomic law of equal areas,* according to which the momentum of velocity for orbits described by particles subjected to some central force is constant. Thus, if the tangential velocity of an oil particle somewhere in the whirl chamber at a diameter of 5 mm is equal to 10 cm/sec., its velocity must be *increased ten times* if it is moved to a diameter of 0.5 mm in order for the momentum of velocity in both cases to be equal to 2.5 sq.cm/sec.

Although this simple equation, applying to *frictionless liquids,* must be replaced by much more complicated ones for the case of the flow of *"real" liquids* (internal and external friction), this general conclusion still holds good in principle for these cases too. Thus BARGEBOER succeeded in solving the mathematics connected with this kinematic problem and found that for values of REYNOLD's number of his theory of dynamical similarity lower than 2000, i.e. for *laminary flow* (thick fuels), the simple equation of KEPLER's law: **r.v = R.V** (v = the velocity of an oil particle at a radius r, and V the velocity of the oil entering the whirl chamber at a radius R), had to be replaced by another, viz.:

$$\mathbf{r.v = R.V.} \, e^{\dfrac{12\pi\mu A}{\gamma Q}} \quad \text{in which } A = \int_{r}^{R} \frac{r}{b} \, dr$$

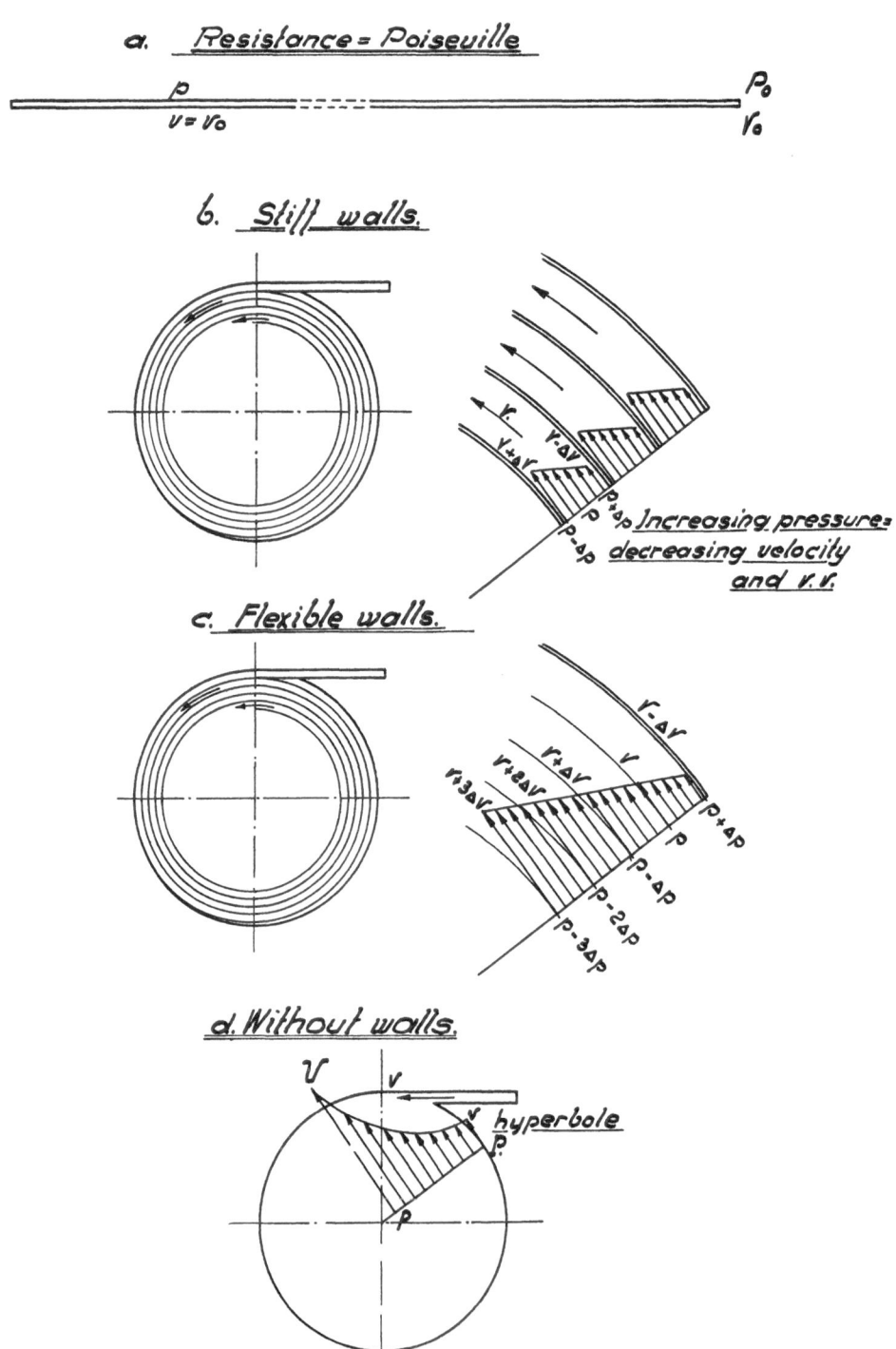

Fig. 169. Imaginary stages of development of whirl chamber flow.

and $\mu$ = absolute viscosity          or $\dfrac{\mu}{\gamma} = \varkappa$ = kinematic viscosity

$\gamma$ = density

b = variable breadth of the whirl chamber = a function of r.

For a *cylindrical* whirl chamber b is a constant and the equation may be transformed into:

$$\text{r.v} = \text{R.V. e}^{\dfrac{-6\,\pi\,\mu\,(R^2 - r^2)}{\gamma\,Q\,b}} \qquad \text{(laminary flow)}$$

If the flow is *turbulent*, which usually occurs in practice, (viz., if REYNOLD's number exceeds the value 2000), the resulting equation proves to be independent of the breadth b and may be written:

$$\text{r.v} = \text{R.V} \dfrac{1}{1 + \dfrac{4\,\pi\,\lambda}{Q\,\gamma}\,\text{R.V}\,(R - r)} \qquad \text{(turbulent flow)}$$

in which $\lambda$ = friction constant of liquid and wall.

In order to give a clear understanding of the important function of the whirl chamber without any mathematics, one might imagine the spiral motion of the oil as being established by means of four successive stages of development, viz.:

*First stage*: If a stream of oil having a *ribbon-like uniform cross-section* is supposed to flow between parallel walls as shown in Fig. 169a, the velocity will likewise be uniform over the entire length of the ribbon, and the pressure difference between the beginning and the end required to maintain this velocity will follow from POISEUILLE's law: $\qquad P_0 - p = c\,\eta\,v\,L\,.\,\varphi\,(b, \tau)$

where    c = constant

$\eta$ = viscosity

v = velocity

L = length of the ribbon

$\varphi$ = a function of the ribbon dimensions b and $\tau$

b = breadth and $\tau$ = thickness of the ribbon,

provided the gravitation (i.e. difference in height) may be disregarded, and laminary flow occurs.

*Second stage*: It might be imagined as a *second stage* that this ribbon-like flow of oil is wound up into a *spiral form* as shown in fig. 169 (b), and if in that case the walls, after being bent, are supposed to be *perfectly stiff* again, nothing will be changed in the conditions of flow, as the centrifugal action only causes a slightly increased pressure at the outer wall and a slightly decreased pressure at the inner wall of the ribbon, which may be disregarded with respect to the flow conditions inside the cross-section of the ribbon.

*Third stage*: If, however, as a *third stage*, the walls are supposed to be *perfectly flexible* (fig. 169c), these small pressure differences caused by the centrifugal action are allowed to *build up a larger total pressure difference between the centre and the outermost wall of the spirally wound ribbon*, which pressure, being opposed to the initial pressure, acts as an additional resistance, tending to reduce the quantity of oil flowing through. In order to force the original quantity of oil through the ribbon with flexible walls, a higher initial pressure, and consequently more energy, will be required than was necessary with the rigid walls. This *increased pressure includes an increase in energy initially applied to the oil*, which, however, not being required to make up for friction losses (for this purpose the above mentioned "POISEUILLE's pressure" is sufficient), must necessarily be converted into kinetic energy according to

BERNOUILLI's theorem and equation: $p + \dfrac{v^2\,d}{2g} = $ constant.

*Fourth stage*: If, as a *fourth stage*, the imaginary flexible walls are supposed to be removed (fig. 169d), the only change in the conditions of flow will be that the *external*

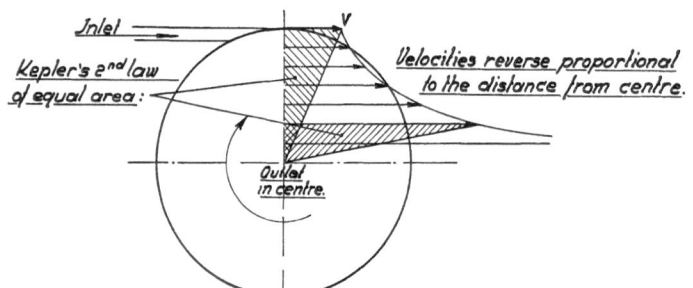

Liquid revolved by internal means. (whirl chamber)

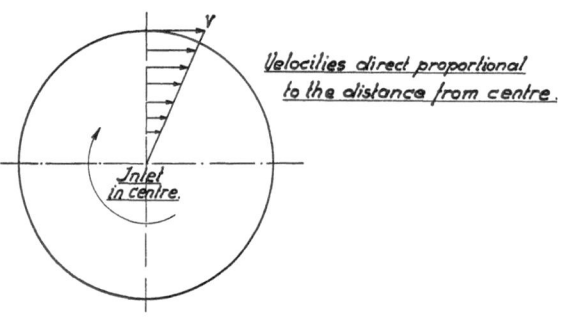

Liquid revolved by external means (revolving plate)

Fig. 170. Difference between whirl chamber and revolving plate.

friction losses caused by the contact of liquid with the walls of the ribbon will be replaced by the *internal* friction losses of the liquid caused by the velocity differences of neighbouring layers, and the only consequence of this change will be that the "POISEUILLE pressure" will be replaced by another value representing the "internal friction pressure."

Although the distribution of the tangential velocities of the oil along the radius of a whirl chamber differ largely from those of a cup filled with oil revolved by external forces — for the latter case the velocities, after reaching the state of equilibrium, are directly proportional to the radius and *decrease* towards the centre, whereas in the former case the velocities *increase* towards the centre (see fig. 170) in both cases an *artificial field of gravitation* is created, due to centrifugal action, and it will be clear that oil particles, if moved against this gravitation, i.e. towards the centre, will obtain a higher potentiality and consequently will receive a "storage of kinetic energy", so to speak, which may be utilized for pulverization as soon as the oil is set free.

c. *Functions of plug, whirl chamber and tip.*

This analysis shows that, whereas it is the function of the *plug* to impart an initial *tangential velocity* $v_0$ to the oil, the *whirl chamber* has to convert as much as possible of the initial pressure $P_0$ into velocity by *forcing the oil*, against the action of the centrifugal force, *towards smaller radii*, which method consequently comprises means of obtaining velocities near the centre which are several times larger than the initial velocity $v_0$ caused by the rotation plug alone. Thus *the action of a whirl chamber might be compared to that of a gear-box of an automobile*, both in fact being transmission devices for "converting pressure into speed", or vice versa.

The function of the *tip* is *setting the oil free at a very small radius*. The *diameter of the tip governs mainly the quantity of oil* which leaves the burner at a certain pressure. The capacity of pressure-atomizing burners only can be regulated for a small range by means of the pressure. For larger variations of capacity it is necessary to change the tip diameter, which is usually done by fitting another burner. Means for avoiding this round-about way of regulation will be considered further on.

Moreover this analysis shows again how very important a factor *viscosity* is, not only with respect to the resistance of the oil in the holes of the plug but even more so as regards the *internal friction* of the layers of oil moving in the whirling chamber, as this latter factor determines how much pressure there will be available from the initial pressure $P_0$ for conversion into velocity.

d. *Some erroneous ideas about the paths of oil particles after leaving the orifice.*

It has been explained that with a pressure-atomizing burner tip, provided it is based on the *regular rotation* principle, the oil, instead of entirely filling the area of the orifice, may also form a hollow whirlpool, thus creeping along the wall of the orifice as a thin film at an extremely rapid rotary speed. As soon as this oil film is no longer supported by the edges of the hole, it breaks up into oil drops, as from that moment onwards each oil particle is left free to follow the direction of its velocity, which in fact is the resultant of the tangential velocity at right angles to the burner axis and of a transversal velocity parallel to the axis. It is clear that the ratio of the latter velocities fixes the top-angle of the spray-cone.

In this connection it must be pointed out that the assertion that rotation plugs will cause oil particles to move in *spiral orbits* through the combustion chamber, which version is often found in literature, is *not correct*, as the oil particles in fact will follow *straight lines*, provided the air is motionless. Of course a spiral movement of the combustion *air*, as is often caused by vanes, will take the oil particles with it, but

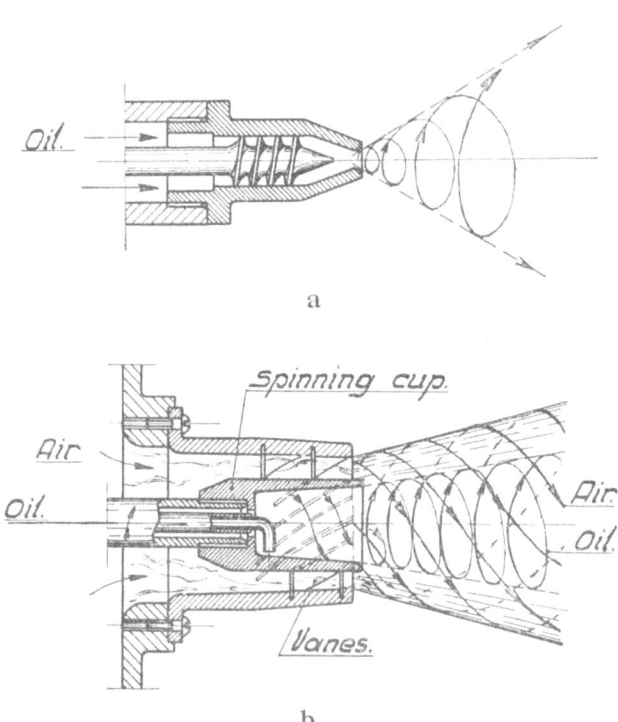

a

b

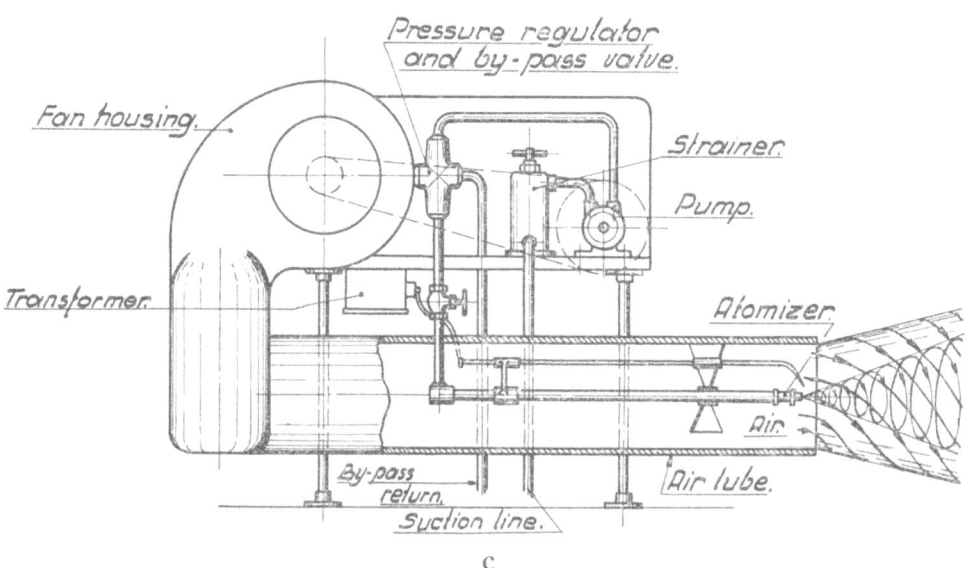

c

Fig. 171. Paths as not followed by oil drops.

propositions of giving the air a rotary motion *opposite* to the rotary motion of the oil in the burner tip, in order to improve mixing, are senseless and not based on clear theoretical views (see fig. 171 and Br. pat. 271, 687, MARKS, 1926, and others).

e. *Some remarks on the design of tips and whirl chambers.*

Some further conclusions may be drawn from the above-mentioned analysis with regard to the dimensions and shape of tips and whirl chambers. First of all the whirl chamber should cause as *little friction* as possible, and therefore the total area of its so-called "wetted walls" per unit of its volume should be as small as possible. From this point of view a spherical whirl chamber would be better than a flat circular box, whilst a *mandril or cone inserted in the centre*, which is sometimes done (see figs. 181*d* and 182*b*), *has a marked adverse influence on the pulverizing effect*, especially because of braking the high rotary speeds occurring in the centre.

As a second point it must be remembered that this increase in rotary speed in the centre is obtained by forcing the oil from larger to smaller radii, and consequently a *whirl chamber must show sufficient differences between inlet and outlet radius* to use this principle in an effective way. In this respect plugs comprising tangential holes in planes *at right angles* to the burner axis may be less favourable than those having holes in planes

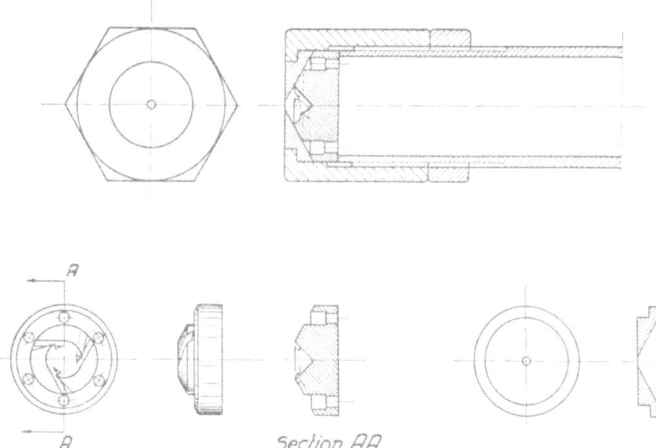

Fig. 172. Conical tip.

*parallel* to the burner axis, although the former type usually shows a lower pressure loss than the latter. A compromise consisting of tangential slots *on a cone* may possess both advantages (See fig. 172).

As regards the construction of the tips, it needs no further explanation that a *knife-edged orifice is better than a cylindrical one*, because the former will cause less friction loss to the rotary motion of the oil film passing through it. It will now be clear what was meant in the previous remark, viz., that any cylindrical part of an orifice causes additional friction without "inducing velocity", and in fact a cylindrical orifice of say 10 mm length will usually absorb enough kinetic energy to spoil badly the pulverization of an ordinary burner.

The only practical objection to the use of *knife-edged orifices* is that they are *more subject to wear* by erosion and corrosion. For this reason tips of domestic pressure-atomizing oil burners are usually made of hardened steel or sometimes of extremely hard alloys (Widia) and even of semi-precious stones, such as sapphire.

### E. *Combinations of Turbulency and pure Rotation.*

The previous discussions were based on the assumption that the rotation took place without any turbulency, which of course never applies completely to actual practice, *every pressure-atomizing oil burner lying somewhere between the two extremes of pure rotation and pure turbulency.*

It is clear that no hollow core can be formed if there is too much irregular motion or turbulency, and consequently the whirl chamber and the area of the orifice of a burner based mainly on the principle of turbulency will as a rule be completely filled with oil. A direct consequence of this is that such a *"turbulency tip"* will usually require smaller orifices for the same delivery of oil than will a *"rotation tip"*.

A "pure rotation burner" might be considered as belonging more or less to the class of *kinetic atomizing* or *centrifugal* oil burners, to be discussed in the following Chapter, which consist of a revolving cup or disc from the edge of which an oil film is projected and pulverized by centrifugal action. It has been pointed out, however, that the velocities of the oil particles in the whirl chamber of a rotary tip are entirely different from those obtained by such external centrifugal action.

The characteristic properties of a pressure-atomozing oil burner depend largely on the *degree of irregularity of its rotary motion, or, in other words, on the orientation between pure rotation and pure turbulency.* Thus, the burner used in Dr. Kühn's above-mentioned experiments, comprising an orifice of 0.53 mm diameter and 2 mm length, was found to produce oil drops the average magnitude of which was practically independent of the presence of the rotation plug; this will now be clear, as such a long cylindrical hole will transform a considerable part of the regular rotation into irregular motion (turbulency), whereas the magnitude of oil drops produced by a "pure turbulency tip" mainly depends on the pressure and less on the rotation.

In this connection it must be noted, however, that two oil mists, although having the same *average* magnitude of oil drops, may differ considerably in combustion quality, viz. if one consists of a mixture of very fine and very coarse drops, while the other mist contains drops of approximately equal size.

For oil fires burning *residues*, useful information on the size of the oil drops may be obtained from the magnitude of the ash particles contained in the flue gases. This ash may be caught by a cyclone and analysed under a microscope (See fig. 156).

As a rule pure rotation tips are able to produce drops differing less in size at lower pressures than turbulency tips, which is only obvious, since in the former case the oil is first drawn out into a film and then torn apart into drops, this being, therefore, almost a *two-dimensional* process with considerably less difference in speed in the oil film than in the latter case, where the oil is directly broken up from a massive stream, this being a *three-dimensional* phenomenon.


## F. *Forms of Construction*

Table V is a classification of pressure-atomizing oil burners divided, horizontally, according to the *constructive forms of the tips* (orifices) and, vertically, according to the *construction of the plugs* (rotation or turbulency elements).

It has been seen that tips may be divided into:

1) *Knife-edged conical orifices* (convergent or divergent),
2) *Sharp-edged cylindrical orifices,*
3) *Round-edged or stream-lined orifices,*

and we will now proceed to discuss the constructive form of the plugs.

First of all a plug may tend to impart either a *pure rotation* or a *turbulent motion* to the oil, this depending on whether there is *co-operation* causing a regular rotation or an *opposed action* of the oil streams leaving the various holes, channels, slots or screw-windings of the plug.

This effect again may be obtained either by using *immovable vane blades* or by forcing the oil through *straight circular holes, straight or bent channels, slots of constant or variable cross section or screw-windings*, which may all be arranged, either in a *flat plane at right angles* to the burner axis, or on a divergently or convergently *conical* or a *cylindrical* surface having the same axis as the burner, as has been indicated in Table V. Consequently, there are so many variations that it is impossible to mention every conceivable burner type of this class here, but one final remark must be added with regard to the efficiency of slots or holes.

It is the task of the rotation plug to transform the oil pressure into a high tangential speed given to the oil particles at the moment they enter into the whirl chamber. As high speeds are accompanied by high resistances and high pressure losses, it is advisable, from the point of view of *pressure economy*, to wait until the very last moment before causing this high speed. This means that, theoretically, the bores or slots of the plugs should be designed with decreasing (convergent) cross sections ending in a *narrow throat* where the oil enters into the whirl chamber. With respect to this point it will be clear that a *long cylindrical* bore is not an efficient element for rotation plugs as, for instance, a quarter of its length would have been entirely sufficient to cause the very same speed, three quarters merely causing resistance and useless pressure loss.

If the cylindrical holes of such a burner are re-drilled with conical drills or reamers, it will be found that the same capacity and pulverization may be obtained at considerably lower oil pressures or, if the same pressure is used, that oils of higher viscosities than before may be perfectly pulverized.

The same consideration applies to rotation plugs comprising *screw thread* as a rotation element. If the screw has a *constant pitch* and leaves a *constant cross section* for the oil passage, it is not efficient from an economy point of view. Therefore, it would be better to construct oil passages similar to the grooves of a rifle, having a *variable decreasing pitch* and, if possible, a *decreasing cross-section* too. As it is not possible to drill bent bores of variable cross-section, such slots are usually cut from the outside on a cylinder or conical surface, an example of which is shown in fig. 172 representing a very efficient design of an industrial oil burner.

The burner tips and plugs are usually subjected to serious wear, especially if residue fuels are burnt. They are therefore generally made of *hardened steel* (e.g. high speed steel) and *preferably polished*, as polished surfaces are less subject to erosion than rough ones. Moreover, the former will cause less friction losses.

As has been explained before, the function of the whirl chamber is to convert pressure into velocity by moving the rotating oil from larger to smaller radii against the centrifugal forces; it follows that, in order to utilize this action to the full, the oil should be admitted into the whirling chamber at distances as far as possible from the centre. Consequently, a plug with bores arranged on a cylindrical surface will generally ensure a better accelerative action of the whirl chamber than a plug with bores arranged on a plane at right angles to the burner axis.


### G.  *Oil Pumps*


As pressure-atomizing oil burners require oil pumps to bring the oil under pressure, a few words will be devoted in this chapter to the various types of pumps.

The first pumps used for this purpose were *plunger pumps*, as may be readily understood, since oil burning required devices with comparatively small deliveries at rather high pressures, usually ranging from 5 to 20 atmospheres. A drawback attaching to crankshaft-driven plunger pumps for continuous oil burning is their *pulsatory* delivery, which may lead to unsteady atomization and may have an unfavourable influence on the combustion and its efficiency. This objection was partly overcome by providing air chambers for levelling the fluctuations of the oil pressure and partly by using pumps with more cylinders. It is interesting to note that the delivery of shaftless steam pumps of the *duplex type* is remarkably constant even at low speeds. This type of oil pump is almost universally used on board of ships.

A disagreeable feature about *air chambers* is that they often *rapidly lose their air*, this being readily dissolved in oil, especially under pressure, as, according to HENRY's law, the quantities of a gas dissolved are directly proportional to its partial pressure, and the more so because the *solubility of air in hydrocarbons is about 6 to 7 times larger than that in water* at equal pressures and temperatures, as may be seen from the following table:

Solubilities

(expressed in BUNSEN units, viz. *cc. gas per cc. liquid* when the partial pressure is 760 mm mercury and the gas volumes are reduced to 0° C.)

| Gas | at 10° C. | | | at 30° C. | | |
|-----|-------|---------|-------|-------|---------|-------|
|     | Water | Gas oil | Ratio | Water | Gas oil | Ratio |
| Oxygen . . . | 0.0380 | 0.229 | 1 : 6.0 | 0.0262 | 0.135 | 1 : 5.2 |
| Nitrogen . . | 0.0186 | 0.135 | 1 : 7.3 | 0.0134 | 0.100 | 1 : 7.5 |
| Air . . . . . | 0.0225 | 0.154 | 1 : 6.8 | 0.0159 | 0.107 | 1 : 6.6 |

Consequently, *multiple plunger pumps* are generally preferred for *industrial*

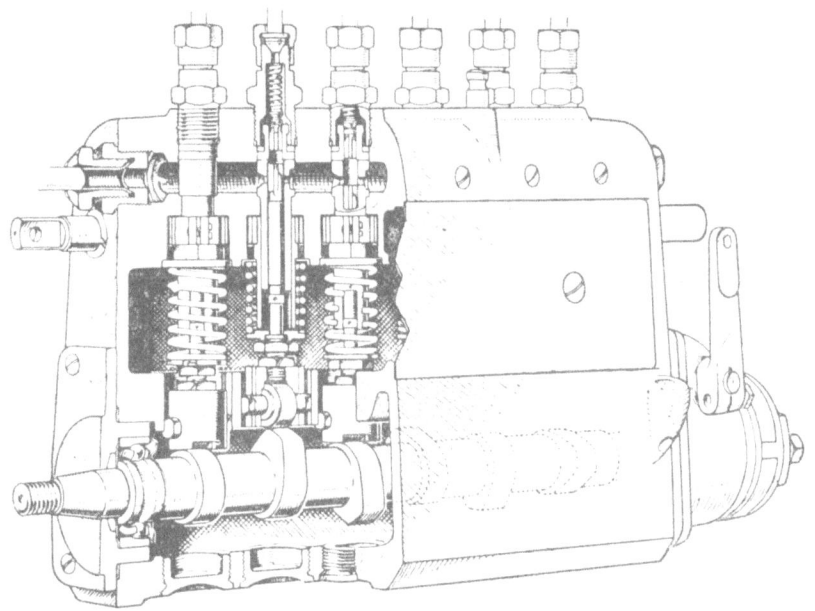

Fig. 173. BOSCH fuel pump for Diesel engines.

purposes, as multiple *centrifugal* pumps, although having perfectly constant deliveries, operate with too low efficiencies for the small quantities and high pressures required in oil burning practice.

*Domestic* oil burning requires such small deliveries (about 1 gallon per hour and even less) that the pumps have to be even smaller than the well-known pumps for high-speed Diesel engines (see fig. 173), so that, if multiple plunger pumps were used, they would have to be of a watch-like construction.

Inventions have been made from time to time for the application of this kind of pump to domestic oil firing, but, although excellent atomization and perfect combustion may be readily obtained with the discontinuous jets of oil mist (these pumps usually operate with pressures from 100 up to even 500 atm.) without any special rotary tips being required, *the power per gallon of oil atomized is too high for domestic use*, and the resulting intermittent shot-like combustions produce such a sharp machine gun effect that it is almost impossible to make a silent domestic oil burner with these pumps.

For domestic oil burning *gear pumps* are now generally adopted, which pumps might in fact be regarded as multiple plunger pumps, the space between meshing teeth acting like a cylinder with piston. Years before their application to domestic oil burning these pumps were developed to a fair degree of perfection by designers of

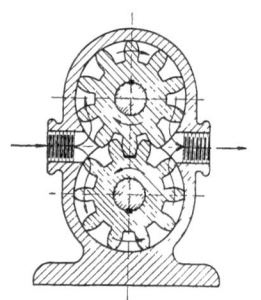

Fig. 174. External gear pump.

automobile engines, who used this type for forced engine lubrication, and still more so by the artifical silk industry where these pumps were used for pumping the "viscose" (= alkaline solution of cellulose) through a number of fine holes for spinning threads. This last application is of special interest, because the irregularities of discharge from gear pumps were then studied closely and found to reach astonishingly high values, particularly because they are apt to be aggravated by interference. Clothes woven from artificial silk were often found to be worthless after being dyed, owing to such fatal interference in the periods of the pumps used for spinning coinciding with the periods of the weaving or knitting machine. The same interference may take place in oil burning practice, sometimes causing a rapid vibration of the atomized oil stream, and often a *noisy, squeaking pump*. This evil, which is very serious indeed for domestic installations, often disappears as soon as the speed of the pump is changed.

The first gear pumps comprised two pinions or spur gears in mesh with each other as shown by fig. 174. The driving shaft operates one gear which causes the other to rotate in the opposite direction. This type is called the *external gear pump* and may give fairly satisfactory results as long as the gears and bearings are not worn too much. A decided disadvantage lies in the fact that the better the gears fit the quicker they are worn, because, as the apex of the tooth tries to come into complete mesh on the discharge side, there is a *tendency to trap* some of the oil at this point. This oil, having no means whatever of escape, is compressed to extremely high pressures of several hundreds of atmospheres, and thus puts a terrific strain on the bearings, which gradually become worn, and when this occurs the pump no longer makes a close fit and its suction characteristic falls off rapidly.

Many remedies have been tried with more or less success, one of them comprising a groove on the face plate at the discharge side, which arrangement, however, has the following deficiencies: If the groove is small it is inadequate to relieve the trapped oil, especially if it is viscous; if the groove is large it will act as a by-pass thus decreasing the volumetric efficiency and suction capacity of the pump.

A second type, the *internal gear pump*, is shown in fig. 175, from which it will be noted that the master rotor or driving element carries internal teeth projecting from the face of the circular disc from which they were originally cut. A stud shaft projecting from the face plate and set eccentric to the centre of the master gear carries a smaller pinion in mesh with the internal teeth of the other gear. A half-moon shaped partition, also part of the face plate, closes the space between the two gears, thus separating the suction and discharge ports. As the master gear rotates, the small pinion will rotate around its fixed shaft in the same direction, and as the teeth come in and out of mesh the displacing action is created.

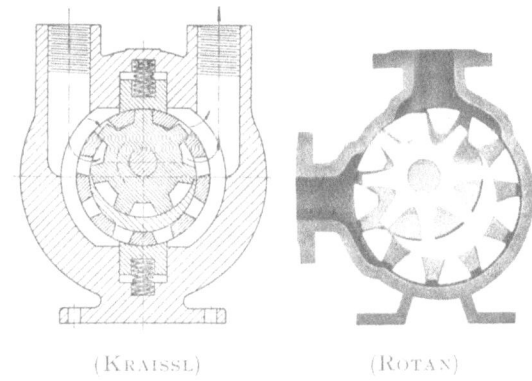

(KRAISSL)          (ROTAN)

Fig. 175. Internal gear pumps.

This pump, which is described by KRAISSL in the "Fuel Oil Journal" of October 1929, possesses not only all the advantages of the external gear pump but, moreover, a very important addition in the elimination of the above-mentioned *trapping effect*, for, since the teeth of the master gear project from the face of the disc, the space between the teeth is open. As the teeth of the pinion come in mesh with these internal teeth the oil is forced out through these spaces in radial direction. Consequently these pumps can be made close-fitting without being subjected to severe wear. Their good priming qualities due to high vacuum can be maintained during a long period, which is especially important for domestic oil burning installations on upstair floors where the oil has to be drawn from underground tanks. A similar design is the ROTAN manufactured by STORK (Hengelo, Holland).

In order to be independent of the suction capacities of the pumps, *booster pumps* are often installed, which are special feed pumps to booster the oil from the underground tank to a small container near the oil burner unit, from which container the oil is drawn by the atomizer pump. Such auxiliary booster pumps are shown in fig. 176 and are commonly used for rotary burners, which are usually fed by gravity. Such pumps are mostly operated intermittently, keeping the oil level between an upper and a lower limit.

Gear pumps, especially when run too fast, may be very noisy due to the oil's being unable to follow the movement of the teeth, resulting in the occurrence of cavitation. In this connection a number of valuable remarks are made by KRAISSL in his above-mentioned article; another interesting publication on gear pumps may be found in "The Engineer" of 15 Feb. 1929 (MERRITT), to which the reader is referred.

A typical form of the internal gear pump type is shown in fig. 177, in which the half-moon shaped partition of fig. 175 is avoided by giving the *master rotor only one tooth more than the pinion*, whilst the tops of the teeth of the former exactly touch those of the latter. This construction ("Gerotor") has been adopted in the QUIET MAY oil burner. Yet another example is shown in fig. 178 representing the STOTHERT & PITT pumps which are especially suitable for pumping heavy viscous oils like asphalt.

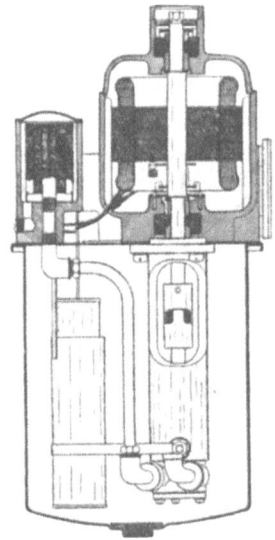

Cross-section COOK pump.

Years ago the artificial silk manufactory ENKA (Holland) developed another type of gear pump, viz. by intermeshing *two parallel helical screw windings*, thus causing the transportation of the oil to take place in axial direction without any periodical interruption. A recent development by the Aktiebolaget IMO-INDUSTRI (Stockholm) comprises *three parallel screws*,

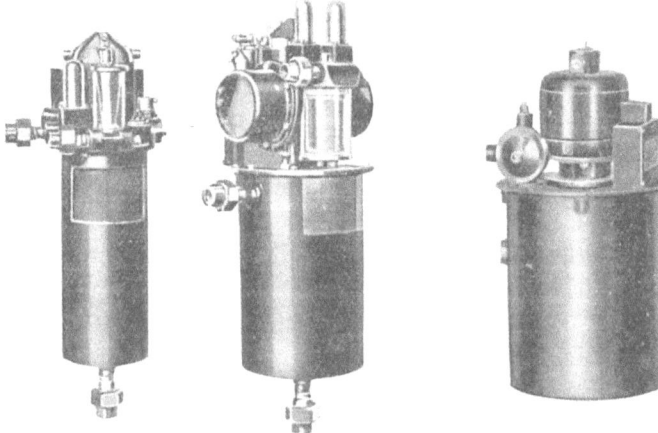

Fig. 176. Booster pumps (COOK and MONROE).

such as shown by fig. 179, which construction is known as the "IMO" pump.

The central or driving rotor operates as a continuously acting piston similar to the Archimedean screw formerly used for polder draining. The other two smaller side rotors, which are located symmetrically on either side of the driving rotor and with their threads intermeshing with the central rotor, only act as rotating sealing slides.

Special advantages of this type are, firstly, that leakage can be reduced to any limit, viz. by choosing the correct length of screws, secondly, that the delivery of the oil is perfectly continuous and, thirdly, that even at high speeds the oil suffers little disturbance, the result being high efficiency and noiseless operation. It is worth men-

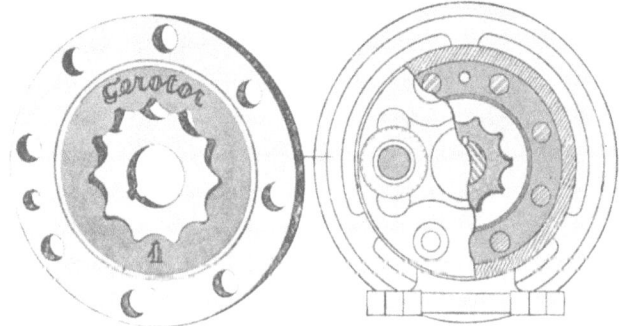

Fig. 177. GEROTOR internal gear pump (QUIET MAY).

tioning that, from the standpoint of construction, the master rotor is perfectly balanced, the thrust of the oil pressure in axial direction being compensated by transferring the oil pressure through a small bore to the other side of the rotor (see fig. 179). As the side rotors are propelled by the pressure of the fluid and not by the driving rotor, there is no metallic contact and hardly any wear.

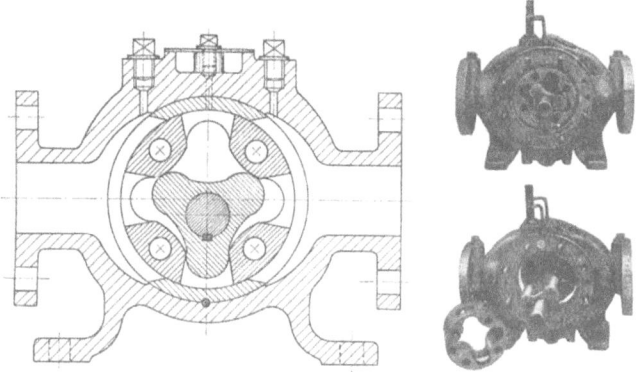

Fig. 178. STOTHERT & PITT pump.

Another type of fuel pump, known as the *"thermo-pump"* has been discussed on pag. 172, whereas fig. 180 shows the LEYLAND pump which needs no further explanation.

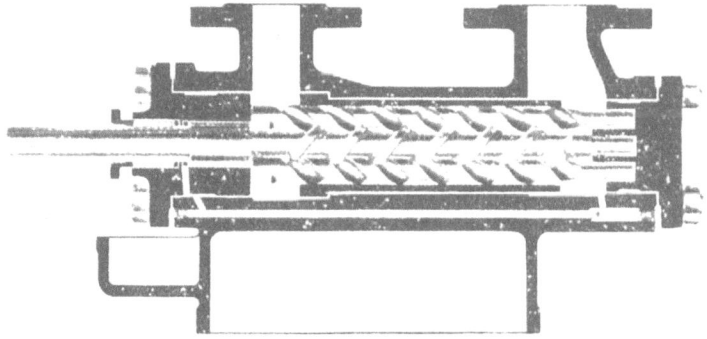

Fig. 179. The IMO pump.

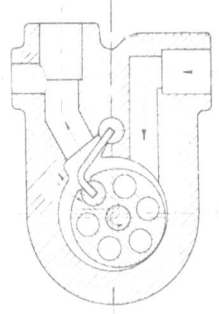

Fig. 180. LEYLAND pump.

The *delivery* of pressure-atomizing oil burners is usually regulated by adjusting the pressure at the burner tip, which is generally done by means of partial circulation: A part of the oil leaving the pump passes through a *relief valve* and is returned to the tank, the valve being automatically adjusted by means of a spring to maintain constant pressure; by varying the spring pressure, the oil pressure and the delivery may be varied (see fig. 171 c).

Another mode of what may be called *"Wide range regulation"*, which offers many advantages over that just mentioned, consists in an extension of the above partial circulation principle, viz. also *through the whirl chamber* of the burner, such as is embodied, for instance, in the PEABODY-FISHER and BARGEBOER devices, which will be discussed further on in this chapter in more detail.

Some examples of the pressure-atomizing class of burners are given in figures 181 and 182.

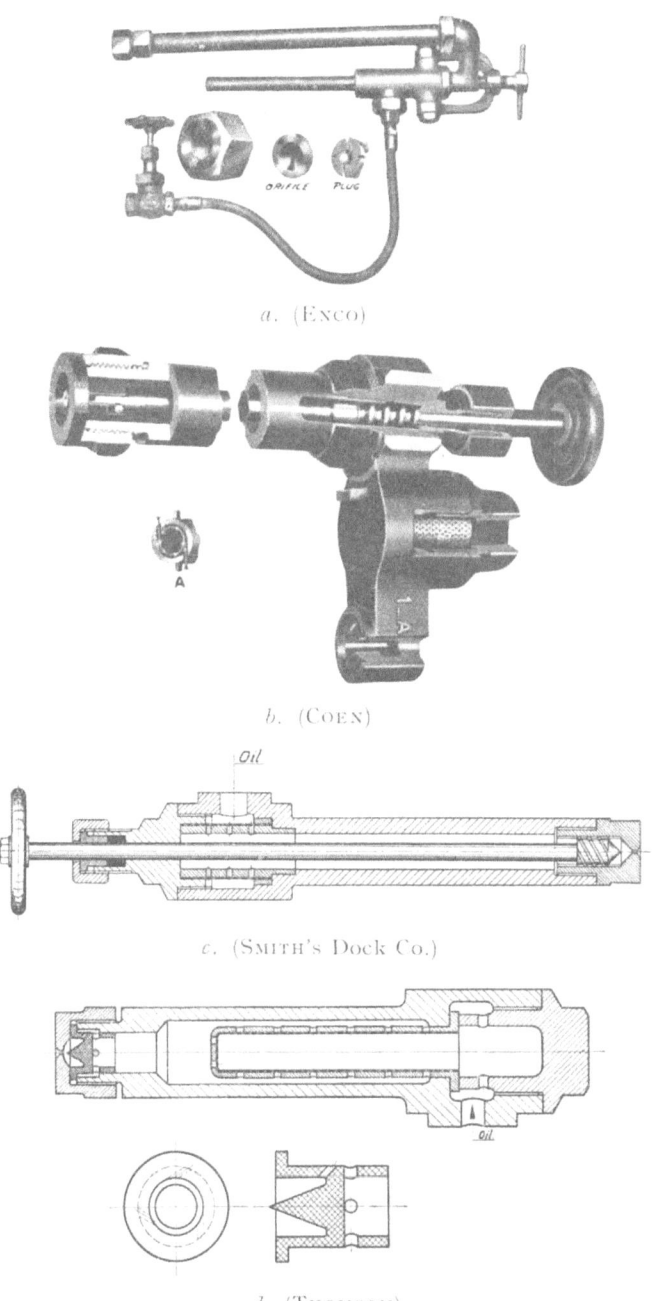

a. (ENCO)

b. (COEN)

c. (SMITH's Dock Co.)

d. (THOMSON)

Fig. 181. Pressure-atomizing burners.

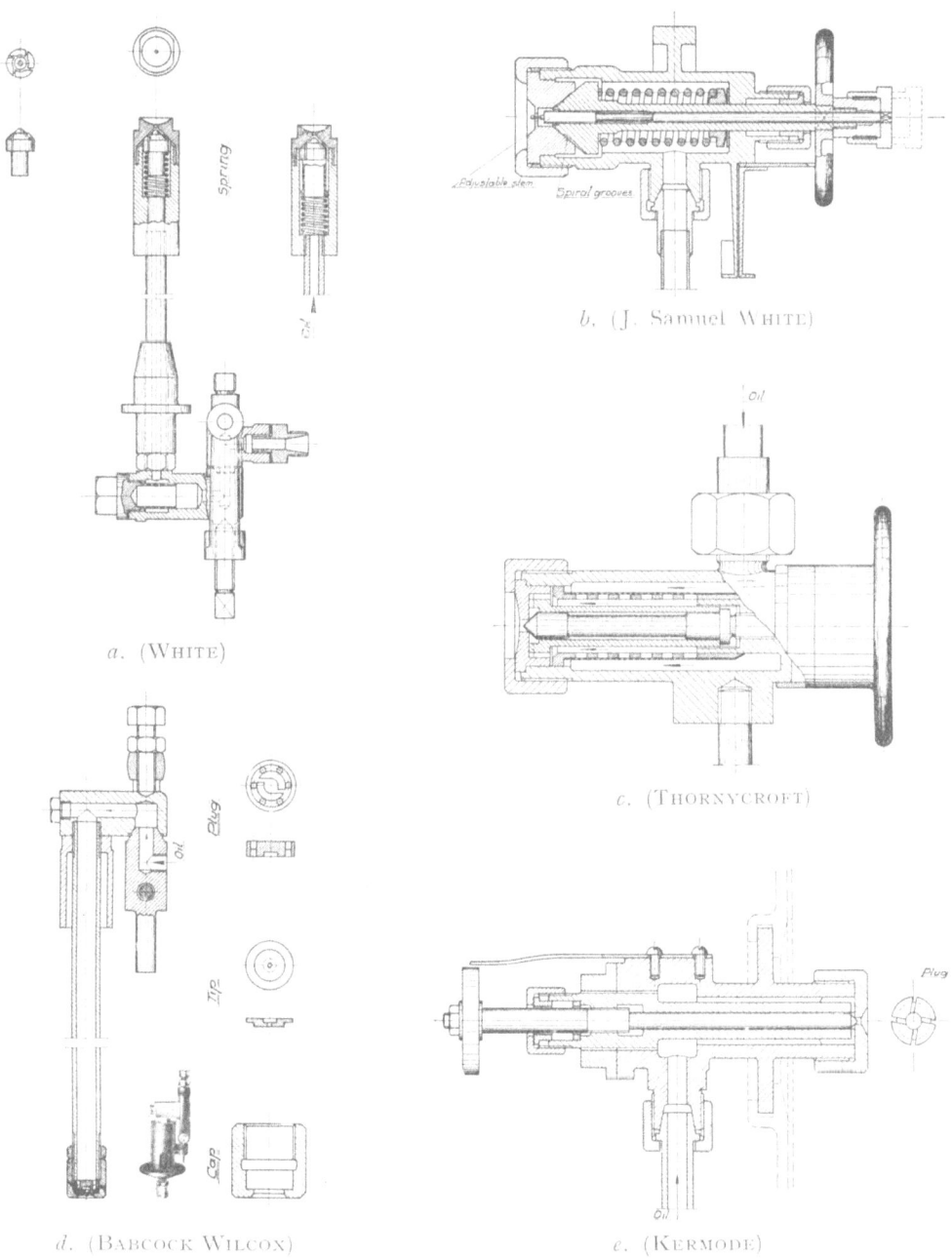

Fig. 182. Pressure-atomizing burners.

### H. *The Target and Air-oil Principles*

There are two more types of pressure atomizing oil burners which will be discussed separately because they do not fit well in the above-mentioned system, viz.

a) The *Target principle* (SWORD burner),

b) The *Air-oil or foam principle* (OIL-O-MATIC).

#### a) *The Target principle of the* SWORD *burner.*

In the beginning of this chapter it was mentioned that almost simultaneously

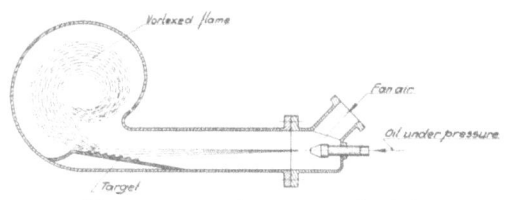

The target and vortex principle.

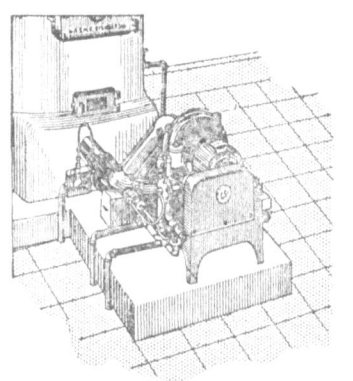

Fig. 183. The SWORD burner.

with KÖRTING's first pressure-atomizing oil burner on the principle of breaking up the oil by *internal* forces, SWENSSON developed a burner on pulverization by means of *external* forces, projecting a solid flow of oil onto a sharp metal cutter. Whereas the first principle has been found to be very fruitful, the second proved to be less successful. One of the few later developments is represented by fig. 183, which is known as the *target principle* and is applied in the SWORD burner for domestic use. A stream of oil is shot onto a rough corrugated surface ("target"), where it is shattered into small particles by the force of the impact. This finely divided fuel is then mixed with air and burnt in a vortexed flame, which by its radiation heats at the same time the target, to assist vaporization.

#### b) *The Air-oil principle.*

This principle, which is used in the WILLIAMS OIL-O-MATIC domestic oil burners, comprises a substitute for atomization by high pressure air. A special pump (see fig. 184) produces a mixture of air and oil, both under the comparatively low pressure of 1 to 5 lbs/sq.in. The *emulsion* or *foam* thus formed is forced through an ordinary atomizing tip, and the oil is pulverized not only by the internal velocities of the liquid but also by the expansion of the air.

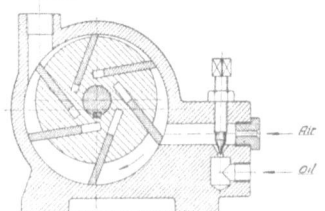

Fig. 184. Air-oil principle.

The above-mentioned analysis of the action of the whirl chamber makes it clear, however, that a "pure rotation" burner tip would be less suitable for this purpose, as its centrifugal action would tend to cause a separation of the air out of the oil before the mixture leaves the orifice. Consequently, when using this air-oil principle it is better to fit a "turbulent" burner tip, whilst moreover every opportunity for the air to settle out, such as may result from wide containers in the line from pump to burner tip, must be avoided.

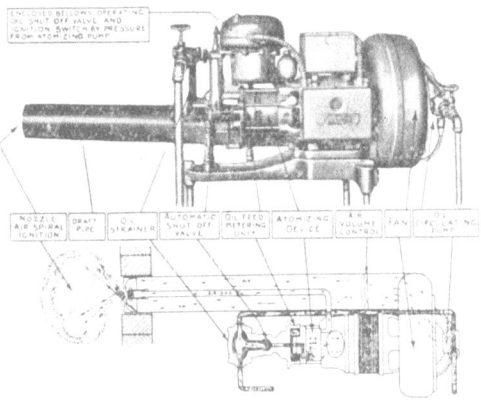

Fig. 185. Scheme of OIL-O-MATIC domestic oil burner (air-oil principle).

A scheme of operation of such an OIL-O-MATIC outfit is given in fig. 185. Another representative of this air-oil principle is the PARWINAC burner described in "Engineering", 28th Oct. 1927 (fig. 99).

*A more recent development by the* WILLIAMS OIL-O-MATIC Heating Corporation of the air-oil principle is shown by fig. 186. The burner unit comprises an additional part, viz., the so-called *"diffusor"* (a better name would be "separator"), which consists of a

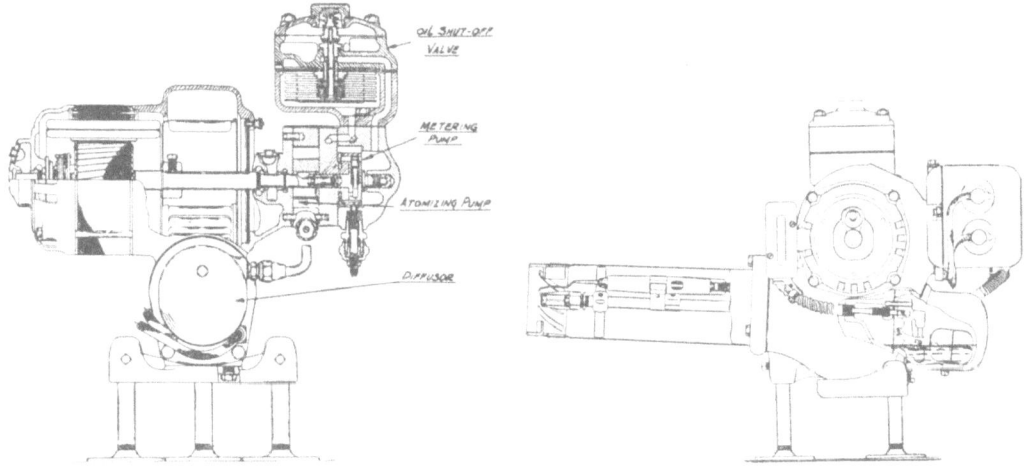

Fig. 186. The OIL-O-MATIC K-1.5 burner unit.

container in which the air-oil foam, after being produced by the *"foam pump"* or — as it is called by the OIL-O-MATIC designers — "atomizing pump" of fig. 184, is collected and allowed to separate. A float device, which is clearly visible in fig. 187, maintains a constant oil level by regulating the oil outlet. A small *medium-pressure air-atomizing tip* effects atomization at air and oil pressures of 1 to 4 lbs, all according to the adjustment.

It is clear that this *new feature is not based on the previously mentioned principle of foam atomization,* as, in fact, it involves just the opposite action, viz. a *re-separation* of the air from the oil. Of course, the air volume thus supplied as medium-pressure air for atomization is not enough to complete the combustion of the oil, and therefore the balance of the air is furnished by the fan, just as is done with the

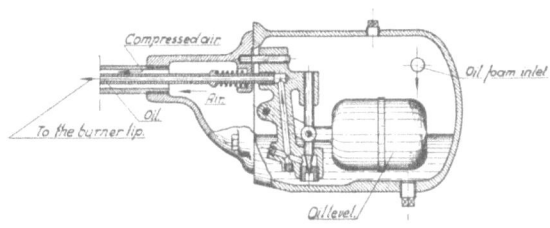

Fig. 187. THE OIL-O-MATIC diffusor.

ordinary type of pressure-atomizing burners, and is blown through the air tube which surrounds the electrodes and nozzle assemblies. The quantity of oil delivered by the metering pump, the pressure produced by the foam pump and the volume of low-pressure air delivered by the fan may be adjusted separately, so that efficient combustion may be obtained at various rates within a sufficiently wide range for each model.

### I. *"Wide-Range" Burners of the Circulation Principle* (PEABODY-FISHER and BARGEBOER)

a. *Drawbacks of plain pressure-atomizing burners. Narrow range of regulation.*

It has often been felt as one of the most serious drawbacks of pressure-atomizing oil burners that the range of regulation is rather narrow, for if the pressure is reduced in order to diminish the quantity of oil burnt, the pulverization gets rapidly worse and consequently a rather narrow limit has to be imposed upon the pressure variation.

The *capacity* of the mechanical atomizing oil burner may be *lowered by decreasing the diameter of the orifice,* and superficially it would seem to be a simple matter to make *burner tips with adjustable holes,* e.g. by inserting a tapered needle in the centre. However, nearly all experiments carried out in this direction have been unsuccessful, because:

1) Of the extremely *small dimensions* and *careful adjustments* required.

2) The needles caused so *much friction* to the rapidly rotating oil in the centre that it spoiled the pulverization, and

3) To some extent it is *theoretically wrong,* as it is not the *area* of the orifice which has to be decreased but its *circumference,* anyhow for the rotary type, which comprises more than 90% of the mechanically atomizing burner tips used.

The graph of fig. 188 shows the *capacities of a well-known industrial pressure-atomizing rotation burner for different sizes of orifices* plotted against the pressure, and

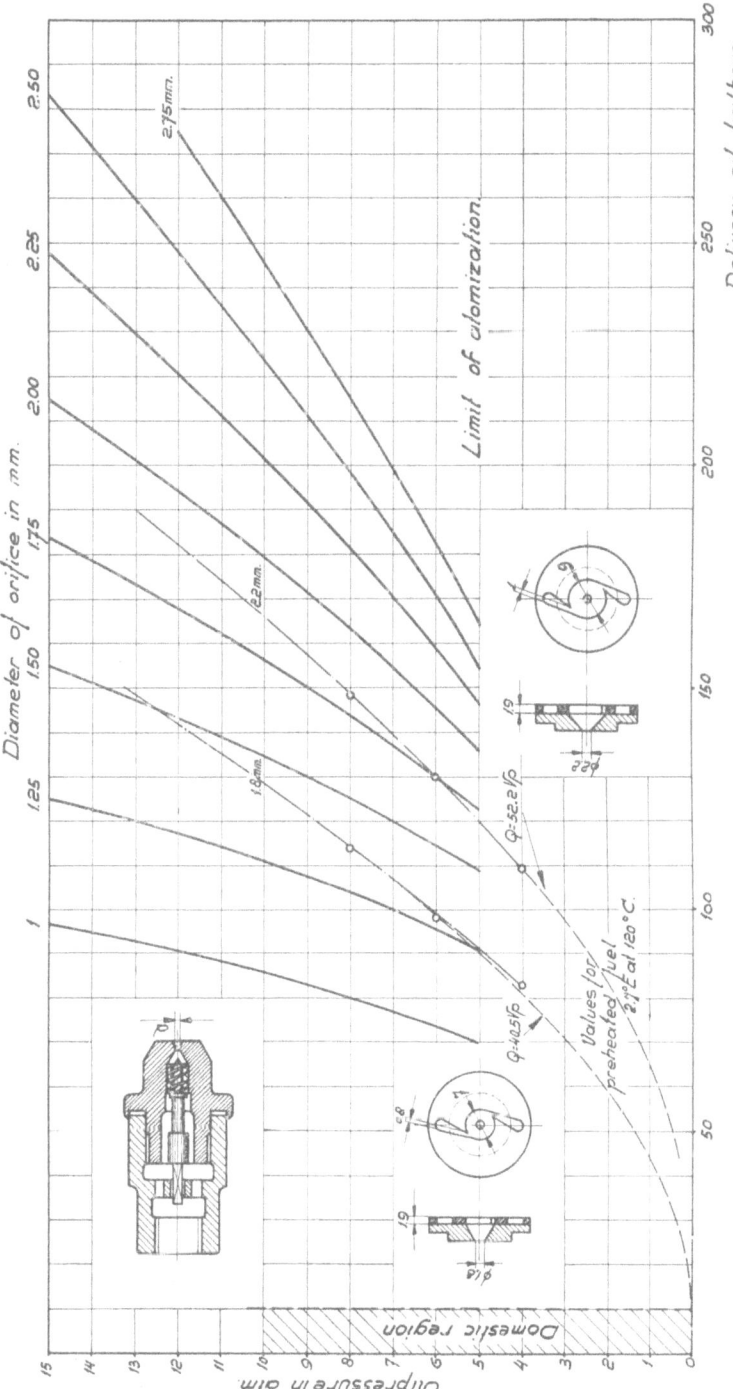

Fig. 188

Capacities of some industrial pressure-atomizing burners.

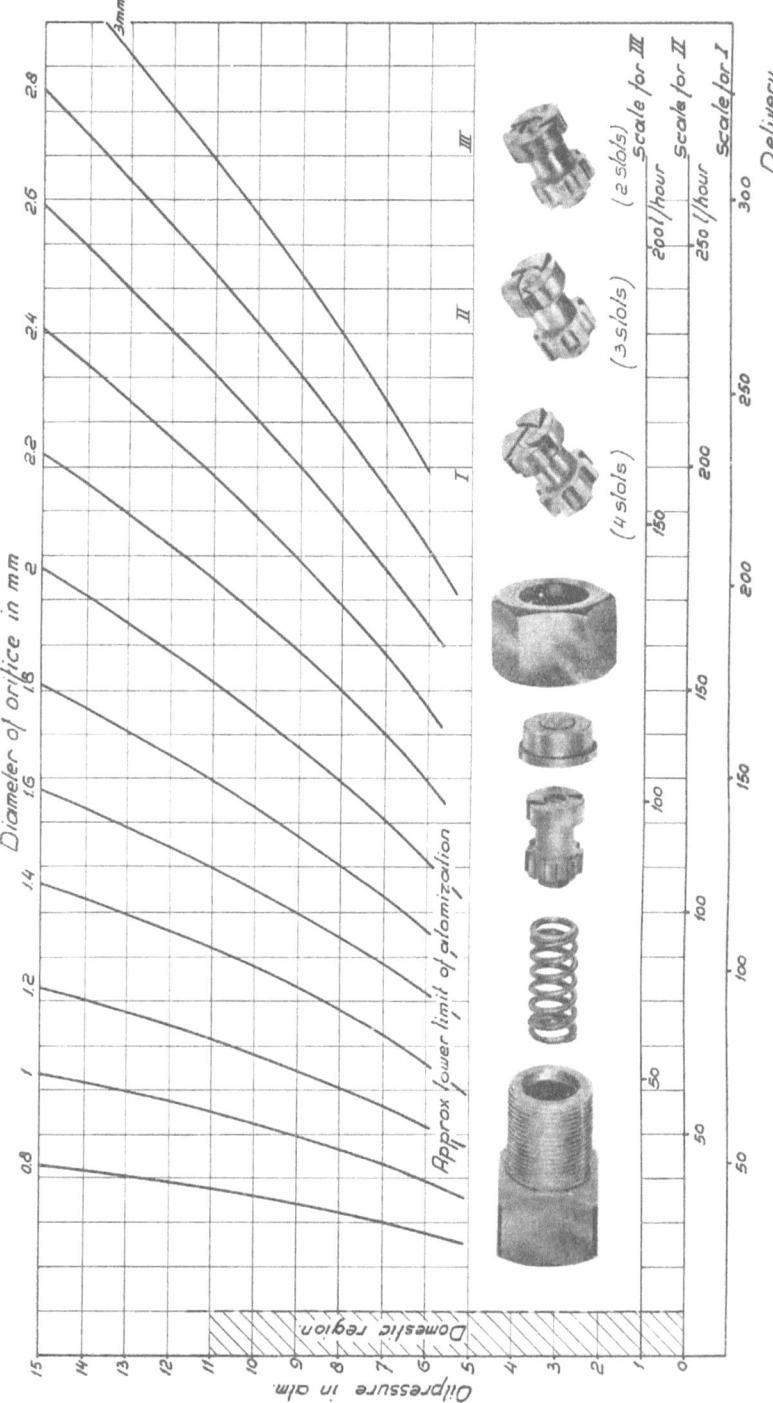

Fig. 189.

Capacities of PILLARD-DC industrial oil burners.

it will be seen that a regulation from 14 atm down to 5 atm, being the lower limit of pulverization, involves a comparatively small decrease in capacity (e.g. for the 2 mm orifice a reduction down to not lower than 65% of the normal capacity), especially for the smaller sizes.

Another example is given in fig. 189 referring to the French PILLARD-DC burners (not to be confounded with the PILLARD-DX, which is a "wide-range" burner). The construction of this burner type comprises a cylindrical flat whirl chamber to which the oil is fed through 2, 3 or 4 tangential slots. For a burner with 3 slots (type II) and 2 mm orifice diameter the maximum capacity at 15 atm amounts to 178 litres oil per hour, which may be reduced to only 112 litres per hour, or 63% of the maximum capacity, by lowering the pressure to 6 atm, which is the lower limit for good pulverization.

For 1 mm orifice diameter the maximum capacity at 15 atm. is 63 litres per hour, which may be reduced to 40 litres per hour by lowering the pressure to 6 atm. These figures show, too, why this problem became particularly urgent for *domestic* oil burning calling for a range of regulation from 1 to 10 litres hourly oil consumption, first of all because such wide variations of the capacity are impossible with the ordinary type of continuously burning pressure-atomizing burner, and secondly, because of the extremely small orifice diameters required. The practical solution was found by adopting *intermittent operation at full capacity* (*"on-off" system*) which will be discussed later on in section IV.

In fact, this *intermittent* operation had also been used for industrial purposes, as the fire under a boiler was generally regulated by the *number of burners put in operation*, and by changing tips.

b. *The* PEABODY-FISHER *Wide-range burner.*

About 1921 an American Naval Officer, FISHER, proposed a new method for regulation which was developed on an industrial scale by the PEABODY Corporation and is described, inter alia, in British Patents Nos. 187,387 (1921) and 380,117

Fig. 190. PEABODY FISHER wide-range oil burner.

(1931). Fig. 37 shows the original diagrammatic sketch of the principle and fig. 190 the design and air register of what is called the PEABODY-FISHER "wide-range" burner. It consists of an orifice with a conical whirl chamber into which the oil is fed as usual in a tangential direction, but a new feature is that *a part of the oil admitted into the chamber is allowed to flow back to the reservoir*. The advantage of this arrangement is that the *rotation velocity* of the oil inside the whirl chamber is *independent of the quantity of oil burnt*. In fact, the case might be imagined of putting the whirl chamber into operation without any oil leaving the orifice, for instance by applying sufficient vacuum to the return tube. By pinching the by-pass valve, the quantity of return oil may be reduced and the quantity of oil burnt gradually increased.

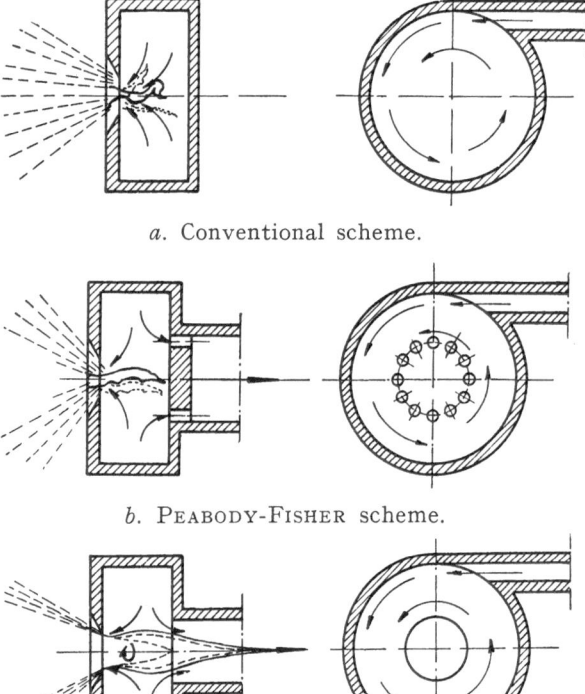

*a.* Conventional scheme.

*b.* PEABODY-FISHER scheme.

*c.* BARGEBOER scheme.

Fig. 191. Whirlpool formation in whirl chambers.

Consequently, the *atomizing power of this burner remains at its maximum throughout the entire capacity range*, and there is practically no change in the working oil pressure.

The fact that it necessitates continuous working of the oil pump at full load, so that the power required per gallon of oil burnt is a little higher, is a disadvantage of minor importance when compared with the great advantages obtained by the perfect regulation. As an additional advantage it may be mentioned that it is possible to join up the return lines of several burners, and it is on record that power plants have been arranged with 56 PEABODY-FISHER burners regulated by means of one single valve or by one thermostat or pressure-stat.

c. *The* BARGEBOER *wide-range burner*.

Independently of the foregoing, BARGEBOER, a consulting engineer (The Hague, Holland), was the first to develop the principle of reinforced circulation through the whirl chamber by returning a part of the oil for *domestic heating* and, in order to arrive at satisfactory results for the smaller loads required, he found it to be most important to *stabilize the conditions of flow in the whirl chamber*.

Theoretically speaking, the oil in a whirl chamber will be able to form a hollow core or whirlpool, no matter whether the oil is recirculated through the whirl chamber or not. Practice, however, has shown that if there is no recirculation and, consequently, all the oil fed to the whirl chamber leaves the burner through the orifice, the velocity towards the orifice will be so large that the hollow pool is almost "washed out", so to speak, and, as its formation is so unstable and disturbed by wall friction, it can hardly be said that there is any core at all (see fig. 191a).

Obviously the conditions for whirlpool formation will become more favourable if a part of the oil fed to the burner is returned as in fig. 191b, as this will make the velocity towards the orifice smaller, whilst the rotary speed may attain to very high values independently of the load of the burner. This condition occurs, for example, with the PEABODY-FISHER burner and is represented schematically in fig. 191b.

BARGEBOER, however, found that in order to get a perfectly stabilized whirlpool it was also necessary to *return the oil through a co-axial orifice having a diameter larger than the tip orifice*. This permits the whirlpool to have its full swing, if necessary even within the return tube, and consequently it will no longer disturb the conditions of flow at the orifice edges (fig. 191c).

Moreover, the oil particles, having extremely high velocities at the inner surface of the whirlpool (according to KEPLER's second law of equal areas the velocities increase in inverse ratio to the radius, or hyperbolically), will undergo a considerable braking effect if they are forced to touch walls, such as in the examples of figs. 191a and 191b. With

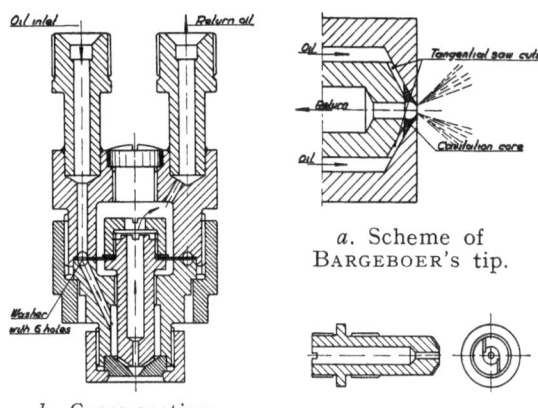

a. Scheme of
BARGEBOER's tip.

b. Cross-section.

Fig. 192. BARGEBOER's domestic burner.

BARGEBOERS' device, however, the high inner velocities are kept as far as possible away from the walls, this resulting in a fine stable development of the cavitation core (see fig. 191c).

By adopting this type of construction, BARGEBOER was able *to "pour" the oil film over the edge of the orifice* (see fig. 192a) and thus atomize the oil into *drops of a magnitude found to be practically independent of the load of the burner.* Moreover, he observed that the *oil pressure in the return line was nearly proportional to the quantity of oil burnt,* which may be lowered to about 20 % of full capacity. This simple relation on the one hand offered a fine opportunity for automatic adjustment of the air quantity by making this pressure move a damper of the air register and, on the other, made it possible to realize a thermostatic or pressure-static regulation of the flame capacity by controlling the return oil pressure automatically by means of a thermometer or manometer working on a valve in the return-line. Fig. 193 shows the results of such automatic regulation.

Other features of the remarkable *N. F. domestic oil burner*, which is the trade name of BARGEBOER's product, will be discussed later on in section IV.

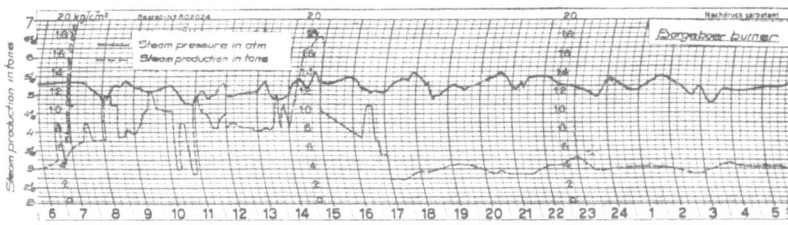

Manually controlled steam pressure.

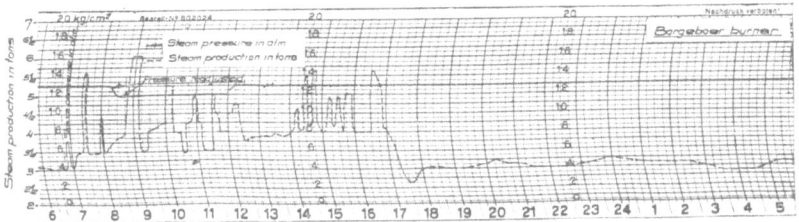

Fig. 193. Automatically controlled steam pressure (BARGEBOER).

Another, but far less important method of regulation by recirculating the oil through the burner is described in British Patent No. 322,337 (1928) granted to TARAGNO (Soc. Anon. per l'Impiego Razionale degli Olii Combustibili IGNEA).

### J. *Wide-range Burners based on automatic Adjustment of Slots or Depths of Whirl chambers*

The French burner manufacturers PILLARD Frères (Marseilles) followed another idea in solving the "wide-range" problem for pressure-atomizing oil burners, viz. *by automatic adjustment of the tangential slots through which the oil is fed to the whirl chamber the depth of which is adjustable too.*

They reasoned as follows: Of course it is possible to construct a pressure-atomizing oil burner capable of delivering the *smallest* load required for the (industrial) purpose in question, but such a burner would need too high pressures for the higher loads, owing to its too narrow passages for the oil. Consequently some means must be provided to enlarge the *cross-section of the passages for the higher loads automatically*, thus reducing the pressure, without, however, reducing the pulverizing effect too much.

Fig. 194 gives a scheme to explain the principle. The burner consists of an orifice plate (1), which is pressed on the tube (2), which contains two or more tangential slots cut on its face and holes through which the oil is admitted. The bottom of the whirl chamber is formed by the face of the movable piston (3) and it is clear that by moving this piston backwards the whirl chamber will be enlarged and the free area of the oil passages will be increased, resulting in less resistance. The piston (3) is moved by the oil pressure itself, being pressed upon the orifice plate by means of a spring. As soon as the oil enters into the whirl chamber, its pressure will push the piston in a backward direction and the higher the pressure of the oil in the tube (pump pressure) is, the higher will be the resistance of the increased quantity of oil leaving

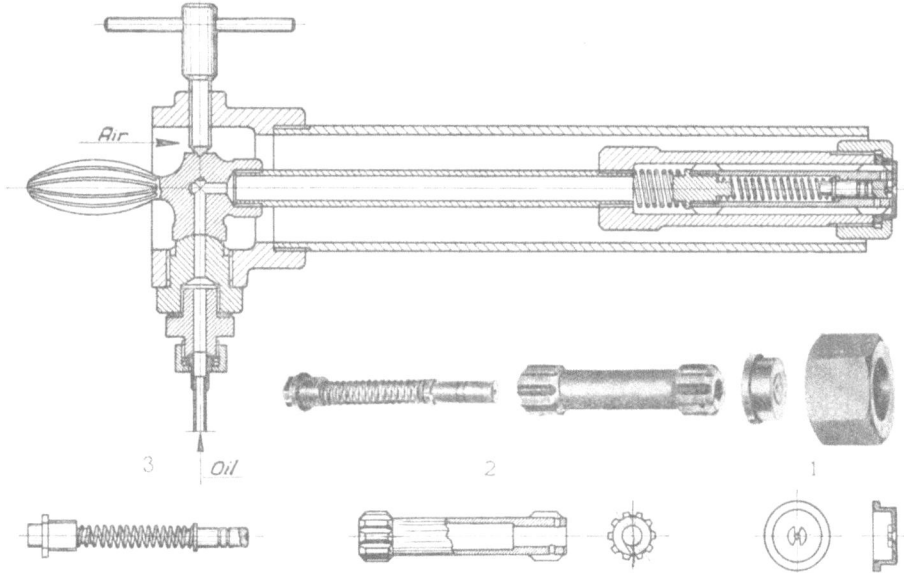

Fig. 194. The PILLARD-DX wide-range oil burners.

*the orifice*, resulting in a higher pressure in the whirl chamber which moves the piston still further back.

It is clear that only a part of the total resistance of the burner tip will be decreased in this way, viz. that of the whirl chamber and of the innermost part of the tangential slots. The other passages are constant and consequently the resistances of these parts will increase approximately with the *second power of the quantity* (parabolic law), just as with an ordinary burner. However, the result of the designed burner of fig. 194 will be that the curve showing the increase in total resistance will be *less steep than the parabolic curves* of the conventional type of burner and consequently larger quantities of oil may be burnt with this burner before the pressure limit of the pump is reached. On the other hand, the *lower limit of this burner is also smaller* than with the conventional burners, because at low capacity the free area of the tangential slots is pinched automatically, resulting in an increased rotation of the oil and improved pulverization even at low loads.

Fig. 194 shows the PILLARD-DX burner based on the above-mentioned principle and the various parts may be easily distinguished. The second spring (8) has nothing to do with the operation of the burner, as it only presses the tube containing the piston (4) and spring (5) against the orifice cap (2).

Similar designs may be found in British Patent No. 246,692 (1925) of KINGSTON and POWELL and in BUTLER's "Fuel Oil" (1914, page 101) and, moreover, in the WHITE oil burners of fig. 181, although here the principle is not applied in its most efficient form. As a recent development the Dutch patent 39,446 of BROWN-BOVERI (Switzerland) may be mentioned.

## K. *Wide-range Burners comprising Multiple Rotation Devices*

Yet another means of regulating an oil fire of pressure-atomizing oil burners over a wide range consists in using burners each of which is composed of *two or more nozzles*, which may be put into operation separately, either by hand, or automatically by the oil pressure itself, or sometimes by means of electrically operated valves. Such equipments, however, are usually rather complicated and up to the present have found but little application, for instance, in domestic oil burning, (see fig. 243).

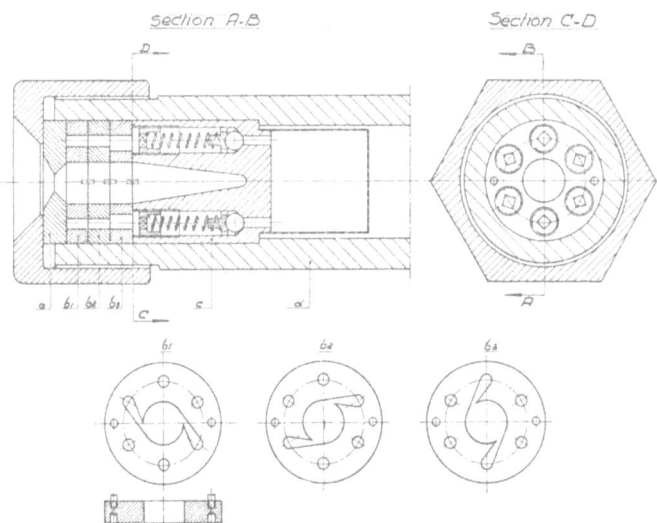

Fig. 195. Wide-range burner with multiple rotation devices.

Another type is suggested by fig. 195. The burner comprises an orifice plate (a) and three discs (b$_1$, b$_2$ and b$_3$) with tangential slots packed together in a position which is fixed by six adjustment pins. Every disc contains one central hole and six small holes, only two of which communicate with the central hole. Each disc has been turned over 60 degrees with respect to its neighbour. Now, the oil admitted into tube (d) must first open a spring-loaded ball-valve before it can enter into one of the six holes and reach the whirl chamber. The springs of the ball-valves may be adjusted in such a way that at low oil pressure (say 1 atm.) two bores are opened up admitting oil into the whirl chamber of plate ($b_1$). Suppose increase of the pressure up to, say, 3 atm. will be sufficient to put the first whirl chamber ($b_1$) into full operation. The next two ball-valves may be adjusted so that they open at, say, 3 atm., thus gradually putting the second whirl chamber ($b_2$) into operation and so on until at, say, 10 atm. all three whirl chambers are put into operation. After reaching this point the pressure may be raised up to the limit of the pump, thus increasing the capacity of the burner in the usual way.

This principle may be worked out in several different ways, e.g. by using whirl plates of different thickness or giving them different central holes (decreasing diameter for the front plates). It offers, moreover, the advantage that at low loads the friction caused by the bottom of the whirl chamber is reduced to the minimum, only the first whirl plate being in operation and causing the development of a fine hollow core in the oil content of the other whirl plates. This effect may be promoted further by giving piece (c) a conical bore as shown in fig. 195.

This type of burner may be compared with a steam-turbine, the capacity of which is regulated by the number of steam-nozzles put into operation. One might also call it: "an oil burner equipped with a three-speed gear-box".

A similar burner design is mentioned by GUILLERMIC in his book, "Le Chauffage par les Combustibles Liquides" (see page 85), comprising two whirl chambers in the burner tip which may be put into operation separately, either by hand or automatically by the pressure.

---

CHAPTER VI

MECHANICAL PULVERIZATION ON THE KINETIC PRINCIPLE
(Centrifugal Atomizing Burners)

A. *General Remarks*

Although, as a matter of fact, the pressure-atomizing oil burners referred to in the previous chapter are also based on pulverization by means of kinetic energy, viz., either by regular or by irregular rotary motion, it has been found advisable to make a separate class of those burners which use *velocity as the primary property* of the oil to effect pulverization, as contrasted with the pressure-atomizing burners in which *pressure* is the primary form and velocity the secondary form (see Table V).

In burners of this kinetic type the oil, which is supplied practically without any pressure, is spread out into *a film which is torn apart into drops by centrifugal forces applied by some external mechanical means.*

From a kinematical point of view the *difference between the dynamic-atomizing* (or pressure-atomizing) *burners and these kinetic-atomizing* (or centrifugal-atomizing) *burners is enormous*, as may be shown by fig. 196.

We have seen that the velocity of the oil in the whirl chamber of a pressure atomizing burner increases hyperbolically towards the centre as the oil is accelerated by being forced from larger to smaller radii against the field of centrifugal gravitation (see fig. 170).

In a cup or on a disc of a centrifugal burner the oil has a velocity which increases with the radius, which increase would be directly proportional to the radius if there were no "slip" between oil and plate (fig. 196*b*), the slip, however, causing the increase to be less than proportional (fig. 196*c*).

The principle of centrifugal pulverization had the attention of several inventors long before dynamic-centrifugal pulverization was discovered, but it was only after the development of the electric motor that practical applications were found. In section I some early developments were mentioned, viz., the AETNA burner for domestic use (1912) (fig. 56), and the WILLIAMS steam-driven centrifugal oil burner (fig. 35) which was described in the report of the U.S. Navy Liquid Fuel Board (1904). In this and several other early types a *revolving flat disc* was used, onto which the oil was fed from the centre, but the pulverizing effect was rather disappointing, firstly because *the oil rotated much more slowly than the disc* and, secondly, because of the oil tending to be *unevenly distributed* (waves) on the plate.

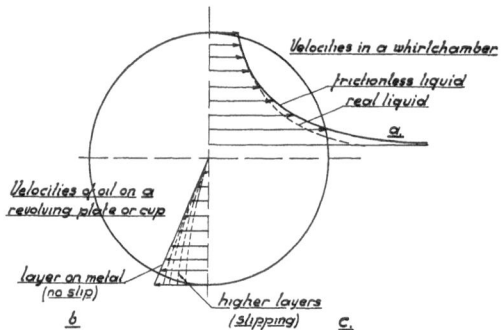

Fig. 196. Velocities centrifugal spraying.

Better results were obtained as soon as the *revolving cup* was adopted as a fundamental form for centrifugal pulverization and, in order to explain this improvement, it is necessary to analyse the movement of the oil.

If a *cylindrical* cup filled with a liquid is revolved with a gradually increasing velocity until a certain speed is reached, from which moment onwards the rotation is supposed to be constant, the liquid will gradually follow the movement, although with a certain delay, and after a while it will have obtained practically the same speed as the cup and its level will show the well-known parabolic form. This behaviour is in the first place caused by the friction of the liquid against the wall and furthermore by the internal friction of the liquid (viscosity), and as a consequence *an imaginary liquid with zero viscosity would not be able to receive any velocity from the revolving cup.*

Instead of proceeding to make a mathematical analysis of this kinetic problem, it may be remarked that perhaps it would not be so difficult to calculate the path followed by a liquid particle on the surface of a flat revolving disc if that particle were put in the centre with a definite, known radial velocity and it were acted upon by a force caused by friction supposed to be directly proportional to the difference in velocity between that particle and the disc surface. This, however, would not bring us very much nearer to the solution of the problem of centrifugal spraying, as, after solving the equations thus obtained, it would apply only to an *infinitely thin layer* of liquid having no differences of velocity in a vertical sense.

In fact, this case does not apply at all to the practical problem, as the *thickness of the oil film is one of the most important factors to be considered* or, in other words, there is a considerable difference between the velocity of the layer in direct contact with the disc surface and that of the upper layers. Thus, instead of taking into consideration the movement of one oil particle at a definite spot of the disc, having a definite velocity and a definite friction caused by the velocity difference of oil particle and plate, it is necessary to consider a large number of oil particles at that same spot, each of which is defined by its distance x above the disc surface and has its own velocity and friction, both depending on x.

From this it will be clear that centrifugal spraying even by such simple means as a flat revolving disc is no easy mathematical problem and that, for the sake of simplicity, it is better to confine ourselves to some general discussions, than to lose ourselves in a labyrinth of mathematical analysis with elaborate differential equations.

## B. *Influence of Viscosity*

*A most important factor in centrifugal spraying is the fact that its effect depends entirely on the viscosity, but in just the opposite way to the effect of viscosity on pressure-atomizing oil burners.* Whereas viscosity opposes atomization by means of pressure, and consequently in that case the best results would be obtained with *a liquid of zero viscosity*, most centrifugal atomizing devices would not be able to pulverize such an imaginary liquid at all, there being no way of imparting velocity to it. It is clear that this fundamental difference makes it still more logical to draw a sharp division between dynamic and kinetic atomization.

The main difficulty experienced by designers of oil burners of this centrifugal type has been *to ensure "enough contact" between oil and disc surface for the oil to be accelerated*. Difficulties were especially encountered, not only when using thin liquids of low viscosities, but also with *flat horizontal plates*, as the oil, being kept down only by its weight, was inclined to be *unevenly distributed* on the surface and apt to form local thick streams (waves) with a comparatively small area of contact per unit of weight, with the result that there was *considerable slip* and bad pulverization. Moreover, when a flat plate was used, the *time of contact* was too short, the centrifugal force being entirely utilized for accelerating and projecting the oil particle in radial direction (see fig. 197a). In this respect, using the outer surfaces of cylinders or plates, such as that in figs. 198 and 199 taken from Br. Patents Nos. 178,931 and 189,908 (1921), was even worse and did not give satisfactory results at all. The same may be said of the British Patent No. 170,760 of VAN DEN HONERT.

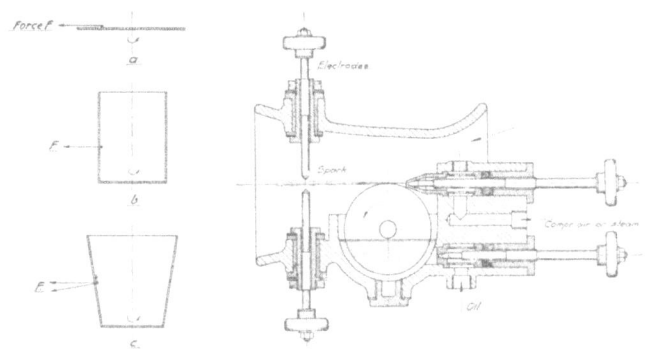

Fig. 197. Centrifugal force.

Fig. 198. MAINA rotary oil burner.

In a *cylindrical cup* (fig. 197b) the oil particle, theoretically, would not be able to move towards the edges, and it is clear that by using a *conical cup* (fig. 197c) the centrifugal force may be split up, as follows:

a) *One component perpendicular to the surface of the cup*, which tends to keep the oil "down" and ensures an *even distribution* of the oil film and a *"thorough contact"* with the cup surface, and:

b) *Another component parallel to the surface*, which determines the axial velocity and consequently the *time of contact* between oil and wall, i.e. the time available for acceleration.

This consideration makes it clear why the majority of modern devices for centrifugal spraying all comprise that same *typical conical rotation cup* of a steepness which has been found by experience to be the most effective.

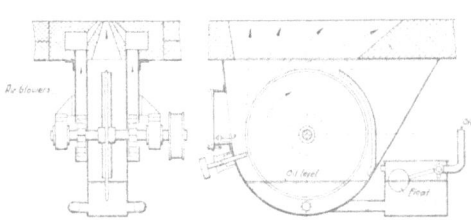

Fig. 199. LINKÉ's rotary oil burner from Br. Pat. 189,908.

Whereas a *high* viscosity (internal friction) is favourable to an efficient acceleration of the oil (and from this point of view the oil should *not* be preheated if it is to be pulverized in burners of the kinetic class), the conditions of the oil film, as soon as it has left the cup, are just the same as those of an oil film of a dynamic rotary burner and consequently require *low* viscosity. It follows that it is not an easy question to decide for kinetic oil burners whether there shall be preheating or not, or to what extent.

## C. *Constructive Principles*

If the disc or cup is strongly heated by radiation or conduction, the oil may be badly accelerated and for this reason a flat plate offering a large surface to flame radiation is inferior to a cup. Consequently, a number of designers protect the rotation devices against radiation, as is described, for instance, in Br. Patents 175,673; 29,298 and 12,821.

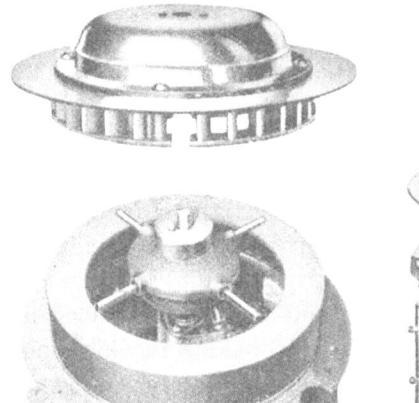
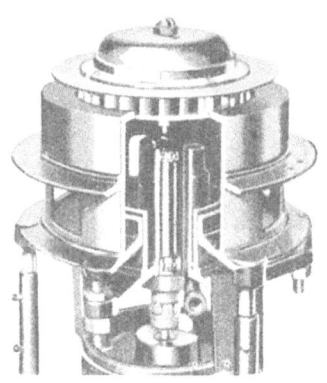

*a*. Atomizing head of the PETRO-NOKOL burner.

In order to avoid the troubles of bad acceleration, a number of types have been designed in which the oil is not accelerated by friction but by means of *tubes*, an example of which is given in fig. 200*a* representing the PETRO-NOKOL domestic oil burner, comprising four centrifugal tubes provided with nozzles at the ends, a device which permits of thin or preheated oils being used. Another example shows fig. 200*b* (FLUID HEAT). Sometimes the oil is fed to the tubes from a constant oil level by means of a device similar to the

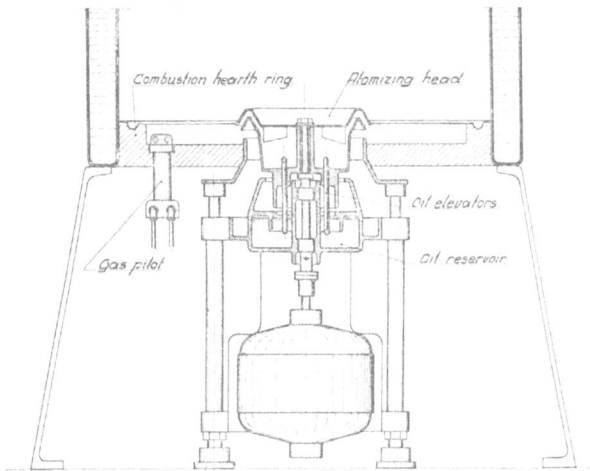

*b*. FLUID HEAT vertical rotary unit with atomizing tubes (see *a*).

*c*. Vertical rotary burner unit with oil elevators.

Fig. 200. Direct-driven vertical rotary burners.

provided with nozzles at the ends, a device which permits of thin or preheated oils being used. Another example shows fig. 200*b* (FLUID HEAT). Sometimes the oil is fed to the tubes from a constant oil level by means of a device similar to the well-known RAMSBOTTOM feed water supply for running locomotives (see fig. 200c).

The arrangements may be vertical or horizontal; directly driven or by means of transmission; driven by an electric motor or by *low pressure air*. Some designs of the

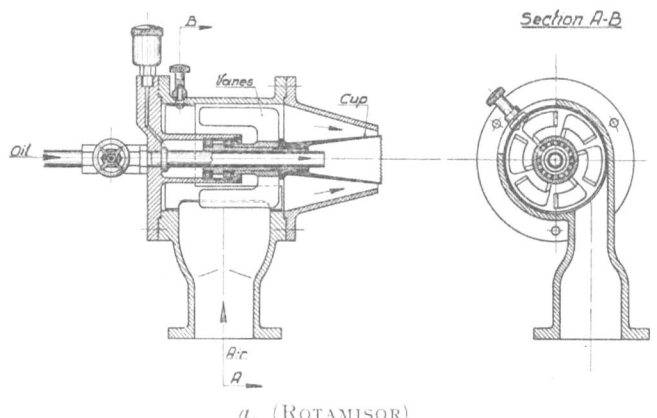

a. (ROTAMISOR)

last type are shown by fig. 201a representing the ROTAMISOR burner (Br. Pat. 203,840) (1922) and fig. 201b showing the LILLIBRIDGE burner, which are particularly useful for small industrial purposes, such as bread-baking, in which case a number of burners may be operated by one single central blower. This type, in which the cup is spun by

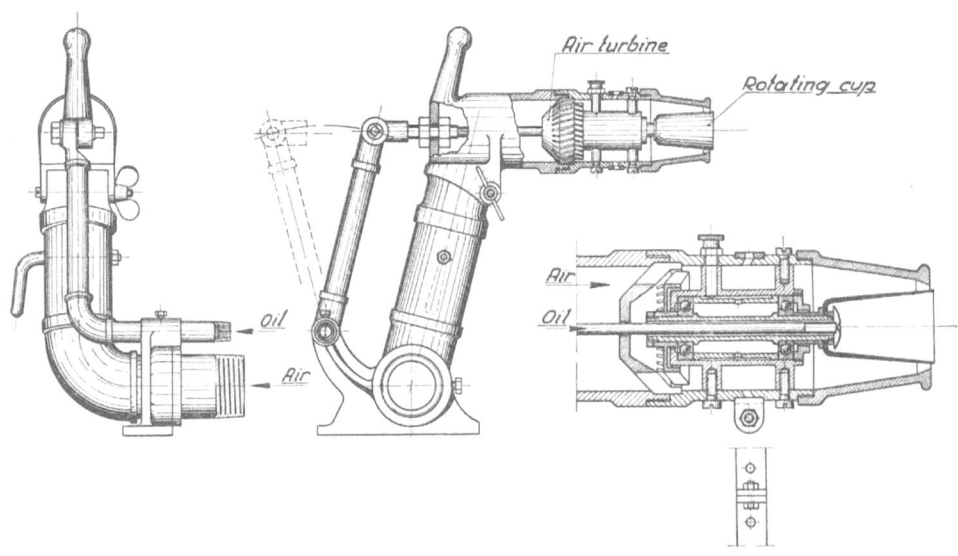

b. (LILLIBRIDGE)

Fig. 201. Air turbine burners.

an impeller operated by means of the air supply to the burner, is often called the *"air-blown rotary type"*.

Figs. 202a and 202b show cross-sections of the JOHNSON and the RAY burner, both of which are *electrically* driven. Fig. 203 shows the *belt-driven* French AUTOPROGRESSIF domestic oil burner (Boutillon-Suresnes). Fig. 204a shows the air-

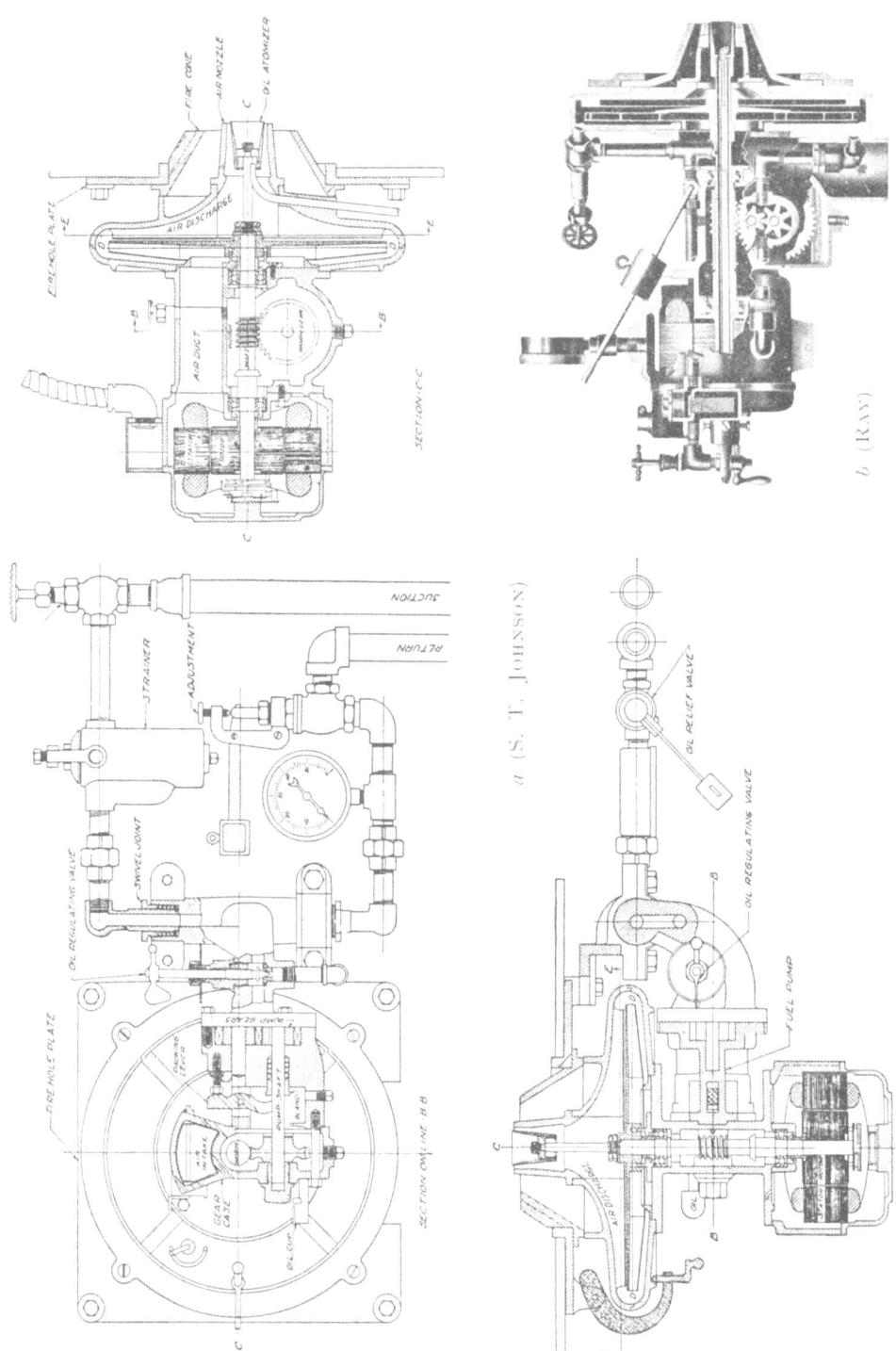

Fig. 202. The S. T. Johnson and Ray horizontal rotary oil burners.

or *steam-driven* RADIOIL burner, and fig. 204*b* the *gear-driven* HARDINGE burner. Sometimes the atomizing cups are provided with saw-tooth edges, apparently

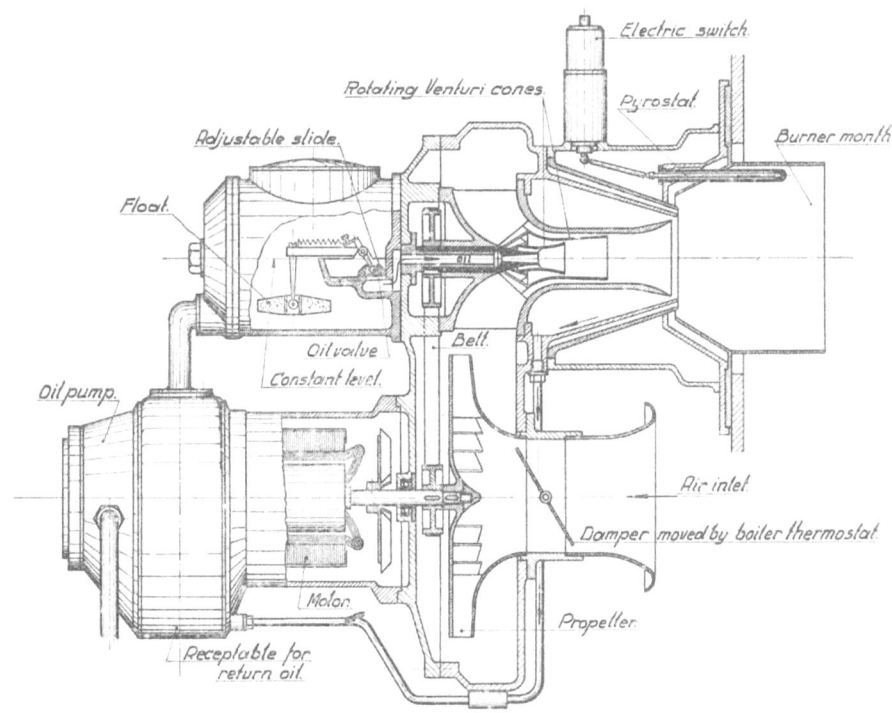

Fig. 203. The French AUTOPROGRESSIF automatic domestic oil burner.

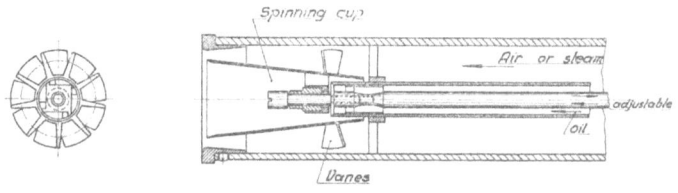

Fig. 204*a*. RADIOIL air- or steam-driven rotary type.

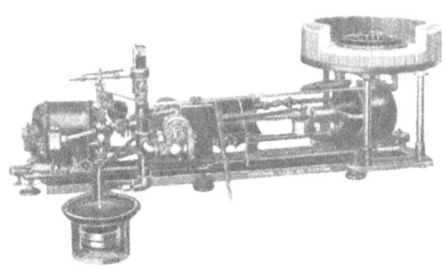

204*b*. Gear-driven HARDINGE burner.

with the intention of improving atomization, but from a theoretical point of view this detail cannot have any great influence.

The horizontal burners usually produce oil flames of the same shape as that produced by pressure-atomizing burners, but the vertical type is often used to obtain a *flat flower-like flame* as shown by fig. 205, which is especially suitable for domestic boilers with circular combustion chambers. For this type of burners it has been found useful to arrange a *circular ring of refractory material* along the wall, onto which any

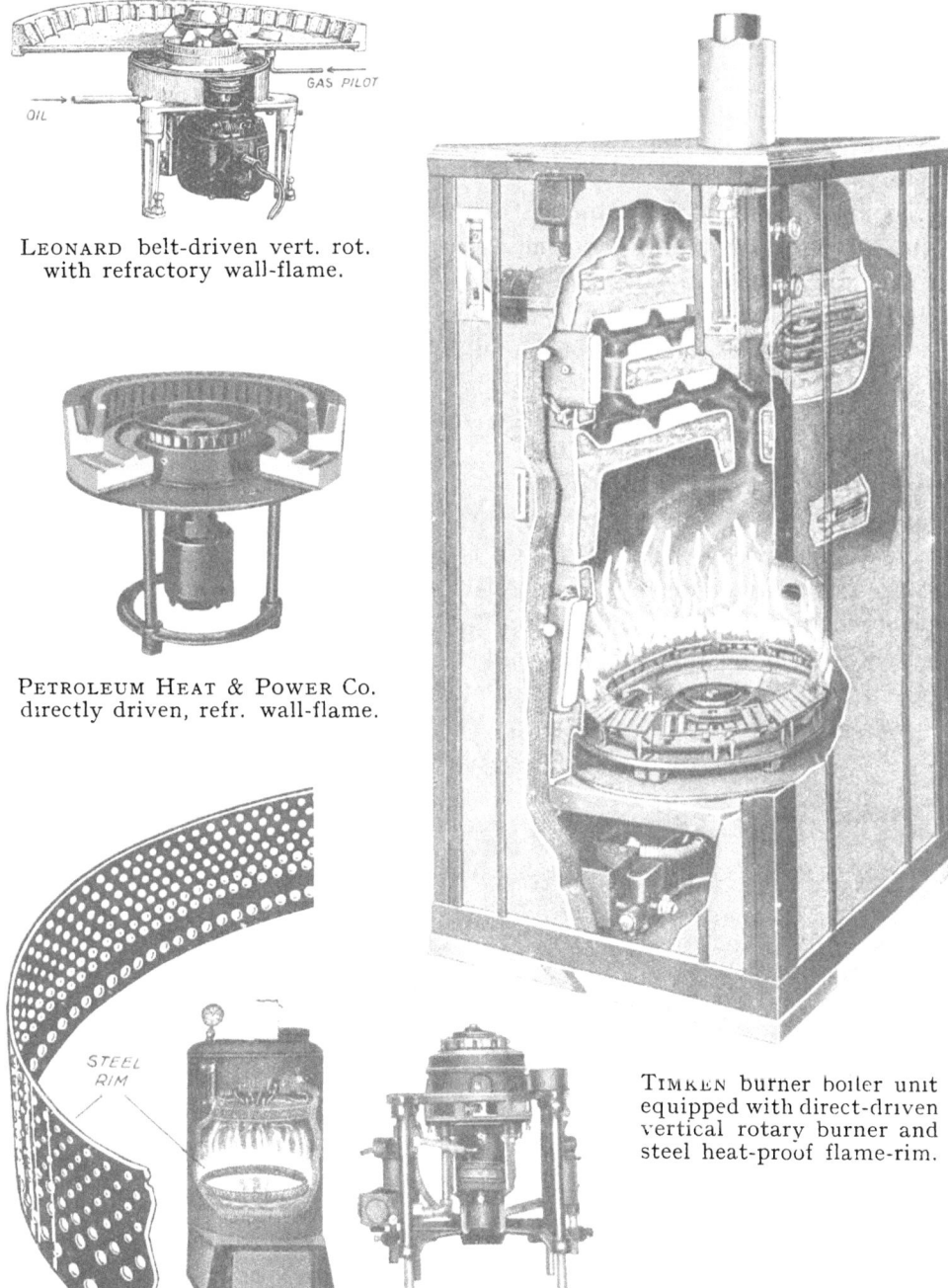

LEONARD belt-driven vert. rot. with refractory wall-flame.

PETROLEUM HEAT & POWER CO. directly driven, refr. wall-flame.

TIMKEN burner boiler unit equipped with direct-driven vertical rotary burner and steel heat-proof flame-rim.

KLEEN-HEET directly driven unit with steel flame-rim.

Fig. 206. Various examples of the wall-flame principle.

coarse oil drops may be projected, gasified and burnt without causing incomplete combustion. This is the so-called *"refractory wall flame"* fig. 206, which principle shows that the atomizing effect of rotary devices may sometimes require some additional improvement. Another way often followed is to use a combination of rotary spraying with low pressure air atomization; this method will be discussed further on.

·The most pronounced advantages of these *kinetic-atomizing burners* over the pressure-atomizing type are that the former have a *large range of regulation* and are *able to burn heavy oils at low preheating temperatures*. Disadvantages lie in the very sensitive moving parts, which may cause much noise and vibration if not perfectly balanced. A flat disc, for instance, may be easily warped out of shape by uneven application

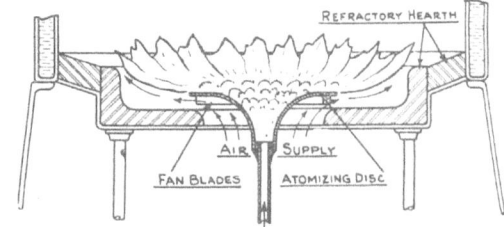

Fig. 205. Flower-like shape of flame (TAPP).

of heat. A high-speed gear transmission is very difficult to keep silent and consequently direct-driven devices are preferred. Furthermore, the initial cost of kinetic-atomizing burners per burner unit is usually considerably higher than that of pressure-atomizing burners, which point is especially important for industrial equipments of several burner sets. On the other hand, this higher cost often may be compensated by lower cost of preheating. The air-turbine type of kinetic atomizing burner, in particular, facilitates industrial arrangements with low initial costs per burner unit.

---

# CHAPTER VII

### PREPARING THE OIL BY MEANS OF STEAM DIRECTLY GENERATED BY THE OIL BURNER ITSELF

## A. *General Remarks*

It has been explained in previous chapters that the use of steam as a pulverizing agent has many advantages, not only because of its large specific vapour volumes, which allow of high velocities with comparatively small weights of steam, and its preheating effect, but mainly because of the *soot-preventing chemical action* of steam, it being possible, provided the temperature is high enough, to convert carbon and water vapour into carbon monoxide and hydrogen ("water gas equilibrium"). Heavy fuels of low hydrogen content such as asphalts may easily be burnt without smoke if they are atomized by steam, which fact was successfully experienced during the early development of oil burning in Russia (see section I).

For industrial purposes in shore installations steam atomization, however, has gone out of fashion mainly because of the inevitable extra chimney loss due to the latent heat of the steam carried away with the flue gases; for modern installations the steam consumption amounts to about 3 to 5 per cent by weight of the steam raised by the oil-fired boiler, a disadvantage which in some cases may be balanced on the one hand by a more efficient clean combustion when using heavier and sometimes cheaper fuels and, on the other, by the power saved by elimination of oil pumps or air blowers. According to the figures of HASLAM & RUSSELL in "Fuels and their Combustion", *the cost of steam-atomization per gallon of oil burnt is about twice that of pressure-atomization.*

For domestic use, where no steam boilers are available, oil burner designers have made several attempts to develop oil burning devices which generate steam for its own atomizing and which are thus operated as self-contained units. The early designs usually comprised actual steam boilers copied on a small scale, but while it is by no means an easy task to keep the pressure of *large* industrial boilers automatically constant, this is still more difficult with very small, toy-like boilers, which are apt to be cooled down so rapidly by supplying a comparatively small quantity of water that at some moments the boiler is completely filled with cold water whilst at others it is quite dry again and filled with superheated steam. *A typical fault of operation in this type of burners was the occurrence of a discontinuous flow of steam (pulsations).*

## B. *Pulsation Troubles*

Most types subsequently developed comprised special means to prevent these pulsations of the steam flow. It was found to be a most important factor to *superheat the steam,* because the weight of water leaving an orifice in a unit of time can be kept more easily constant as a gas than as a wet vapour of variable liquid content. Moreover this superheating has a beneficial influence on the oil flame, which otherwise tended to be extinguished if the steam produced was occasionally too wet.

Another way of preventing pulsations was to make the system less "lazy" by *reducing the water content of the boiler* to a minimum, in which case the water supply followed the consumption more closely. Of course the return valves admitting the water into the boilers must be very sensitive in operation.

A drawback to all these systems is that a drop in steam pressure may be caused, not only by a shortage of water supply, but also by a drop in temperature. If in the latter case the water supply is automatically increased, which is done as a rule by a

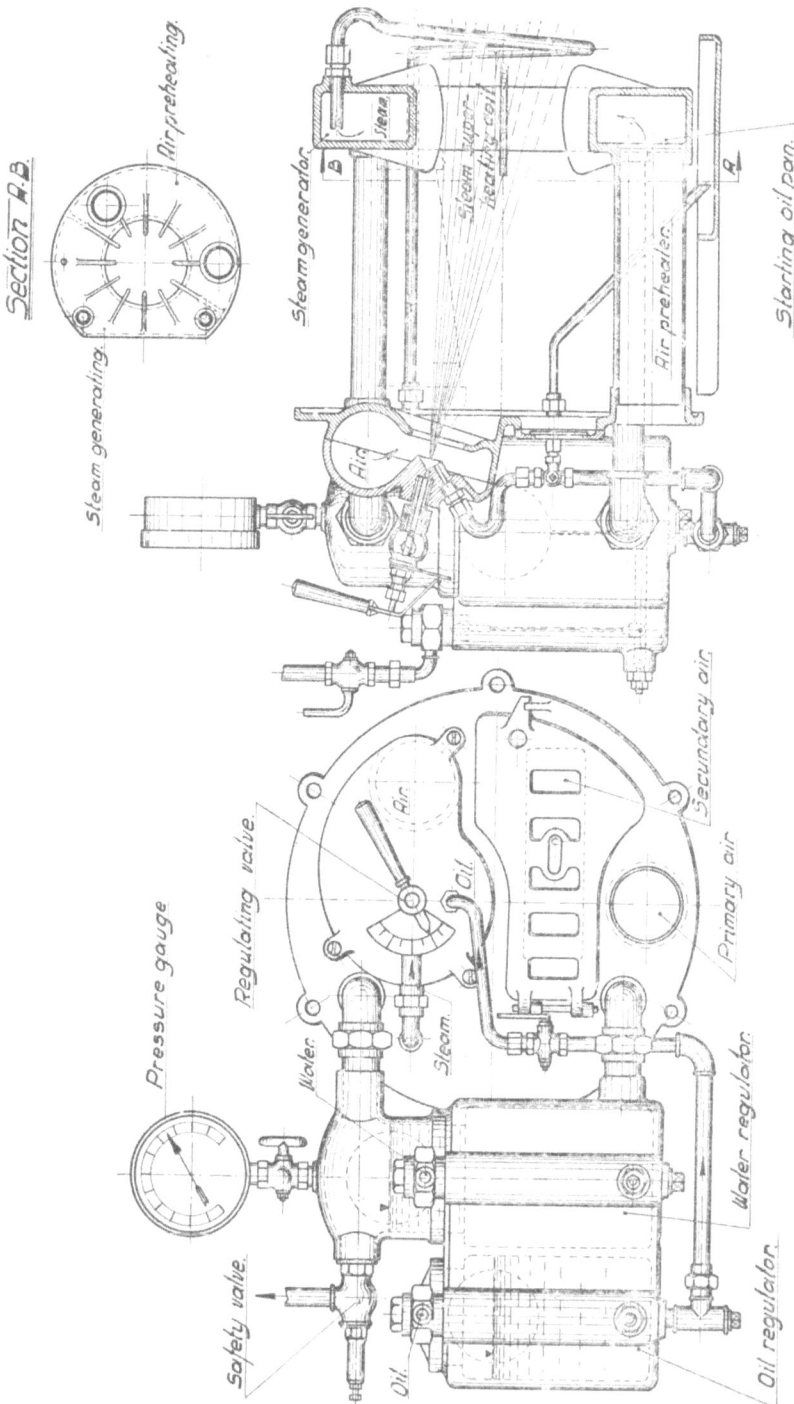

Fig. 207. Tacchella steam-atomizing domestic oil burner (Butler).

return valve, the pressure will even drop still further, until finally the boiler will be filled with cold water. This thermal lability may be prevented to some extent by installing another valve in the water supply line, which valve is shut by means of a bimetallic element as soon as the boiler temperature drops below a certain limit, a system which in fact comprises a *thermostatic* and a *pressure-static* device.

Yet another method is to keep the temperature of the steam constant by means of a *boiling bath of constant temperature*, such as, for instance, diphenylamin (300° C.). This allows, moreover, of preheating the fuel to that temperature, thus ensuring perfect pulverization at very low pressures of steam and fuel. Details of this principle will be further discussed in the Fourth section.

## C. *Constructive Forms*

The fact that these so-called *self-generating steam-oil burners* have found but little practical application proves that their operation is not yet sufficiently reliable. The

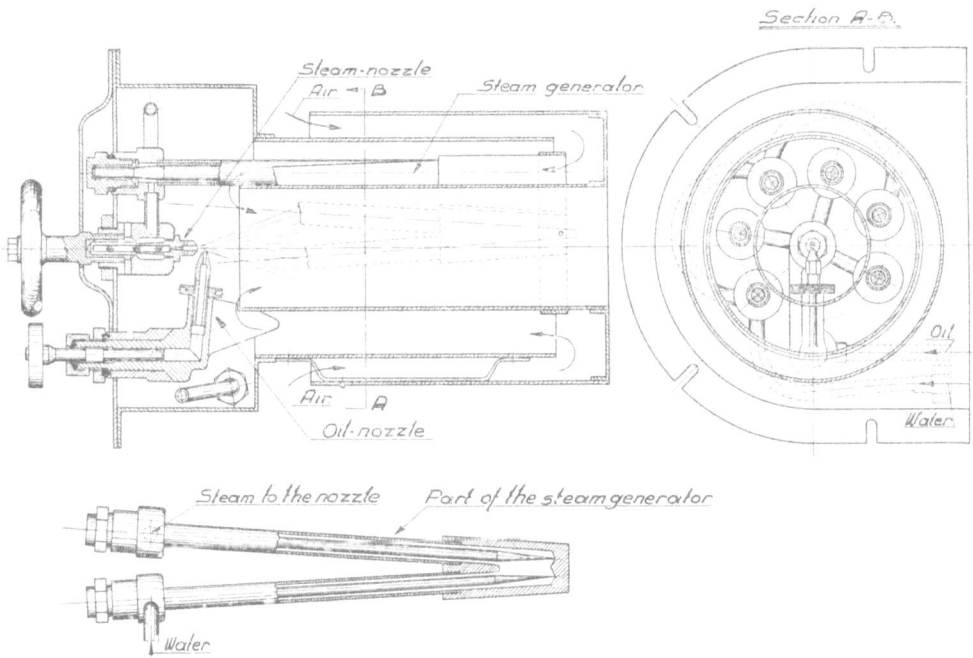

Fig. 208. RUTHERFORD's steam-jet oil burner.

flash-boilers often used for this purpose comprise generally an annular space with tubes exposed to the flame. There are many others, but it will suffice to quote only the following:

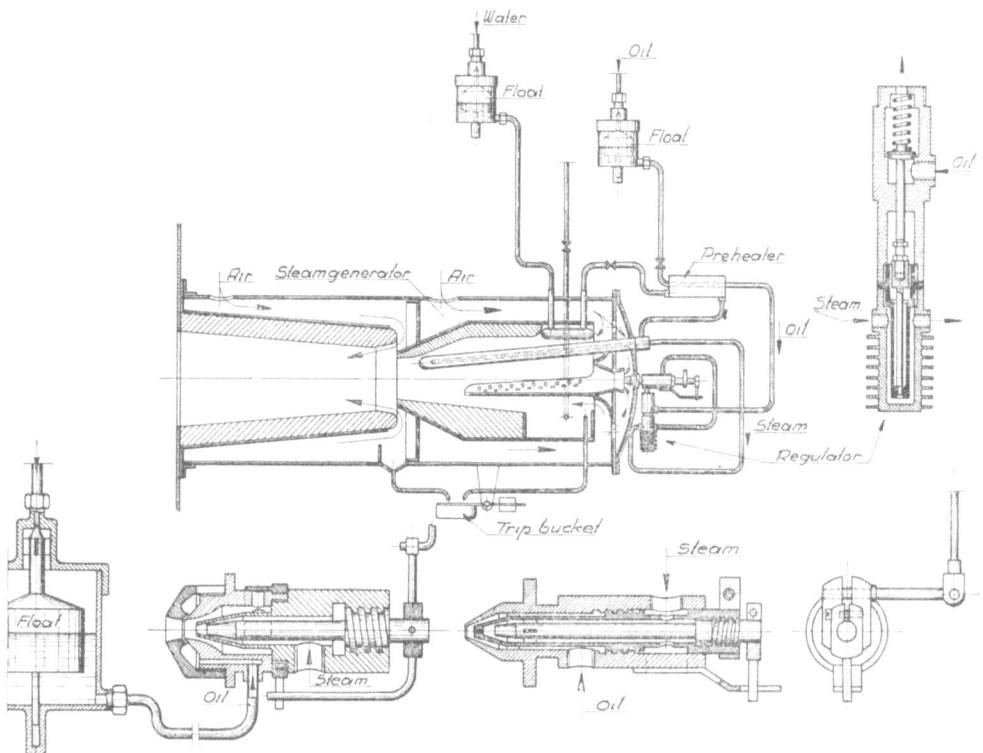

Fig. 209. Brûloil steam-jet oil burner.

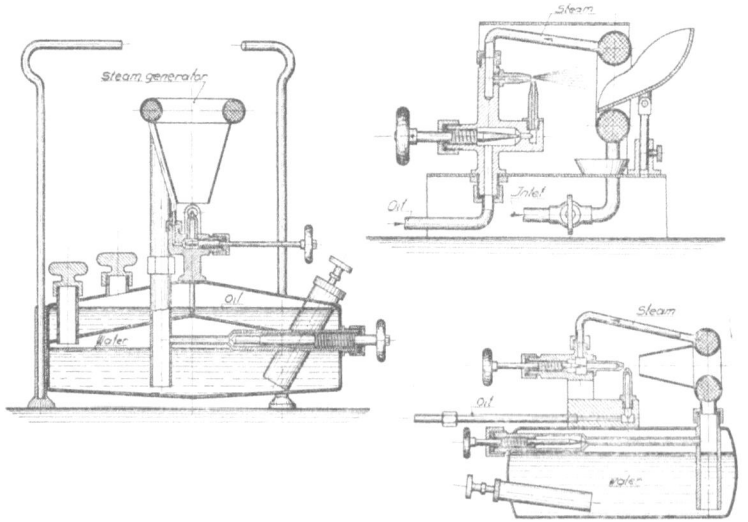

Fig. 210. Primus type steam-jet burners of Androvic & Smith.
(Br. patent No. 279,239).

TACCHELLA (see fig. 207); RUTHERFORD (see fig. 208), who developed a portable unit as described in British Patents 237,203, 237,204, 238,266 and 383,016; THOMPSON (Br. Pat. 161,831 — 1920); BRÛLOIL (fig. 209) (French Patent 724,027 — 1932); SCARAB (Br. Pat. 162,549 — 1920); DAVILA (Br. Pat. 163,186 — 1920); CASTELLAZZI (Br. Pat. 203,363 — 1922); ANDROVIC &

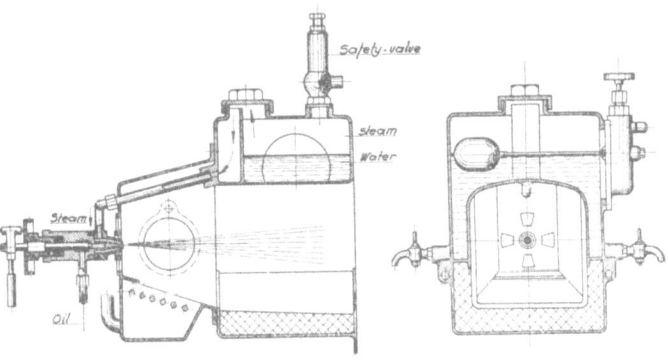

Fig. 211. The SIMPLEX steam-atomizing domestic oil burner. (Austria).

SMITH (fig. 210) (Br. Pat. 279,239 — 1926), who developed a portable apparatus of the Primus type; PERFECTION STOVE Co. (MARKS), (Br. Pat. 328,546 — 1928) and the Austrian SIMPLEX burner (fig. 211) (also see Table VI).

In order to avoid pulsations, it has been recommended to generate the steam in coils which are arranged in such a way that steam bubbles are allowed to rise freely, as is described, for instance, in German patent 438,375 and Br. Patent 356,150 granted to UTINASA (FRÖHLICH & HECKL) (fig. 212) and in the Br. Pat. 227,207 (1923) of the FILMA Oil burners Ltd. Br. Pat. 325,953 granted to

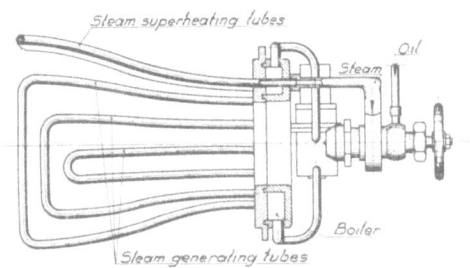

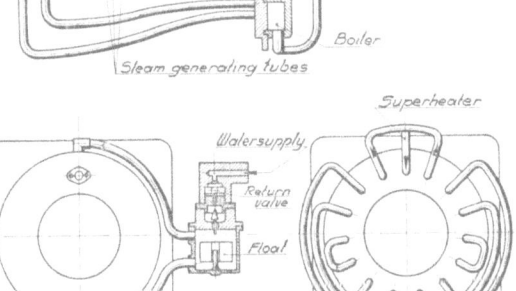

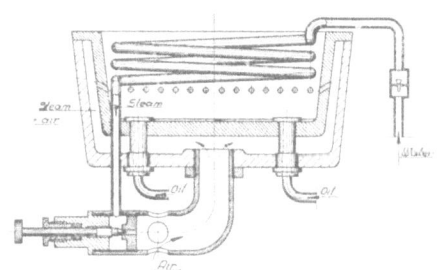

Fig. 212. FROHLICH & HECKL (UTINASA) domestic steam-jet oil burner.

Fig. 213. FITZPATRICK steam-jet burner.

FITZPATRICK (1929) (see fig. 213) uses the steam, not for pulverization of the oil, but for *forced air supply*, the oil being evaporated in an open trough similar to PRIOR'S (Br. Pat. 276,586 and 284,611).

A serious objection to the use of water in a self-generating steam-atomizing burner is the *formation of boiler-scale* in the vaporizing device, which deposit may cause bad heat transfer and may clog the passages. Excellent waters distributed in large cities will nevertheless usually leave some $\frac{1}{4}$ to $\frac{1}{2}$ a gramme of solid deposit per litre of water evaporated, and the rough calculation will show that this will be enough to choke a steam tube of $\frac{1}{2}''$ rapidly.

Thus we see that with this type of burners the *water* vaporizer problem is just as perilous as it is with the burners in which the *oil* is previously vaporized; in both cases it is difficult to prevent a deposit.

---

## CHAPTER VIII

### PREPARING THE OIL FOR COMBUSTION BY MEANS OF STEAM GENERATED BY A FOREIGN SOURCE

### A. *General Remarks*

Both Chapter VII and Chapter VIII deal actually with the same subject, viz., atomization by means of steam, but it has been thought useful to make a separate class of the self-generating type in order to draw more attention to it. In this Chapter the details of steam-atomizing nozzles will be discussed further, although several types have been quoted before in the section on the Historical Development. As a matter of fact, steam-atomization has lost much of its interest and has been superseded to a large extent by mechanical or low pressure air atomization.

This is not astonishing, considering that for industrial oil burning, according to HASLAM & RUSSELL's figures, *the total cost for steam atomization per gallon of oil burnt is about twice that of mechanical or low pressure air atomization*, interest, depreciation, maintenance and repairs included. Moreover, in the case of steam atomization part of the total boiler costs, calculated according to the ratio between steam consumed by burner and the total amount of steam raised, comes to the account of oil burning.

### B. *Types*

The method of classification of the U.S. Liquid Fuel Board (1904) (page 17) has been adopted here too, viz., a vertical division into *outside mixing burners*, comprising "drooling burners" and "shearing burners", and *inside mixing burners*, comprising "chamber burners" and "injection burners" (fig. 32, 214 and Table VI).

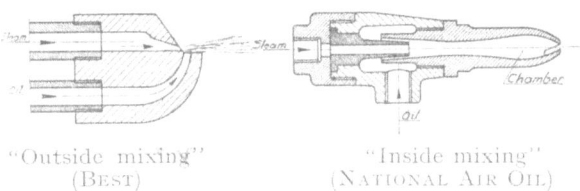

"Outside mixing" (BEST)  "Inside mixing" (NATIONAL AIR OIL)

Fig. 214. Outside mixing versus inside mixing type.

With the *outside mixing type* the oil either dribbles from (drooling type), or is forced through an orifice at comparatively low pressure (shearing type) and just as it issues from this orifice it is struck by a jet of steam blown at high velocity usually from a direction at approx. right angles to the flow of oil. The steam-jet breaks the oil and projects it into the combustion chamber in a fine spray or mist.

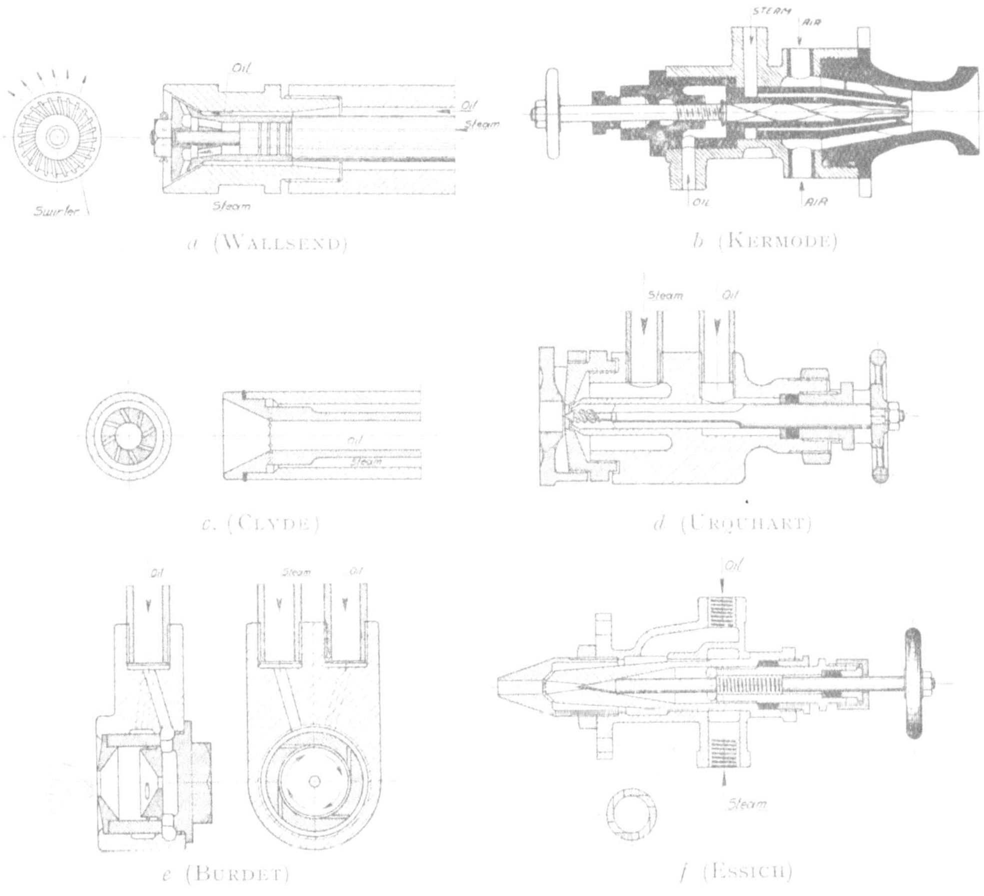

*a* (WALLSEND)  *b* (KERMODE)

*c.* (CLYDE)  *d* (URQUHART)

*e* (BURDET)  *f* (ESSICH)

Fig. 215. Some steam-atomizing oil burners with rotation devices.

The *inside mixing type* comprises a mixing chamber (chamber burners) between the steam orifice and the burner opening. The mixing inside the burner permits of some previous vaporization of the lighter hydrocarbons by the heat of the steam, so that the mixture issuing from the burner opening ignites readily and maintains a steadily burning flame. This burner class may also be equipped with steam injectors

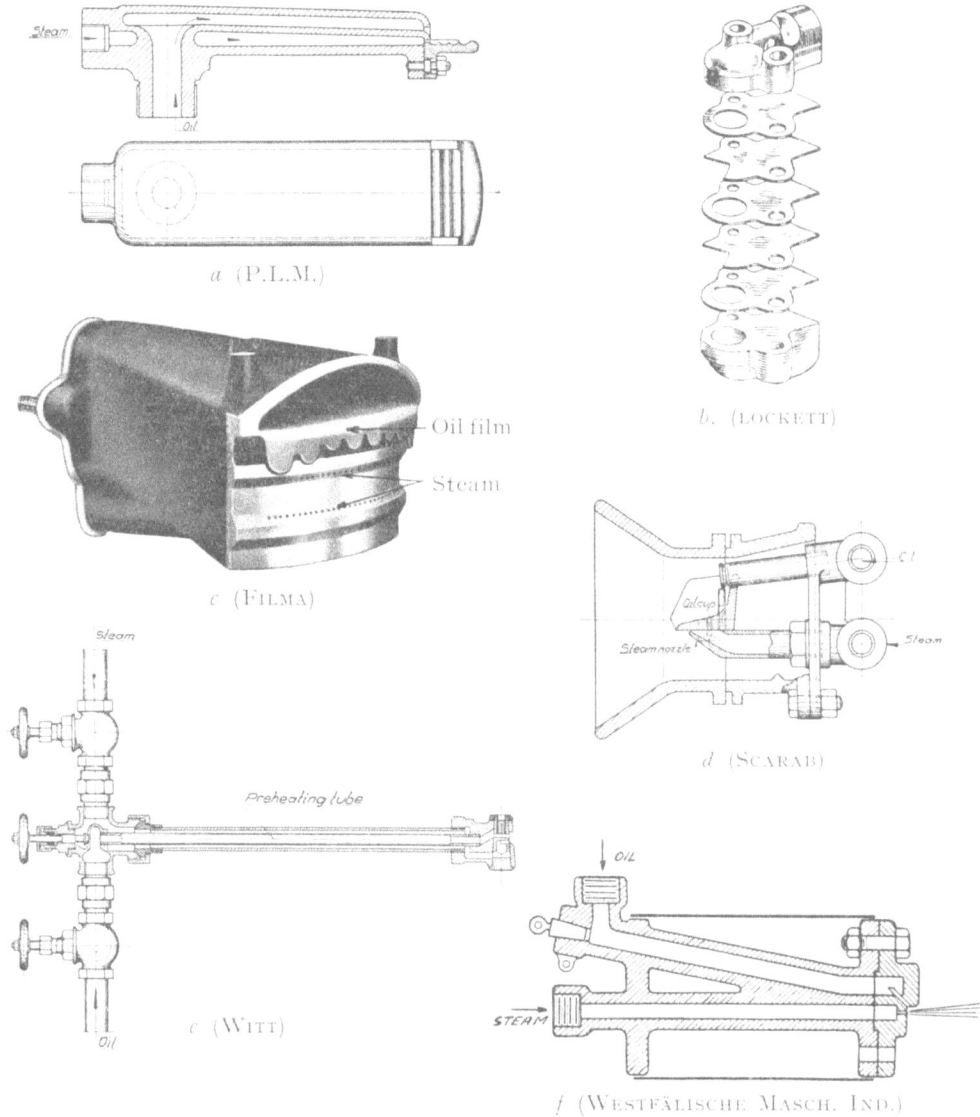

Fig. 216. Some steam-atomizing oil burners of the drooling type.

(injector burners), the steam jet, by blowing through a narrow throat, causing enough vacuum to draw in the oil and sometimes air too.

The former type (outside mixing burners) is usually somewhat less economical

as to steam consumption but is not so sensitive to fluctuations in the steam pressure, as the steam can exert no back-pressure on the flow of oil. The second type (inside mixing burners) causes a better contact between steam and oil, often resulting in better preheating and better pulverization, and, consequently, is as a rule somewhat more economical than the first. On the other hand, irregularities in the steam pressure may influence the flow of oil, and with the inside mixing type of oil burners a constant steam pressure is necessary to obtain steady flames.

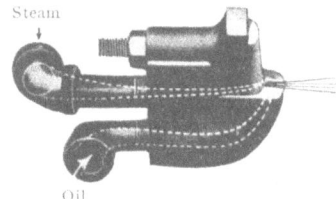

Fig. 217. BEST pulverizing type burner.

*The steam should be dry and superheated*, as then the same velocities may be obtained with lower weights of steam, as is clearly shown by the following figures:

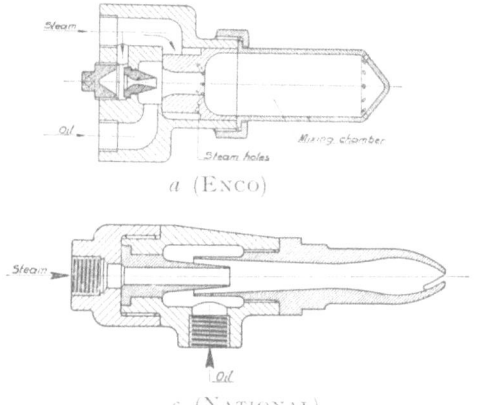

*a* (ENCO)

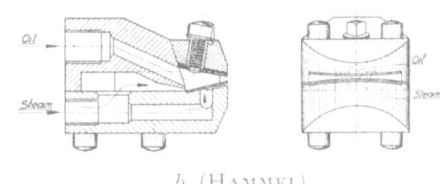

*b* (HAMMEL)

*c* (NATIONAL)

Steam volume of one kilo water at 100° C. and 1 atm. = 1.73 cub.m. against 2.68 cub. m. for the same at 300° C.

Modern designers of steam-atomizing oil burners have taken advantage of the progress of atomization by pressure and directed their new designs to a combination of both principles, thus giving the oil and steam a whirling motion by means of *rotation devices*. One of the earliest types of that school, viz., that of KERMODE (see figs. 29 and 215*b*), has already been mentioned in the Historical section, whilst other examples

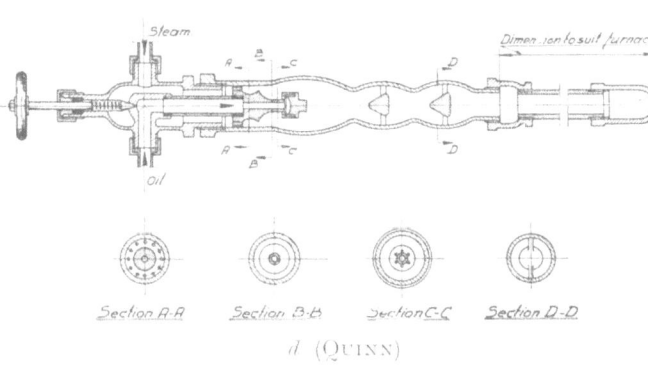

Section A-A    Section B-B    Section C-C    Section D-D

*d* (QUINN)

Fig. 218. Examples of "chamber" burners.

of such designs are: BABCOCK WILCOX (Br. Pat. 289,745; 1928), TODD (Br. Pat. 306,616; 1927), WALLSEND (see fig. 215*a*) comprising tangential grooves on a conical surface, CLYDE (see fig. 215*c*) having tangential bores in a plane at right angles to the burner axis and discharging the steam through a VENTURI throat. Another typical modern steam-atomizing oil burner is that of BURDET shown in fig. 215e, in which the oil is fed through an orifice into a whirl chamber, whilst the steam receives a rapid rotary motion by passing through tangential channels. The resulting mixture of steam and oil leaves the burner through a second sharp-edged orifice.

One of the main advantages of rotation devices for the steam flow is that it reduces the length of path of the resultant oil spray, and thus enables the combustion

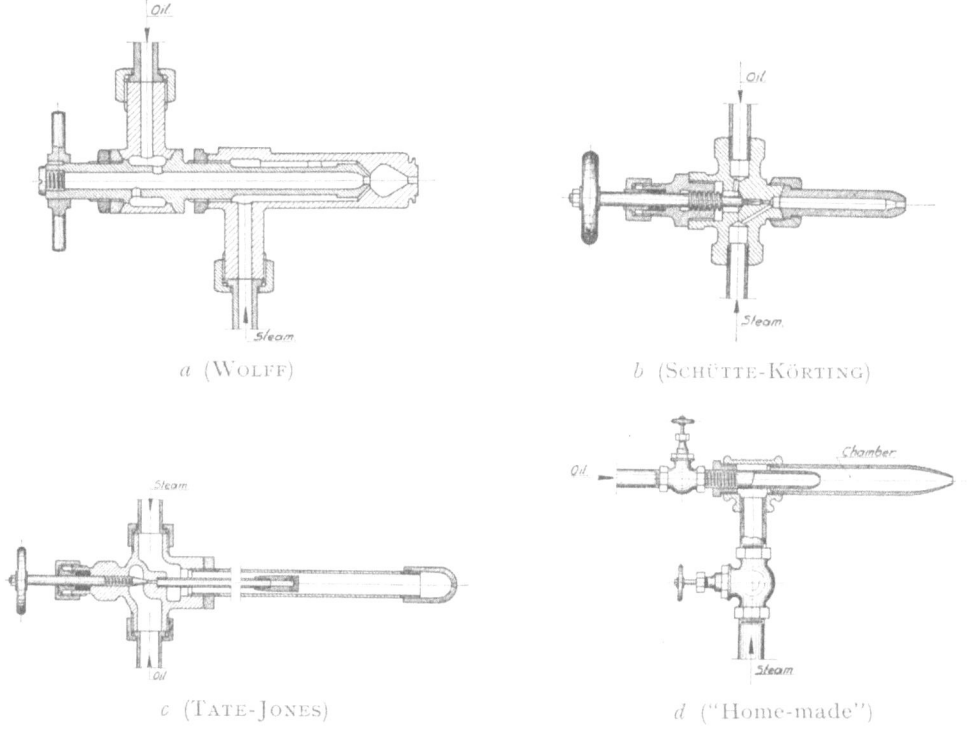

a (WOLFF)          b (SCHÜTTE-KÖRTING)

c (TATE-JONES)          d ("Home-made")

Fig. 219. Other "chamber" burners

to take place with *shorter and wider flames* than those produced by straight shot burners. Upon the whole, adding steam to oil flames has an extremely beneficial influence on the character of the flame, which is made short, clear and brilliant.

A steam-atomizing oil burner usually requires for atomization 1.5 to 5% of the total steam generated by the boiler which is heated by it, 2 to 2.5% being a good average.

The *vertical division* of the classification of Table VI is made according to the *shapes* and *directions* of the streams of steam and oil, which is self-explanatory.

Some types may be quoted:

1) *The Drooling Type of Oil burner.*

"Mexican trough" type, SCARAB fig. 216*d*, P.L.M. fig. 216*a*, BOOTH, FILMA fig. 216*c* (Br. Pat. 219,228; 267,743 and 389,903). This type, which as a rule has large passages for the oil, is especially suitable for burning dirty fuels, such as waste tar oils, acid sludges etc.

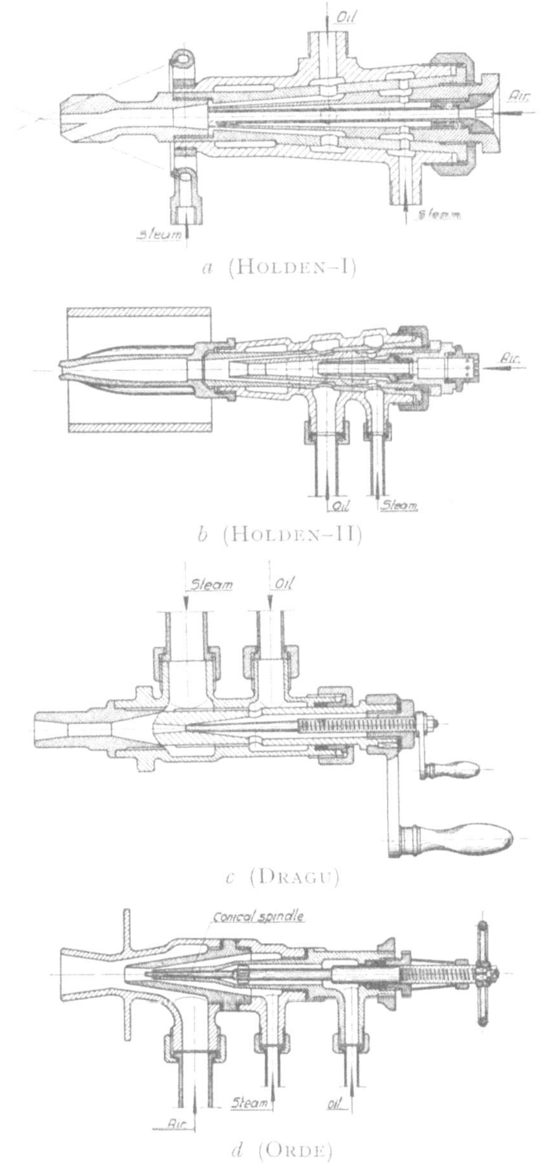

2) *The Pulverizing or Shearing Type of Oil burner*, in which the oil is forced at approximately right angles onto the jet of steam instead of flowing by gravity. GUYOT fig. 20 (French Navy), LOY et AUBÉ, BEST fig. 217 (Br. Pat. 171,173; 1920), COSMOVICI (Br. Pat. 182,501; 1921), DAVIES (Br. Pat. 185,184; 1921), and JONES (Br. Pat. 342,363; 1929), LASSOV (Br. Pat. 757; 1909).

3) *The Chamber burners*, which are widely applied in America: NATIONAL AIR-OIL fig. 218*c*, ENCO fig. 218*a*, TATE-JONES fig. 219*c*, HAMMEL fig. 218*b*, QUINN fig. 218*d* (Br. Pat. 176,138; 1920), THOMPSON (Br. Pat. 186,148; 1921), HAUCK (Br. Pat. 219,965; 1923), GUTHRIE (Br. Pat. 247,855; 1925), SCHÜTTE-KÖRTING (fig. 219*b*), WOLFF (fig. 219*a*) (Br. Pat. 20,122; 1911). An inexpensive "home-made" burner made of standard fittings and gaspipe is shown with fig. 219*d*.

4) *The Injection Type of Oil burner*, in which the *Venturi principle* is used to create vacuum by means of a Venturi throat, this vacuum being used for drawing in the oil or air too. HOLDEN figs. 220*a* and 220*b* (Br. Pat. 21,837; 1910), KERMODE fig. 215*b*, ORDE fig. 220*d*, ESSICH fig. 215*f*, DRAGU fig. 220*c*, the last-mentioned being used by the Roumanian Railways.

*a* (HOLDEN-I)

*b* (HOLDEN-II)

*c* (DRAGU)

*d* (ORDE)

Fig. 220. Examples of injection burners.

## CHAPTER IX

### PREPARING THE OIL FOR COMBUSTION BY HIGH OR MEDIUM PRESSURE AIR

Although it is not, of course, possible to draw a sharp demarcation between high, medium and low pressure air, in the following pages pressures *below 75 cm water gauge*

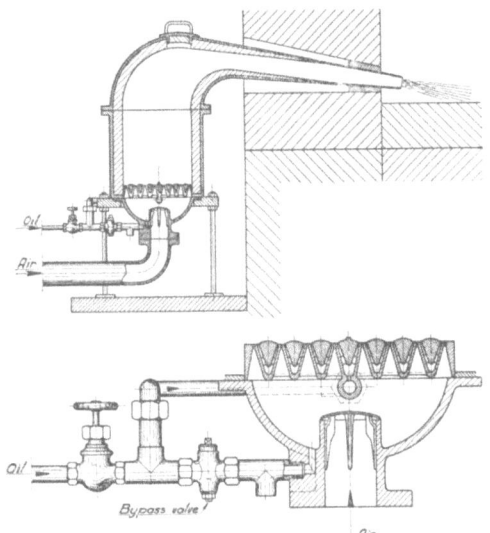

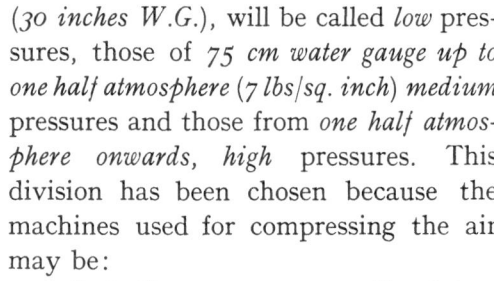

Fig. 221. WESTPHAL's Metallurgical blow-torch burner.

(*30 inches W.G.*), will be called *low* pressures, those of *75 cm water gauge up to one half atmosphere* (*7 lbs/sq. inch*) *medium* pressures and those from *one half atmosphere onwards*, *high* pressures. This division has been chosen because the machines used for compressing the air may be:

1) Ordinary *open* or *centrifugal fans* for low pressure.

2) So-called "*positive pressure blowers*" for medium pressure and,

3) *Piston compressors* or *multiple turbo-compressors* for higher pressures.

As has been explained before, the application of atomization by high pressure air — which direction was followed for a while during the last decennia of the past century in order to avoid the consumption of fresh water by the steam-atomizing oil burners on

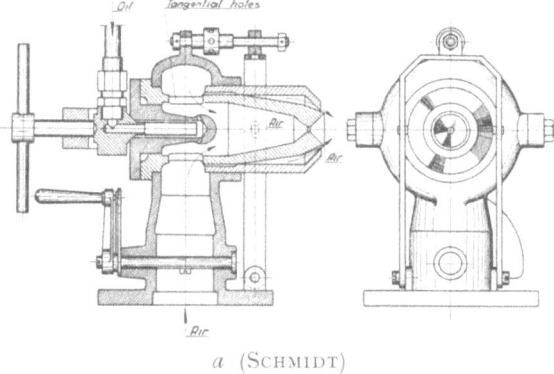

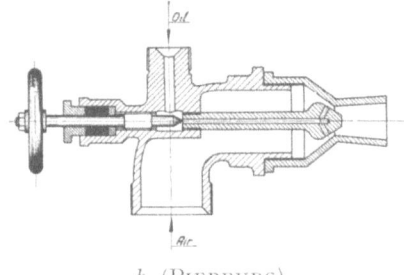

*a* (SCHMIDT)                                        *b* (PIERBURG)

Fig. 222. Medium air pressure oil burners.

board ships — was not very successful, mainly because of the complicated compressor installations consuming a considerable amount of power, requiring skilful attendance and involving high cost of upkeep.

As regards the fires, high pressure air atomization was found to produce very noisy roaring flames of such high temperatures that damage was often caused to the boiler walls by impingement of the flame.

Thus it looked for a long time as if air atomization had no future except for some special metallurgical purposes, such as convertor ovens, blast furnaces etc. (see fig. 221 representing the WESTPHAL burner), until it was gradually recognized that it was not at all necessary to use such high air pressures to get reasonable atomization. A consequence was that during the following period from 1910–1920 several new inventions were made for reducing the air pressure, resulting in the formation of a separate class of what may be called *medium pressure air-atomizing oil burners*. Examples of this type are: the SCHMIDT burner (see fig. 222a), PIERBURG (see fig. 222b), WITFIELD (Br. Pat. 15,520), SELAS (Br. Pat. 12,808), DELANAY-BELLE-VILLE (Br. Pat. 14,964), FILMA (fig. 216c) (Br. Pat. 219,228; 267,743 and 389.903), LOY & AUBÉ (Br. Pat. 148,385; 151,015 and 248,847), CALOROIL (Br. Pat. 295,698 and 295,707) (fig. 223), (also see Table VII).

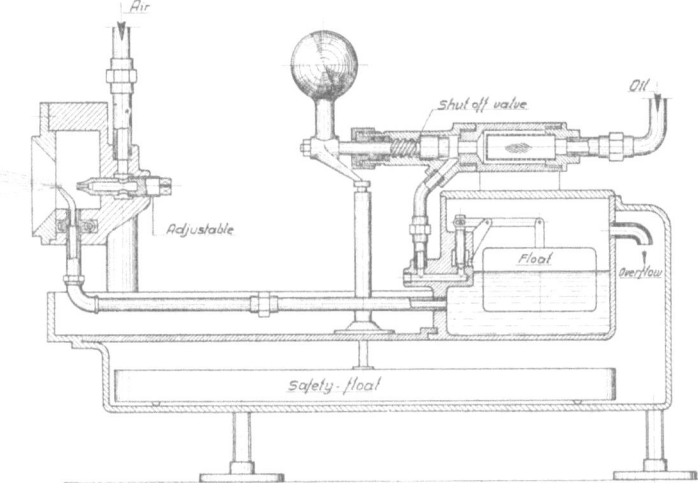

Fig. 223. CALOROIL domestic oil burner using compressed air.

Burners of this class are based on the same constructive principles as the steam atomizing burners described in the previous chapter, to which the reader may be referred.

As to the *cost* of oil burning by high pressure air, it may be mentioned that, according to HASLAM and RUSSELL's figures, this is about 4 times that of pressure-atomizing or low air pressure oil burning. Furthermore it may be mentioned that the amount of *air required for medium pressure atomization may roughly be estimated to equal one half of the weight of the oil burnt*, which figure corresponds to a steam consumption of $2\frac{1}{2}$ to 3% of the total steam generated if the air compressors or positive blowers are steam-driven and if the burner is used for high pressure steam raising.

# CHAPTER X

## PREPARING THE OIL FOR COMBUSTION BY MEANS OF LOW PRESSURE AIR

### A. *General Remarks*

From about 1920 onwards further considerable reduction in the air pressure was obtained for air-atomizing oil burners, which new development was especially stimulated by the young domestic oil burner industry introducing a new demand, viz., *noiseless operation*. Blowers of the so-called *"positive pressure type"*, such as shown in fig. 224a (ROOTS *blower*), were less suitable for domestic use on account of noisy operation and were therefore replaced by *simple or multiple bladed centrifugal fans* (see fig. 224b) delivering air at pressures from 30 to 50 cm water gauge (12 to 20 inches W.G.). A frequently adopted type is that shown in fig. 224c, known as the SIROCCO *type*, delivering air at pressures from about 10 cm water (4 inches W.G.), whilst also the *ordinary fan* comprising four crossed blades, being the least noisy type and producing air at only a few cm of water pressure, found widespread use.

It was not only because of the noise that lower pressures were required, but also because this meant a considerable reduction in the *power* required and consequently lower cost for electricity, which point was found to be especially important for small domestic units.

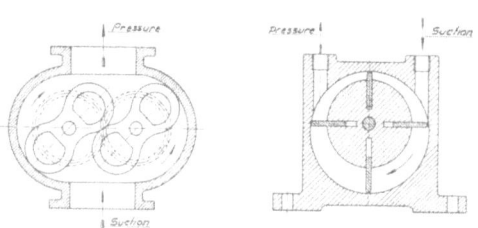

*a* "Positive pressure" blowers.

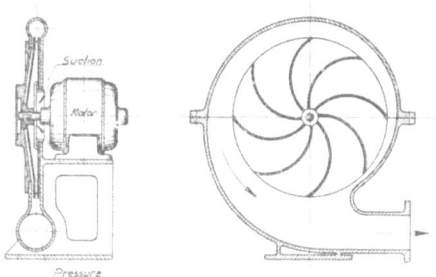

*b* Medium pressure blower.

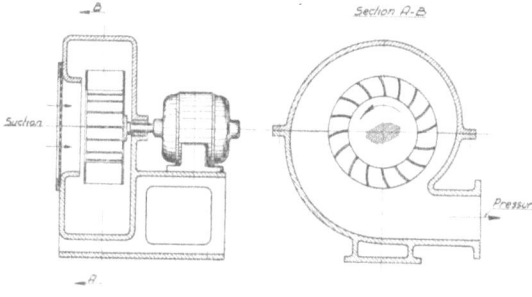

*c* Low pressure blower (SIROCCO).

Fig. 224. Various types of blowers used for oil burning (after TAPP).

B. *The Low Pressure Air divided into two Streams with Separate Functions.*

Mention has already been made of the fact that for high pressure air-atomizing oil burners only a small part of the total air necessary for combustion was compressed (e.g., at 3 atm. pressure about 10 to 15%), the remaining quantity being drawn into the combustion chamber by means of natural or forced draught. When *reducing the air pressure* it was soon found to be necessary *to increase considerably the quantity of compressed air used for atomization,* medium pressure oil burners of 0.1 to 0.2 atm. pressure using about 50 to 60% of the total air for atomization of the oil. This percentage was even raised to *70 or 80% for low*

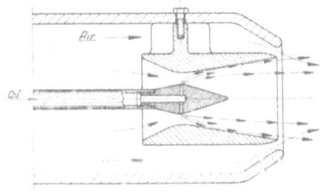

Fig. 225. Air stream of low air-pressure burners divided into two portions.

*pressure air atomizing oil burners,* and of course at such low pressures it was scarcely worth while supplying the remaining 30 or 20% of the air by other means, and thus we see at the present day the fans or blowers of low pressure air installations delivering the *total amount of air necessary for combustion.*

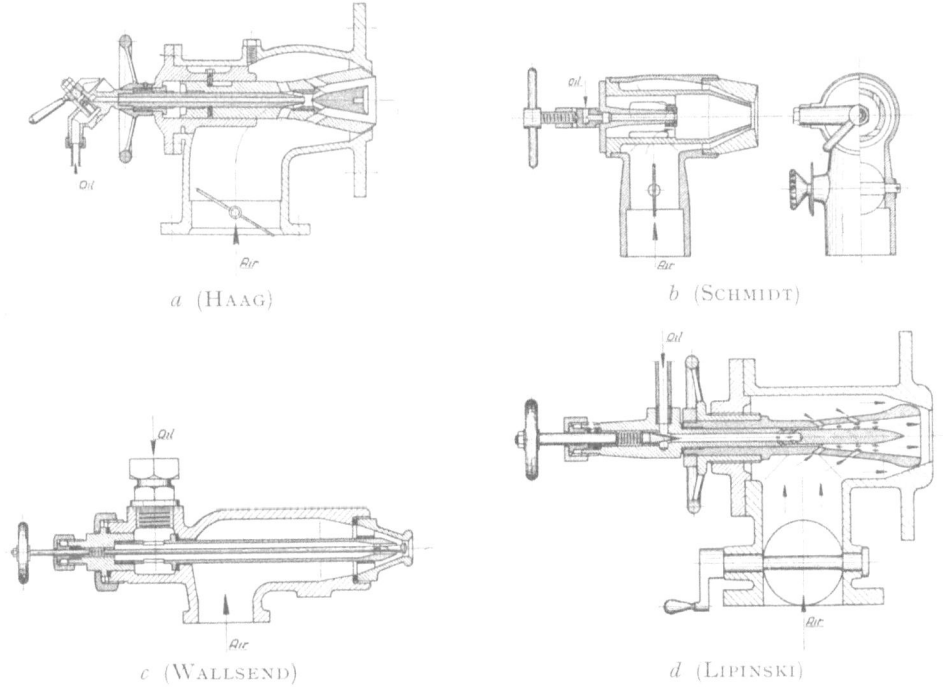

a (HAAG)

b (SCHMIDT)

c (WALLSEND)

d (LIPINSKI)

Fig. 226. Low air pressure burners.

A general feature of this type of burners, however, is still the dividing of the quantity of air into *two separate streams,* one part being used either to atomize the oil itself or to receive the atomized oil, whilst the other stream of air is added to the first mixture a little further on. The first part of the air may be called *atomizing air* and the second *forming air*, the latter having considerable influence upon the shape of the flame. Fig. 225 gives a diagrammatic sketch showing clearly the division into two parts, the atomizing part flowing through a VENTURI throat, producing some kind of a

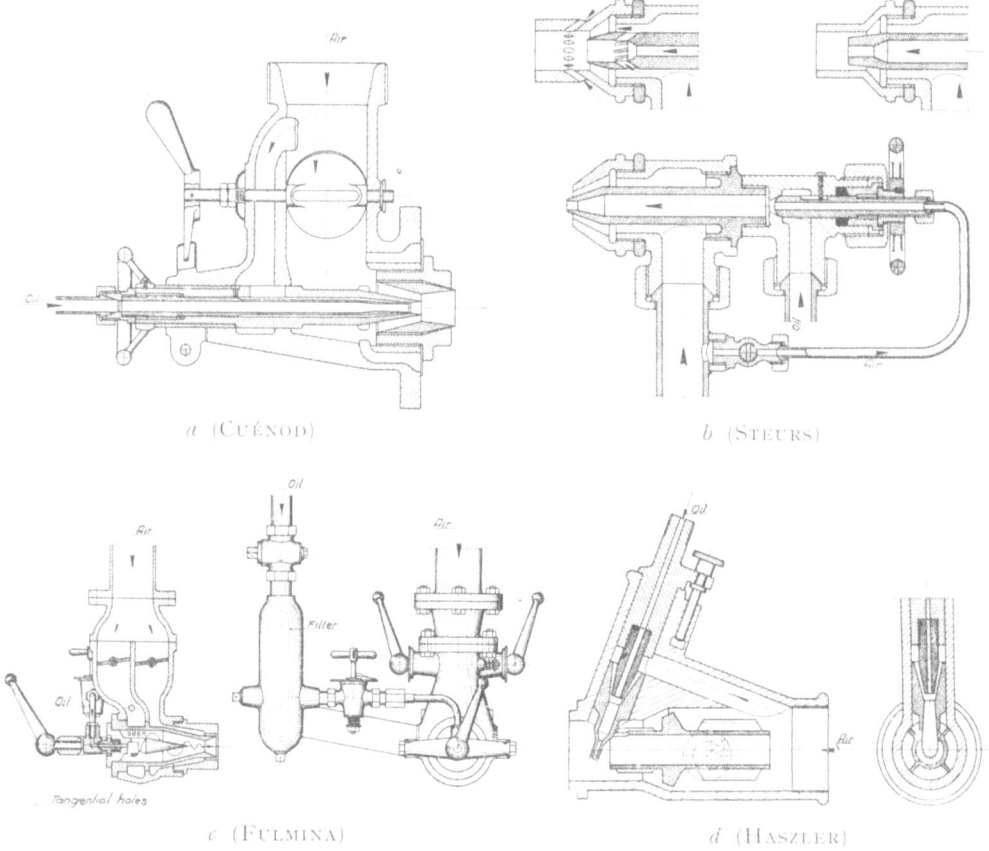

a (CUÉNOD)    b (STEURS)

c (FULMINA)    d (HASZLER)

Fig. 227. Low air pressure burners.

carburettor effect, for which, however, considerable velocities, accompanied by pressure losses, are required.

A frequently used principle is to *blow the oil off from the edges of a cup, plate, cone, needle, sphere or other device* placed in the middle of the air stream. Examples of such types are CUÉNOD fig. 227a (Br. Pat. 201,167) (Cup), SCHMIDT (fig. 226b) (Conical cup), WALLSEND fig. 226c (Br. Pat. 200,860) (tube with central valve stem), LIPINSKI fig. 226d (Br. Pat. 10,040) (cone and cup), HAAG fig. 226a (cup and cone), STEURS fig. 227b (conical tubes), FULMINA fig. 227c (cone), HASZLER fig. 227d (saw-tooth blade), CUSTODIS fig. 228*a* (flat blade), ROLL fig. 228*b* (sphere with bores), WINDLE (Br. Pat. 132,286, sharp-edged tube), (also see Table VII).

For perfect atomization it is important that the above-mentioned obstruction device —which generally receives the oil by gravity — should end with *sharp edges* on to which the oil film should be forced by the friction of the air stream. A *dull* ending such as that in fig. 228*c* (BOYE) gives the oil an opportunity of accumulating

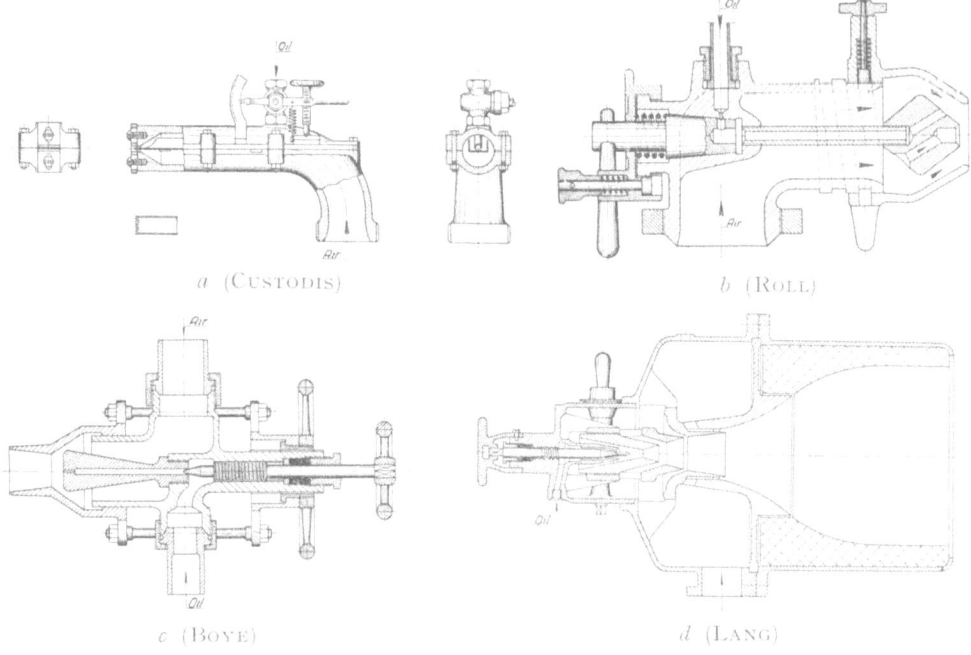

a (CUSTODIS)        b (ROLL)

c (BOYE)        d (LANG)

Fig. 228. Various low air pressure oil burners.

for a moment in the dead corners, which may lead to an irregular oil delivery and to coarse atomization or unsteady flames.

Other patents of this type are: REGNAC PAILLE (Br. Pat. 293,275), SWINNEY (Br. Pat. 377,821), LANG fig. 228*d* (Br. Pat. 237,587), HETSCH fig. 229*f* (Br. Pat. 146,359 and 261,006), BRILLIÉ (Br. Pat. 377,765), LANSER (Br. Pat. 340,858), DIVE (Br. Pat. 351,075).


C. *Low Pressure Air Atomization combined with other Means of Preparing the Oil*

It is common practice to combine the VENTURI principle with the above-mentioned "blowing-off principle", such as has been done with PIERBURG's burner (see fig. 222*b*), the oil being drawn in by a vacuum caused by an increased velocity of air flowing through a VENTURI throat. Furthermore it is possible to combine the low

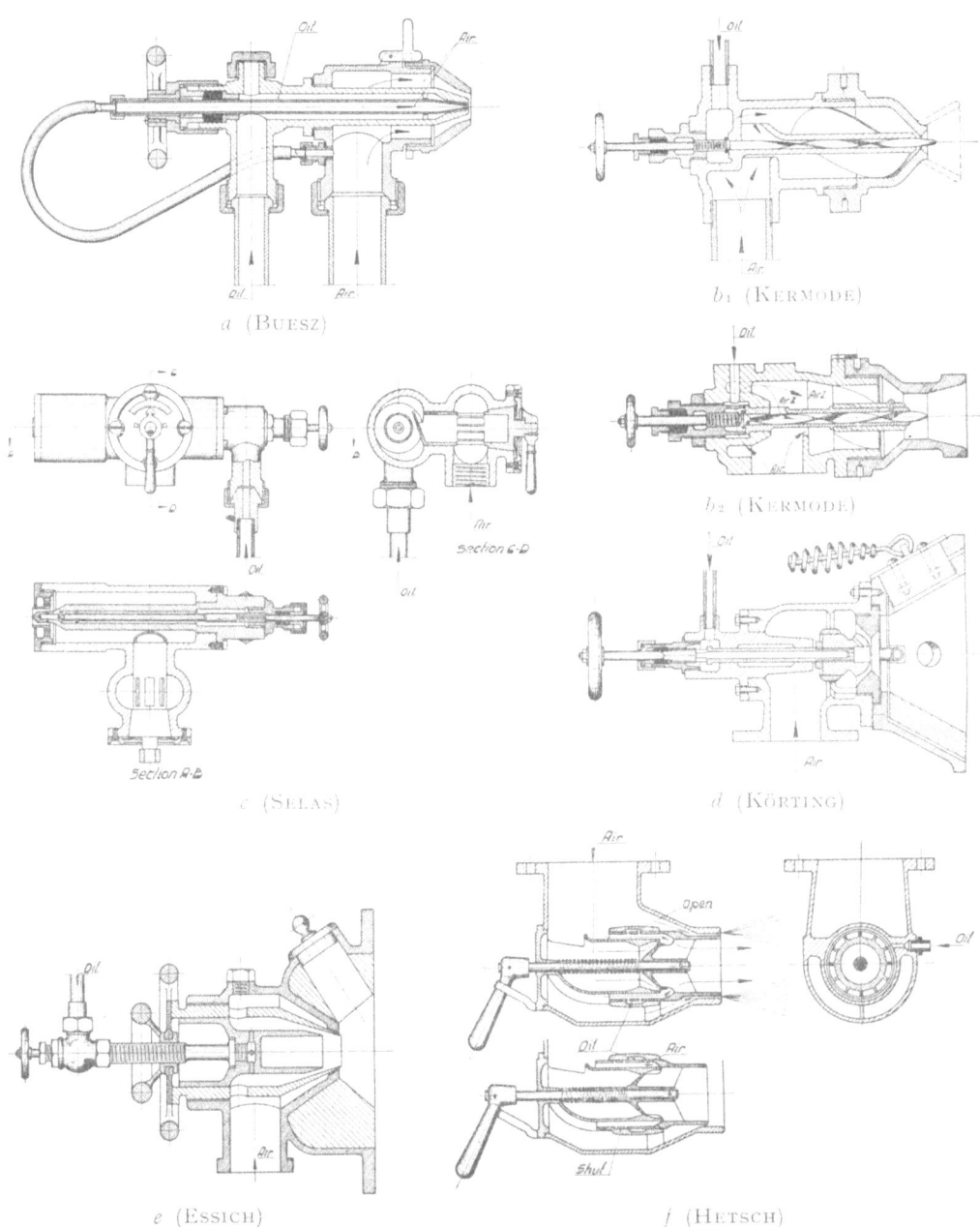

Fig. 229. Various low pressure air-atomizing oil burners.

pressure air atomizing principle with some other means of preparing the oil for combustion, and thus we get combinations such as the PRIOR burner fig. 61 (Br. Pat. 276,586; 284,611), comprising an oil film exposed to radiant heat of the flame while

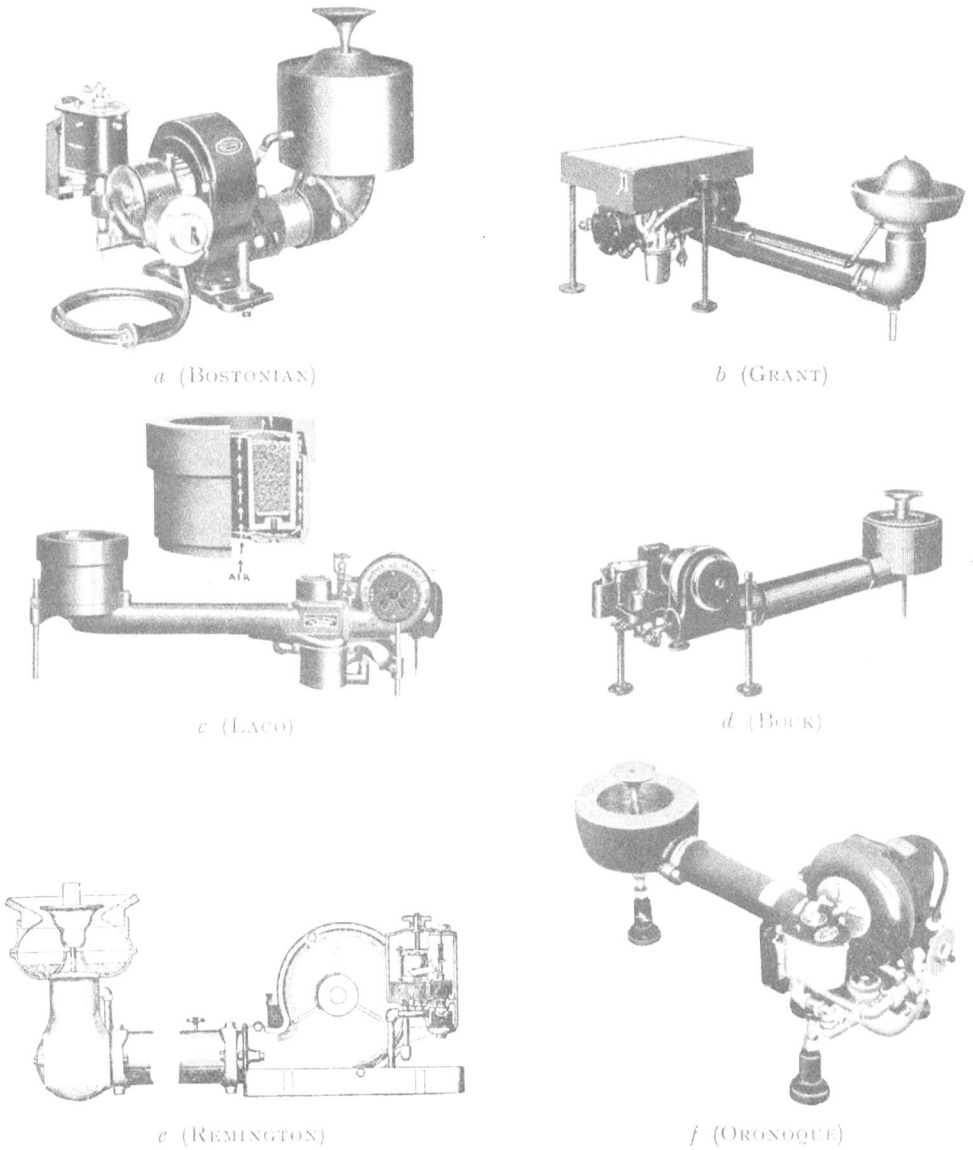

a (BOSTONIAN)                    b (GRANT)

c (LACO)                    d (BOCK)

e (REMINGTON)                    f (ORONOQUE)

Fig. 230. Various domestic oil burners of the fire-pot type.

at the same time a strong current of air is blown on to the oil surface, a principle which is sometimes called: "*mechanically assisted evaporation*". British Patent 377,107 describes a similar burner (DEN BOER), in which the air chamber is divided into two parts and, by applying different air pressures, the flame may be directed to the right or left, a manipulation which is sometimes useful for bread-baking, etc.

A similar type of "mechanical assistance" incorporates a *fire-pot of refractory*

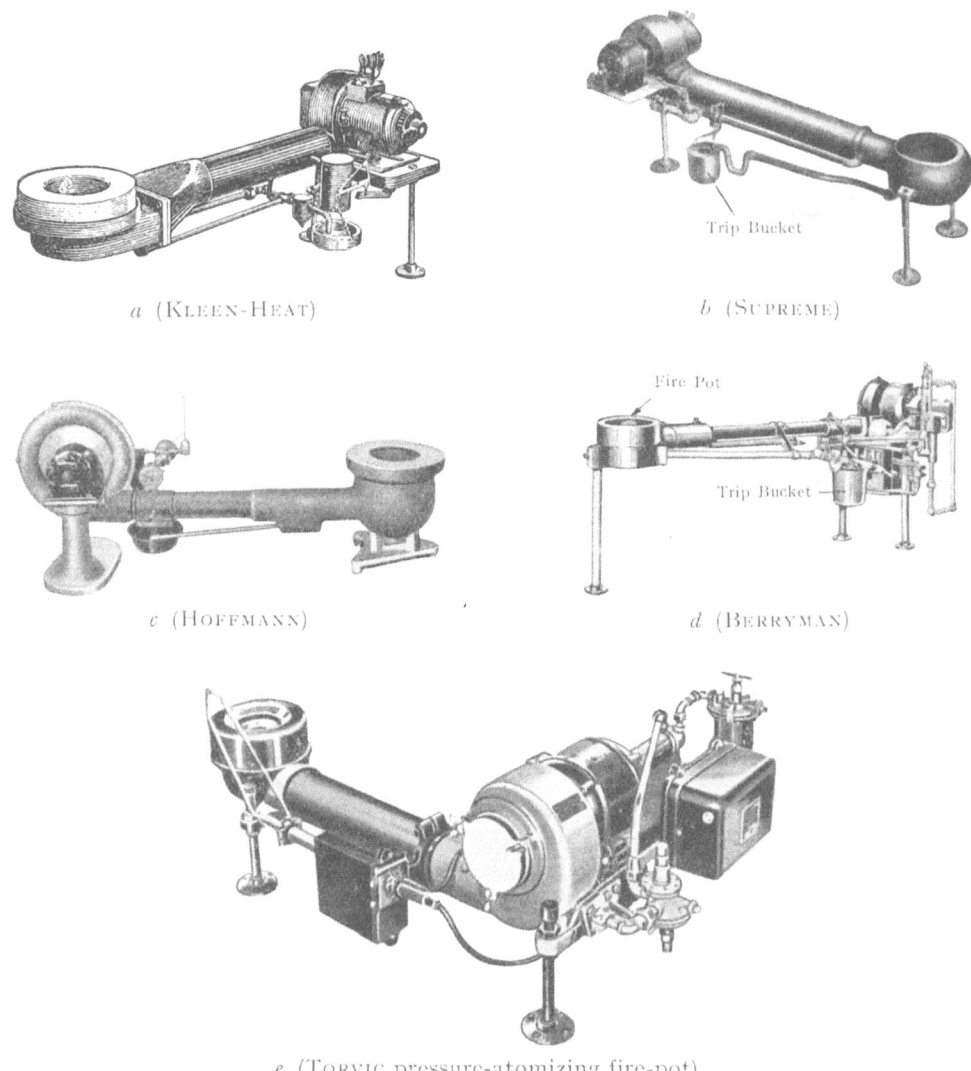

*a* (KLEEN-HEAT)   *b* (SUPREME)

*c* (HOFFMANN)   *d* (BERRYMAN)

*e* (TORVIC pressure-atomizing fire-pot)

Fig. 231. Various domestic oil burners of the fire-pot type.

*material* having such a high temperature that the oil is not only evaporated but also partly cracked, and this principle has found widespread application in America as a cheap form of mechanical domestic oil burner; it is generally called the *vaporizing-pot type* or *fire-pot type*, such as described in, for instance, Br. Patents 269,821 and 281,182 (1927). Of the numerous representatives in the American domestic oil burner market the following may be mentioned: Bostonian fig. 230*a*, Grant fig. 230*b*, Laco fig. 230*c*, Bock fig. 230*d*, Fluid-heat, Kleen-heat fig. 231*a*, Supreme fig. 231*b*, Hoffmann fig. 231*c*, Berryman fig. 231*d*, Northern fig. 59, Oronoque fig. 230*f*, Marr, fig. 141, Nuway. Regarding these pictures it should, however, be borne in mind that the manufacturers often change their designs and also often make other types of mechanically atomizing burners.

Certain pressure-atomizing domestic oil burners, e.g. Torvic (fig. 231*e*), use low pressure air as an additional means of pulverization, and this is especially true of several centrifugal oil burners.

---

# CHAPTER XI

### PREPARING THE OIL FOR COMBUSTION BY MEANS OF FLUE GASES

It is, of course, an old and well-known advantage of economizers to utilize the heat of the flue gases for preheating the incoming air or oil, but the comparatively slight temperature differences and low heat transfer coefficients necessitate the use of considerable heating surfaces, whilst, moreover, the additional resistance for flue gases and air is often rather objectionable.

Another way of utilizing a part of the heat of the flue gases, at the same time avoiding these objections, consists in *mixing a part of the flue gases with the incoming air*, which is a simple method of obtaining good preheating. This principle attracted considerable interest almost from the beginning of oil burning, as it was soon found to offer useful means of improving combustion. The dilution of the mixture of combustibles with inactive gases slows down the combustion speed and consequently *favours the soot-free aldehydeous combustion*. Moreover, it *raises the water vapour content* of the flame gases and thus has a soot-preventing action, as at sufficiently high temperatures soot and water vapour may be transformed into carbon monoxide and hydrogen.

For such *recirculation of the flue gases* it is necessary to provide means for entraining the gases either by steam injectors (Br. Patent 17,302; 1902, Thwaite) or by fans or blowers (Am. Pat. 1,743,205; 1930, Frenier). As a modern application of this principle may be quoted the frequently used method of circulating the hot flue and flame gases in tubular stills, especially of cracking plants (Dubbs), which is often done with a steam-driven special blower of heat-resisting material and is intended, not only to improve combustion, but also to obtain an efficient heat transfer by intensive turbulency of the gases around the tubes.

For circulation of the *flame* gases the troublesome mechanical means may be avoided by using the *vortex principle*, i.e., returning the flame in a swirl by its own kinetic energy and thus obtaining a partial mixing of flame gases and fresh combustibles. This principle is described in several patents, such as: Swiss patent No. 85,701 (1918); D.R.P. 492,520 (1930); British patents 1,786 (1909); 13,000 (1909); 105,431 (1926); 224,511 (1923); 204,951 (1922); French patent 690,455 (1930), the American Patents 1,624,943 (1927) and 1,782,050 (1930) and the Dutch Patent 34,379.

Then again the above-mentioned *vaporizing-pot type*, comprising a fire-pot of refractory material in which oil is evaporated and gasified in the presence of a strong current of whirling air, belongs more or less to this same class.

A further development of the vortex principle will be discussed in the fourth section.

# FOURTH SECTION


## FUTURE DEVELOPMENT OF OIL BURNING

———

# CHAPTER I

Although realizing that it is not only extremely difficult, but even perilous to one's reputation to speak about future developments, the author has felt it as a natural consequence, following upon the first section dealing with the *historical development*, the second and third sections treating of the *present theoretical bases* and the *present forms of construction*, that the fourth section should be devoted to the *future development of oil burning*. Indeed it is hardly feasible to conduct research work in a fruitful manner without having some definite ideas as to the trend of future development and, instead of following the usual line of keeping silent about future and prospects, the author wished to expound his ideas about the subject frankly, though fully appreciating that some may not share his view or that the coming years may belie him.

The author will confine himself to suggesting the following ten points as a species of forecast of the development of oil burning:

## Industrial Oil burning

1. The application of industrial oil burning will continue to increase steadily in the next years, special attention being given to the use of *lower-grade fuels*, i.e. "heavier" oils of lower fluidity or lower C to H ratio than those at present distributed. A more active application of the cracking processes, to meet the requirements of the gasoline and diesel oil markets, will especially foster the use of *cracked residues*.

2. *Mechanical atomization*, either in accordance with the pressure atomizing or the centrifugal principle, will be mainly used for the heavier oils. The latter principle in particular will gradually come to be more appreciated than it is at present because it does not necessitate strong preheating in the burning of heavier oils.

3. The principle of *atomization by means of low pressure air* will be further developed for the medium oils, especially for the smaller types of industrial plants.

4. Industrial oil burning will gain more ground by further development of *automatic regulation of the flame capacity*, by systems activated, for instance, by the steam pressure of a boiler or the outlet temperature of the product to be heated. At

the same time, this will be combined with equipments *keeping the oil to air ratio automatically constant.*

5. The *use of steam or water vapour* in oil burning practice, not for atomization but *for correcting the atmosphere of the combustion chambers* in order to obtain soot-free combustion, especially with heavier oils, will be recognized and applied, particularly for shore installations.

### Domestic Oil burning

6. The use of *kerosenes* (CONRADSON value = zero) will show considerable increase for *cooking, emergency heating of rooms* and for *hot water supply*, mainly on account of better burner designs (the range burner type) and of the lower cost of heating as compared with city gas. Automatic equipments for hot water supply will be successfully employed.

7. Attempts to use *gas oils* (CONRADSON value = not zero) for this purpose will be seriously handicapped and, even if the cost of gas oil may be half that of kerosene, the carbon troubles encountered with the former oils will largely obstruct their use, especially for the automatic devices.

8. *Gas oils* will still be moderately used in *burners of the fire-pot or hot plate type,* either using mechanical or natural draught, for domestic central heating by hot water systems or for commercial heating in large kitchens of hotels, laundries, for bread baking, boiling butcher tubs, dyeing tubs etc. *Mechanically assisted evaporation* will maintain its predominant position over natural draught.

9. *Gas oils* will be more and more required for *automotive oil burning* due to a vivid development of high-speed Diesel engines for commercial traction, especially in Europe. Consequently there will be a *strong tendency to use heavier oils for domestic heating* and more particularly those oils which are less suitable for use in high-speed Diesel engines. The availability of gas oils as low-priced domestic fuels may probably be regarded as a temporary state of transition.

10. The development of *fully automatic mechanical oil burning equipments* will be mainly directed towards the following points:

a) A better *adjustment of the flame intensity* to the requirements of the season. The "on-off" system will be replaced by the "high-low" system or by continuous regulation.

b) *Lower cost of operation* by lower electricity consumption, more continuous operation and by using lower grades of fuel oils.

c) *Harmonious design of burner and boiler*, especially with respect to the combustion chamber, resulting in a more general adoption of *burner-boiler units*. Automatic devices for preheating the heavier oils.

d) Possibly *adding water vapour to the flame* to reduce its tendency to soot.

## CHAPTER II

### BURNERS REQUIRED FOR DOMESTIC OIL BURNING

Owing to the time available, the discussion of the future possibilities in the original Lecture IV had to be restricted to *domestic* oil burning, which subject was specially chosen because this is a young and very promising branch of the business, doubtlessly appealing to most of those concerned with oil burning.

Undoubtedly, when referring to the future developments of domestic oil burning, the thoughts of readers involuntarily incline towards the fully automatic electric equipments. Most perfect, though highly complicated, devices are imagined, ultimately becoming so cheap as to promise a period in which "everyone enjoys the comfort of oil heating". Indeed, a very important branch will be occupied by the fully automatic oil burner and more particularly by the problem of how to make the complicated systems of piping, wiring, automatic switches and valves sufficiently fool-proof *to keep the expenses of service within reasonable limits*. But it should not be forgotten that there are several other types of oil burners which are non- or semi-automatic devices but nevertheless based on sound principles, and these burners certainly also deserve attention.

In considering the numerous possibilities of application for various types of oil burners, discussed in the third section, it is well to remember that:

1. Various *countries* may be divided according to:
   a) The climatic conditions (severe or mild winters).
   b) The prices of fuel oils (taxes, import duties, etc.) compared with other fuels (coal, wood, etc.).
   c) The general state of wealth.
2. *Each country* may be divided into *regions* according to:
   a) Low prices of gas and (or) electricity.
   b) High prices of gas and (or) electricity.
   c) No available gas or electricity.
3. The *population* of each region may be divided into classes:
   a) Well-to-do people.
   b) Middle-class people.
   c) Poorer class people.
4. The *burners* may be divided according to the *purpose of heating*, viz.,
   a) For room heating either direct or indirect (central heating).
   b) For cooking.
   c) For hot water supply.

This rough division, which by no means claims to be comprehensive, gives not less than *81 different conditions for oil burning application*, which shows that it may be taken almost as certain that *there is a field to be found for any sound type of oil burner*. It is wrong to shut one's eyes to anything that is not of the fully automatic type, considering all other types as being practically knocked out of competition.

When considering the above conditions at least six types of domestic oil burners come to the fore, viz.,

1. *The fully automatic electric oil burner of the mechanically atomizing type for central heating*, installed either in a boiler originally built for solid fuel or as a complete *boiler-burner unit;* equipped with all kinds of safety devices, thermostats, etc., and preheating devices for the use of heavier oils and sometimes combined with "air-conditioning systems", thus making the room atmosphere completely independent of the season.

2. *The non- or semi-automatic electric oil burner of the fire-pot type* ("mechanically assisted evaporation") *for central heating* with manual control or thermostatic regulation of the water temperature leaving the boiler (semi-automatic). Such equipments are suitable for middle-class people living in small private houses or in flats, who do not object to attending to the oil burner now and then for adjustment. Such devices may be considerably cheaper than those mentioned under 1, but will as a rule be more "fastidious" with respect to the oil quality than their mechanically atomizing brothers.

3. *The non- or semi-automatic natural draught burner using no electricity* for central or direct heating in regions where electricity is either not available or only available at very high prices. Manual control or semi-automatic regulation by the water temperature at the boiler. Such burners may be suitable for farm houses, hot-houses etc.

4. *The small capacity oil-hearth burner* to be installed in ordinary coal-hearths and stoves, using *small quantities of electricity* (say equal to the consumption of a normal bulb) with automatic ignition and manual control or thermostatic regulation by the room temperature. To be used for direct room heating by people having no central heating plants.

5. *The small capacity oil-hearth burner using no electricity* for direct room heating equipped with manual control and ignition. To be used for direct room heating by the poorer class of people.

6. *The small capacity oil burner of the "range burner" type* to be used for cooking or hot water supply with manual control and ignition or equipped with automatic regulation by the water temperature. To be used by middle and poorer class people in regions where prices of gas and electricity are high.

In the following chapters the most important of these different types of oil burners will be discussed briefly as far as the main points of future development are concerned.

## CHAPTER III

### DOMESTIC OIL BURNERS AND THEIR RANGE OF REGULATION

#### A. *Climatic Conditions*

The range of regulation required for domestic oil burners depends entirely on the climatic conditions of the country in question and through lack of available data the author was obliged to confine the considerations of this subject to the conditions in HOLLAND, which, however, may be taken as very similar to those prevailing in BELGIUM, Northern FRANCE and ENGLAND.

In order to fix the demands for the range of regulation the following calculation has been made, based on the climatic conditions of HOLLAND during the heating season of 1932/33, which may be considered as a fair average for

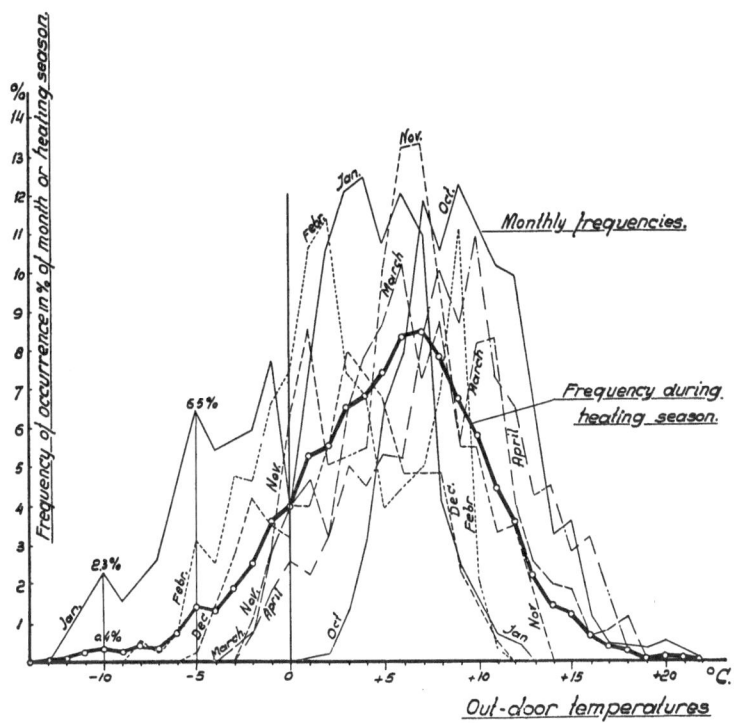

Fig. 232. Frequencies of hour-averaged out-door temperatures

that country. The graph of fig. 232 represents the *frequency of ambient temperatures* measured by the Meteorological Institute at DE BILT, a small town near UTRECHT situated in the centre of HOLLAND. These figures are not based on the values of the ambient temperature averaged for 24 hours but for each hour separately, as an oil burning equipment will be adjusted rather more according to the *hour averages* than to the *24-hour averages*.

The graph of fig. 232 shows, for instance, that during January the frequency of the temperature of —10° C. was 2.3%, that of —5° C. 6.5%, and so on; obviously all the values for one month must add up to 100%. The curves for October 1932 up to

April 1933, which period comprises a heating season of 220 days, give an average curve, indicated by a thick line, which shows that the frequency of the temperature of —10° C. during the whole of that season was only 0.4%. It may be clearly seen that this curve has that typical volcano-like shape sometimes called "chapeau de gendarme", well known from the theory of probabilities, and which may be roughly represented by the formula:

$$P = 9.4\,e^{-0.033(t-5)^2}$$

in which   P = frequency in per cent of the duration of the heating season,

e = base of natural logarithm = 2.718

t = ambient temperature in degrees Centigrade in the middle of Holland from October to April.

This formula was worked out by the author on data covering 16 years (1899–1914, De Bilt).

The next graph of fig. 233 has been compiled on the assumption that:

1) The central heating system is *put into operation as soon as the ambient temperature falls below 15° C.* (59° F.).

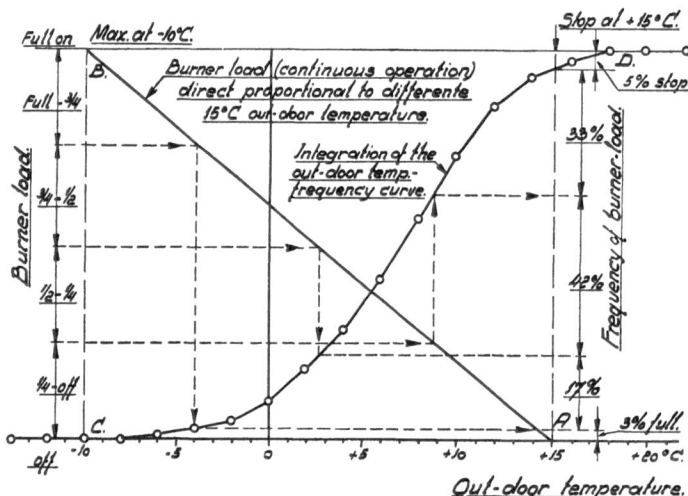

Fig. 233. Burner loads.

2) The central heating system is operating at *maximum load when the ambient temperature is —10° C. (14° F.)*, which figure applies to conditions in Holland.

3) *The load of the system is proportional to the difference between the ambient temperature and 15° C.*, and thus the load may be represented by a straight line running from A (= 15° C. ambient temperature, burner stops) to B (= —10° C., burner at full capacity).

The second curve CD represents the integration of the frequency curve of the first graph, thus at *zero °C. ambient temperature* the sum of the frequencies of all temperatures below 0° C. amounts to 10% of the total sum of the frequencies of all temperatures occurring. By combining the two curves AB and CD in the way shown in fig. 233, the following remarkable inferences may be made (see fig. 234):

| Load of the burner | Time required in % of total duration heating season | Net heat required in that time in % of total net heat delivery per season |
|---|---|---|
| 1. Between full and ¾ load . . . | 3% | 9% |
| 2. Between ¾ and half load . . | 17% | 32% |
| 3. Between half and ¼ load . . | 42% | 47% |
| 4. Between ¼ load and stop . . | 38% | 12% |

These facts, which have been largely confirmed by measurements of oil consumptions of various oil burning installations, show clearly how very important it is that (for conditions like those in Holland) *a domestic oil burner should be very economical at about half capacity*, its consumption at full load being scarcely of any importance with respect to the total season economy, which rule applies more or less to other countries too.

Moreover the graph shows that during a considerable part of autumn and spring not more than about $1/_{10}$ of the maximum heat production will be required, and consequently *the domestic oil burners should have preferably a range of regulation of 1 to 10.*

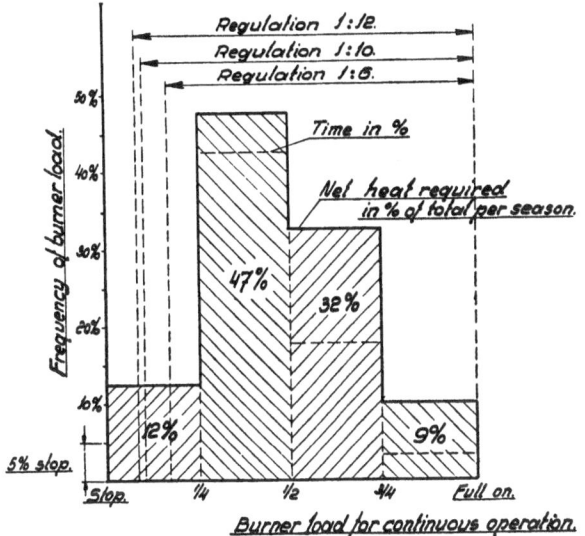

Fig. 234. Frequencies of burner loads and required heat for continuous operation.

Correctly adjusting a combustion process at *one* definite load is one thing; but to secure the efficient adjustment *for all loads between full load and stop* is quite another. It is extremely difficult to do automatically. Numerous attempts have been made to devise a satisfactory so-called *"automatic graduated"* or *continuous regulation*, but they have either failed to operate properly or led to very complicated arrangements, the main trouble being that whilst a thermostat can be made to switch an electric current *on or off*, such a simple device alone is not capable of giving a certain graduation between its two extreme positions.

## B. *"On-off" regulation. Safety Devices*

A simple solution for obtaining a sufficiently wide range of regulation was found by adopting *discontinuous regulation*, which, in fact, might be considered as a special case of the industrial method of regulating pressure-atomizing oil burners by means of the *number of burners in operation*, viz., using alternately *one* or *none*. Of course this so-called *"on-off" regulation* made it necessary to possess a *very reliable automatic ignition system*, and, moreover, *reliable safety devices* which stop the burner as soon as ignition fails or something else goes wrong.

Such devices have been mentioned before in section I, and an example is given in fig. 235 showing a *"vis-a-flame"* (Br. Pat. 277,344; 280,179 and 281,619) based on expansion of a gas volume contained by a bulb heated by radiation of the flame, thus closing an electric contact and shutting off the current as soon as radiation from the flame disappears. There are several other similar *"protectostats"* equipped with metal membranes or bimetallic spiral elements. So-called *"stack-stats"* are similar devices

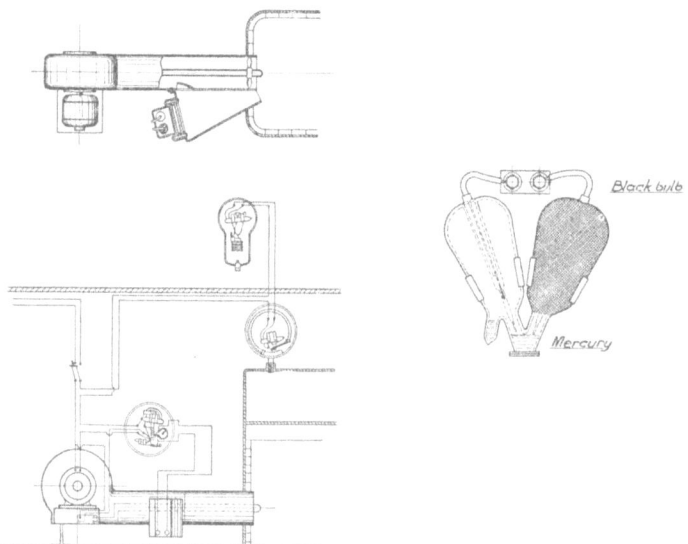

Fig. 235. "Vis-a-flame" combustion control. (McCabe, Br. Pat. 277,344 & 281,619).

acting upon the temperature of the flue gases in the stack. Descriptions of several automatic safety-devices may be found in Tapp's excellent book on domestic oil burning: "Handbook of Oil Burning" Ch. XIV.

Another *"flame guard" principle* developed by Bargeboer is shown in fig. 236 A bimetallic strip (a) is exposed to the radiation of the flame and moves the lever (c) to the end of which is fastened a second bimetallic strip (b) placed behind the first one. The movement of the end (d), being the result of the movements of both metallic strips, operates a vacuum switch (e). A special feature of this arrangement is that its "laziness" is reduced to the minimum, it being unnecessary, as in the case of most other devices, for the bimetallic strips to have considerably cooled down to operate the switch. All that is needed is that (a) shall be colder than (b), which it is as soon as the flame is extinguished.

Another safety measure, which is indispensable with automatic ignition, is that *the fan must be put into operation some moments before atomization and ignition is started* (gas clearing period), in order to remove oil vapours which may have collected in the combustion space and which are thus apt to cause an explosion. Such a delay of operation may be obtained by means of *time switches*, which close the main current flowing through the motor indirectly by means of a bimetallic element heated by a primary current closed by the thermostat switch, which heating takes a certain time, depending on the heat capacity of the element. Such arrangements, however, necessitate either two electric motors, viz., one for the fan and another for the oil pump, or an automatically detachable coupling between fan and pump or an electric-

ally operated oil valve which shuts off the oil supply to the burner during the first moments of operation.

With the Dutch N.F. domestic oil burner, which will be discussed in detail further on, *delayed atomization* is obtained by means of an air chamber in the oil line between pump and burner, which chamber takes some time to be brought under pressure, the oil supply to the burner being shut off in the meantime by an automatic valve which does not open until the correct atomization pressure (18 atm.) is attained.

Another risk of oil-gas explosions lies in the so-called *re-ignition*, i.e., when the spark, after having failed to ignite the atomized oil the first time, is automatically re-lighted for a second attempt. In the meantime considerable

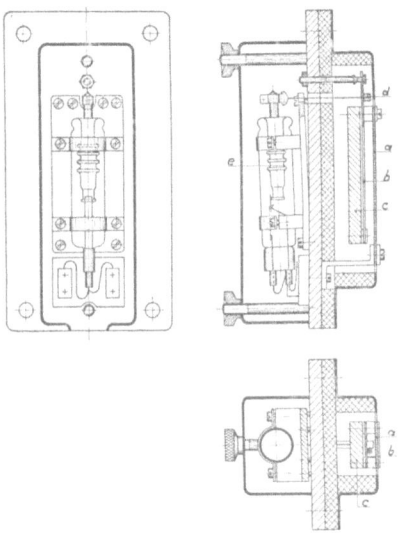

Fig. 236. BARGEBOER's flame guard.

quantities of oil may be vaporized in the hot combustion chamber, and in order to prevent explosion *the oil supply should be automatically shut off as soon as the first attempt at ignition has failed*. The second ignition should be preceded by a gas clearing period too and thus the combustion space should be ventilated by the fan before the spark and oil are started for the second attempt. Moreover, there are burners which repeat these attempts not more than two or three times, and if after that the spark still refuses to start the flame, the burner is definitely stopped. Regarding the danger, which is often exaggerated, it should not be forgotten that even with stopped burners the natural draught of a hot chimney may remove combustible vapours in a very effective way. In order to reduce the risk of oil-gas explosions, *continuous sparking* is often preferred, although this may raise the cost of electricity considerably.

No oil should be allowed to drip from the nozzle of a stopped burner, as this would not only tend to foul the tip but might also produce explosive oil gas-air mixtures in the hot combustion chambers, which might be ignited by a glowing particle of soot. In this respect devices which shut off the air inlet of the fan automatically during the stop periods are not entirely without objection, although the oil consumption will thus be reduced to some extent by preventing cold air from being drawn through the combustion chamber by chimney draught during the "off" periods.

Several patents are known for preventing fouling of the ignition electrodes by carbon deposits, which might cause failing ignition. Usually they are either withdrawn automatically by a bimetallic element, or a strong current of air is blown along them, which, while preventing oil vapours from coming into their neighbourhood, causes the spark to be more or less "projected", so to speak, into the atomized oil. An example of another method is shown by the picture

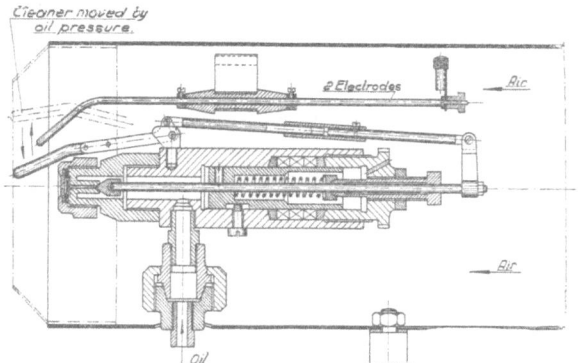

Fig. 237. Electrode-cleaning device of Br. Pat. 341,486. (LUBBOCK & JOYCE).

of fig. 237 (LUBBOCK and JOYCE) comprising a wiper moved by the oil pressure.

All these special precautions and devices are the natural consequence of solving the wide-range problem by means of *intermittent operation* and these requirements have made the fully automatic type of domestic oil burner very complicated units indeed. It will be an important point of research to make these equipments simpler without sacrificing safety.

## C. *Influence of Intermittent Operation on Room Temperatures*

Discontinuously firing a hot water boiler of a central heating plant seemed at first sight to offer no serious objections with respect to the *pulsatory delivery of heat*, as the total heat capacity of the water content, obviously being so large that it takes a considerable time to cool down or heat up, is thought to have a sufficiently equalizing effect on the resulting room temperatures. However, during the last few years the view has won ground that this is not quite true, as it is not the temperature of the *water* in the radiator in the main room that governs the periods of the burner, but the *air* temperature there, and the time necessary for the heat transfer from the water to the air constitutes an important *time lag* resulting in considerable "*laziness of operation of the system*".

This time lag is moreover aggravated by the *lack of sensitiveness of most room thermostats*, a sensitiveness of a few degrees Fahrenheit generally being considered to

be quite sufficient for the purpose. Let us consider the following example: Supposing a room thermostat with an insensitiveness of + or —1° F., it is obvious that, if the *temperature-observing element* or *"feeler"* of that thermostat has cooled down, say, from 70 to 69° F., the surrounding air must be of lower temperature in order to establish a flow of heat away from the feeler. It is clear that the larger the heat capacity of this feeler is, the longer it will take to be cooled down one degree F., and the further the air temperature will fall in the meantime before 69° F. (at which temperature, for instance, the burner is supposed to be put into opera-
tion) is attained. This phenomenon evidently brings about an additional time lag, which makes it very desirable to design temperature-feelers of thermostats of *small heat capacities* (as light as possible) and *very well access-
ible to the air*. In this respect covers may do much harm, and should at least be amply perforated.

Then, yet another lag is caused by the fact that if in due course the temper-
ature-observer of the thermostat has cooled down from 70 to 69° F. and the burner is consequently put into operation, it will take a certain time to heat the water content of the systems, during which time the room temperature

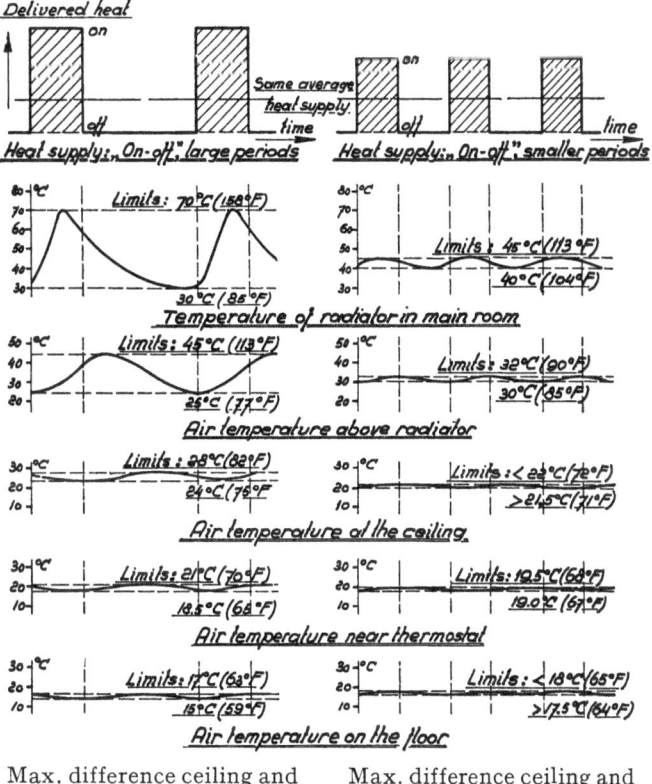

Max. difference ceiling and floor = 13° C.

Max. difference ceiling and floor = 4° C.

Fig. 238. Influence of pulsatory heating on the air temper-
ature in rooms. (Composed from data collected by BARGEBOER, De Ingenieur, 29 March 1935).

will fall still further, depending on the heat capacity of the system as compared with the capacity of the burner, until at last the radiator temperature will be raised sufficiently to compensate for the loss of heat.

A rising temperature will then follow the same process as that described for a falling temperature, as at the moment that the desired air temperature of 70° F. is attained, the temperature of the thermostat-feeler will be still lower and, in order to raise it to 71° F. (at which temperature the burner is supposed to be stopped again), the surrounding air must be raised to temperatures higher than 71° F. in order to procure a flow of heat into the feeler. But if the burner is stopped at 71° F. "feeler temperature", this will not prevent the *air* temperatures from rising even still further beyond 71° F., a large amount of excess heat being stored in the hot-water content at that moment.

Figure 238 may give a good impression of the temperature deviations and fluctuations occurring with "on-off" regulated systems; this figure was composed on the basis of data collected by BARGEBOER. It shows at the same time that intermittent heat supply with *shorter* periods of *less* intensity reduces the fluctuations considerably ("De Ingenieur", 29 March 1935).

Thus we see that *a sensitiveness of 1° F. above and below the desired temperature does not at all mean that the air temperature will move between these two limits*, as in fact big heating systems with large water contents (old-fashioned radiators) equipped with comparatively large, "over-rated" burners and thermostats of large heat capacities may show limits as much as *5° F. above and below the desired temperature.* To keep the room temperature constant within 1° F. it is necessary that *the sensitiveness of the corresponding thermostats should be a few tenths of a degree Fahrenheit.*

There are room thermostats on the market which are claimed to possess that degree of sensitiveness. This claim, however, is often based on a curious physical error, the "sensitiveness" being caused by the electric heat radiated by the current flowing through in the "on" position. The electric current, flowing through mercury and connections, encounters sufficient resistance (sometimes even purposely so) to heat the surroundings, thus causing the bimetallic element to switch off after a certain length of time. In fact the regulator is then a kind of "time switch more or less dependent on the air temperature", as the latter governs the switch periods by the rate of dissipation of the heat electrically supplied. This kind of thermostats may be easily recognized by putting an insulating layer between the mercury switch and the bimetallic element, when they suddenly show a very greatly reduced *true* sensitiveness. The same happens if the thermostat is operated with a relay and only very weak currents flow through.


D. *Air temperatures and Draught in Heated Rooms. The "Cold 70" Phenomenon*


The most common method of room heating — direct heating by stoves excepted — is the central hot water heating system with radiators in the rooms. The name "radiator" is strictly speaking less correct than "convector", because the radiation effect is almost negligible, the major part of the heat being transferred by convection. Convection, however, implies air flow, which is sometimes felt as draught. Consequently modern heating technique shows a tendency to use more radiation and less convection ("panel heating"). As far as the oil burners are concerned, the question arises as to *the degree in which discontinuous oil firing may decrease the comfort of heated rooms*. This question will be discussed now.

First of all it may be said that, although it is very easy to speak of *the* average room temperature, it is very difficult to measure it anywhere, as the air in a heated room is a very complicated complex of different temperatures causing all kinds of fluctuating currents and draughts. As a rule the air layers near the floor are much colder than those at the ceiling, differences of 7 to 10° F. not being at all rare. During the *heating-up period* the radiators, usually placed under the windows, will there cause a strong upward current of hot air, which comes down somewhere in the middle of the room, with the result that the head of anyone sitting in the room is

*Is this your idea*

**of "automatic" regulation in an oil fired job?**

What's wrong with *this* picture? An hour ago the room thermostat hit 72 degrees—its upper limit. The oil burner was then stopped. Now the radiators have cooled and instead of the air currents from the window sweeping up from the radiator and keeping the room comfortable they are dropping down to the floor. The thermostat still records seventy-one, because the cooling air-current haven't yet reached it.

The thermostat will tell you that there is nothing wrong with this picture. But the woman will tell you that the room is chilly and uncomfortable. Both are right!

The room *is* seventy-one degrees, but it is *also* „chilly"—due to the movement of air currents. Temperature can no longer be considered an accurate index to comfort

The hundreds of people who have been complaining about the alternate chilly and stuffy conditions that go with so-called automatic regulation are at last getting recognition The condition has been said by one heating engineer to be the most serious obstacle in the path of oil-burner or gas-burner progress. He refers to this chilly condition as „cold 70!"

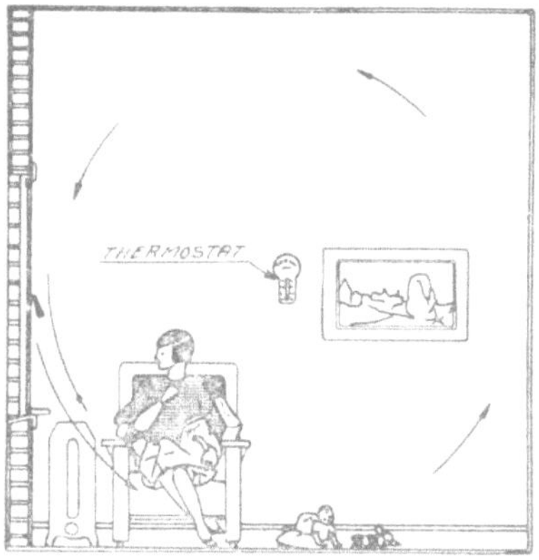


**What is „cold 70"?**

„Cold 70" is actually the aftermath of over-heating! It is due to the lag between radiator heat and its effect on the room thermostat Basically it results from failure to compensate the upper limit of boiler or radiator temperature for outdoor temperature. The „boiler limit" sought to correct this, but even if set weekly, it would be a practical failure for the simple reason that weather won't obey the calendar!

But now comes the Raymond Duo-Stat—the device that can wipe out the complaints that almost all oil burner dealers have received. This device is simple and inexpensive. It is as dependable as it is simple. It is the one way of compensating the heating system automatically and unfailingly throughout the entire heating season.

The Duo-Stat gives the job of regulating temperature to the room thermostat up to the point where weather compensation is necessary. Then it goes into action and stays in action just as long as limitations are necessary for comfort

**F. I. RAYMOND CO., River Forest, Ill.**

Fig. 239. "Cold 70" advertisement of RAYMOND DUO-STAT (Fuel Oil Journal).

exposed to a higher temperature than his feet, which causes a disagreeable *feeling of congestion*. During the first part of the *stop period* there will continue to be a draught in the same direction, due to the "laziness" of the heating system, but after a length of time the *decreasing radiator temperature* will not be able to maintain the upward current of air any longer, resulting in a *draught in the reversed direction coming from the cold windows and covering the floor with a layer of cold air*. This situation, taking place during the second part of the stop periods, causes a disagreeable *shivery feeling*, owing to the occurrence of air currents as indicated by fig. 239, taken from an advertisement of the RAYMOND DUO-STAT (Fuel Oil Journal, July 1928). This phenomenon is often called *"cold 70"* because, while people positively feel uncomfortable, if they look at the thermometer — which usually hangs 5 to 6 feet above the floor — they see that it nevertheless registers about 70° F., their conclusion then being: "Well, I thought there was something wrong with the heating system but I see I was mistaken". If the thermometer were on the floor, however, it would tell a different story.

Fig. 240   Views of BARGEBOER's thermostat for domestic use
(cover and mercury switch have been taken off).

BARGEBOER, the designer of the N.F. burner for domestic use, who has made an exhaustive study of this subject (De Ingenieur, 29 March 1935), arrived at the conclusion that in order to avoid "cold 70" it was not only necessary to adopt *graduated regulation* for oil firing instead of the pulsatory "on-off" system, or at least an "on-off" regulation with a *graduated regulation of the "on"-intensity of the fire*, but also to use room thermostats which are extremely sensitive. In order to increase the sensitiveness of a bimetallic element to be used for observing the room temperature, he developed an ingenious device by applying axial compression to the bimetallic strip by means of a pair of springs which are adjusted just a trifle below the critical collapsing stress of the strip. Thus it is claimed that a change in temperature of only *one-tenth of a degree Fahrenheit* will cause the strip to bend either to one or to the other side, resulting in a considerable momentum quite sufficient to operate an electric mercury switch. Fig. 240 is a photograph of this thermostat.

Another attempt to meet the "cold 70" phenomenon is illustrated by fig. 241 representing the HEIL COMBUSTION

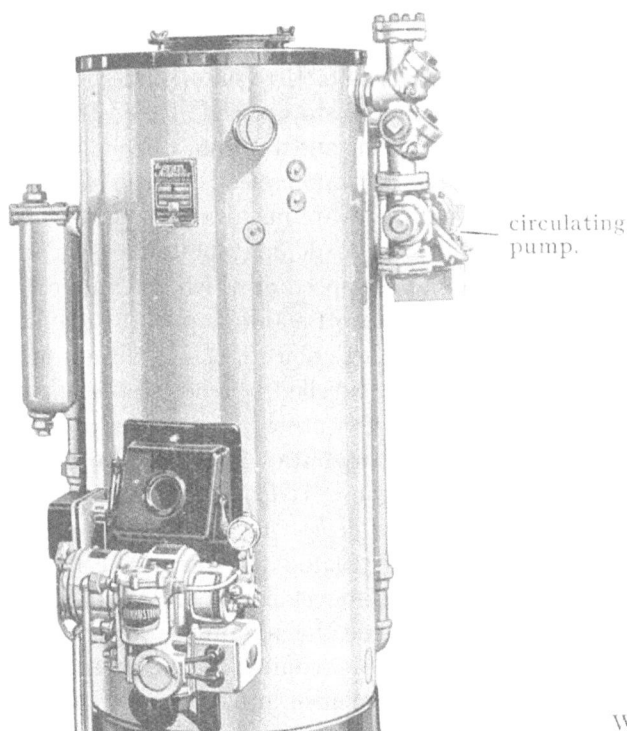

circulating pump.

Motor

Water

Fig. 241. HEIL COMBUSTION burner-boiler unit.

Fig. 242. THRUSH water-circulating pump.

burner-boiler unit, the operation of the burner being governed, not by the room thermostat, but by the temperature of the water leaving the boiler, or sometimes by a combination of the incoming and outgoing water temperatures, whilst the *room thermostat switches on or off an electric pump for circulating the water* through the heating system. This is a feature which reduces the laziness of the system, as a falling room temperature is in the first place immediately counteracted by a more intensive water circulation which moves the stored heat from the boiler practically without delay to the place where it is required. In a second stage the oil burner is put into operation as soon as the water temperature in the boiler drops below a certain limit.

This system may be easily installed, as electric pumps such as shown in fig. 242 (THRUSH circulating pump) are available on the market. It will be clear from the above, however, that while this expedient may do good, it does not attack the evil at its roots.

### E. *"High-low", "High-low-off" and other semi-graduated Regulations*

It should be mentioned, regarding the objection to the "on-off" regulation of domestic oil firing just discussed, that the longer the stop periods lasted the more serious these troubles were; that is to say, during spring and autumn, when only small amounts of heat are required. This is especially objectionable if for the sake of a marked acceleration ("pick-up") over-rated oil burners are installed, which of course further increases the duration of the stop periods. The reader may obtain a clear idea of the matter if he imagines a 50 H.P. motor-car the speed of which is regulated at, say, 10 miles per hour by applying alternately full throttle and stop to the engine.

The step taken by American designers to meet these objections was the adoption of the *"high-low" regulation* which, on the one hand, levelled to a certain extent the fluctuations of the pulsatory heat delivery but, on the other, necessitated efficient adjustment of the combustion process for two capacities instead of only one. This again was not found to be so easy, it being necessary, of course, to adjust the *low fire equal to or lower than the lowest amount of heat required.* Consequently it is not astonishing that the first "high-low" regulated automatic oil burners showed rather high oil consumptions, which led to the development of a third system, called the *"high-low-off" system.* This, although having the advantage of permitting low adjustments *higher* than the lowest amount of heat required, necessitated rather complicated wiring schemes with a considerable number of automatic switches and relays.

In order to solve the problem of efficient combustion both at high and at low load, a modern development may be mentioned which comprises a *double-nozzle unit,* such as made by the PROGRESSIVE OIL BURNER CO. (NITEK), see fig. 243. This unit, which is arranged for "high-low-off" regulation, consists of two burner units built together, one for high and the other for low fire, put into operation by means of electrically operated oil valves. The "low" burner is adjusted at half the capacity of the "high" burner and is put into operation as soon as the desired temperature is reached by means of the "high" burner. The oil supply to the large burner is shut off electrically and that to the small burner opened, while at the same time the speed of the electric motor is reduced from 1400 to 700 revolutions per minute, which reduces the quantity of air delivered by the fan and the quantity of oil delivered by the pumps roughly to half. Consequently, the small burner will operate with about the

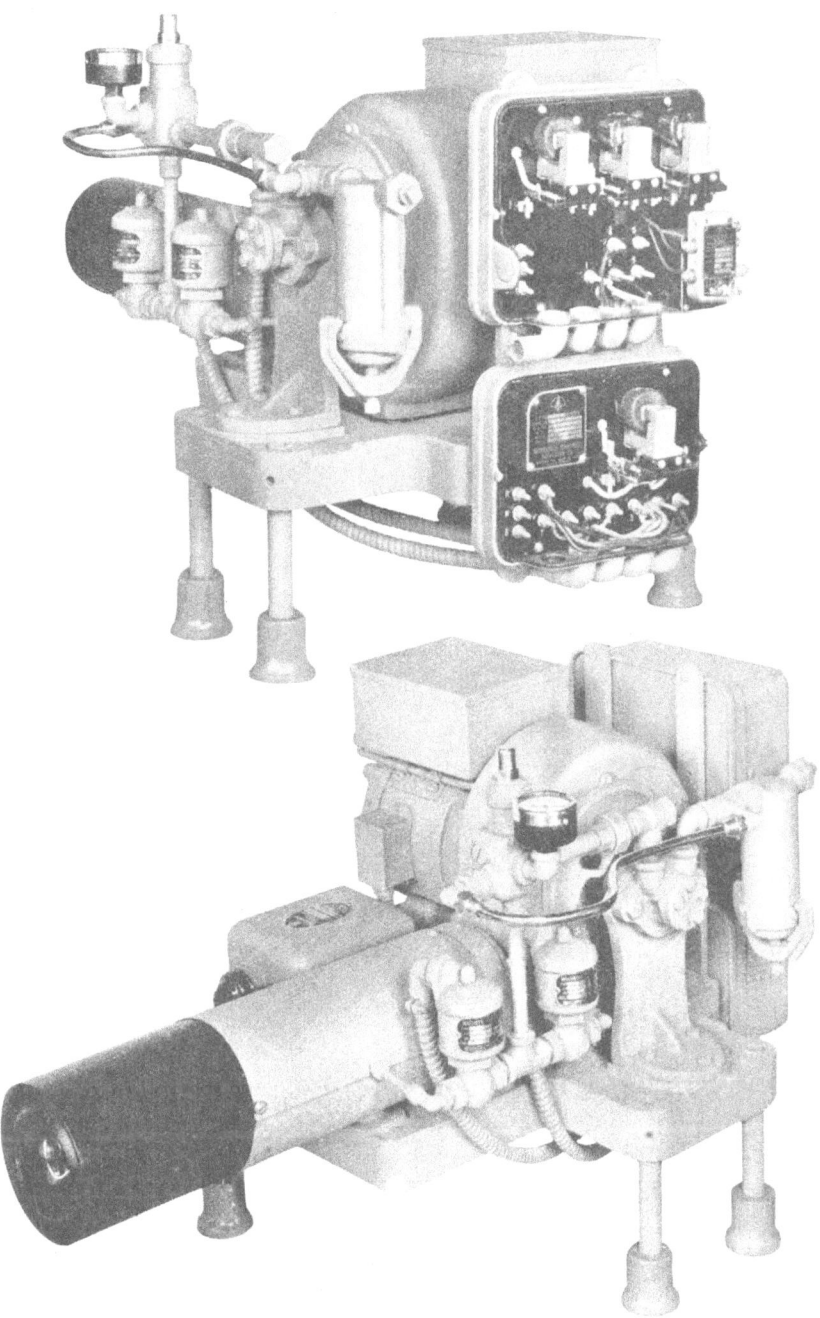

Fig. 243. The PROGRESSIVE (NITEK) double-nozzle unit.

same oil to air ratio as the large one, which solves the problem of efficiency in a conclusive way. On the other hand, it need scarcely be said that this is a rather

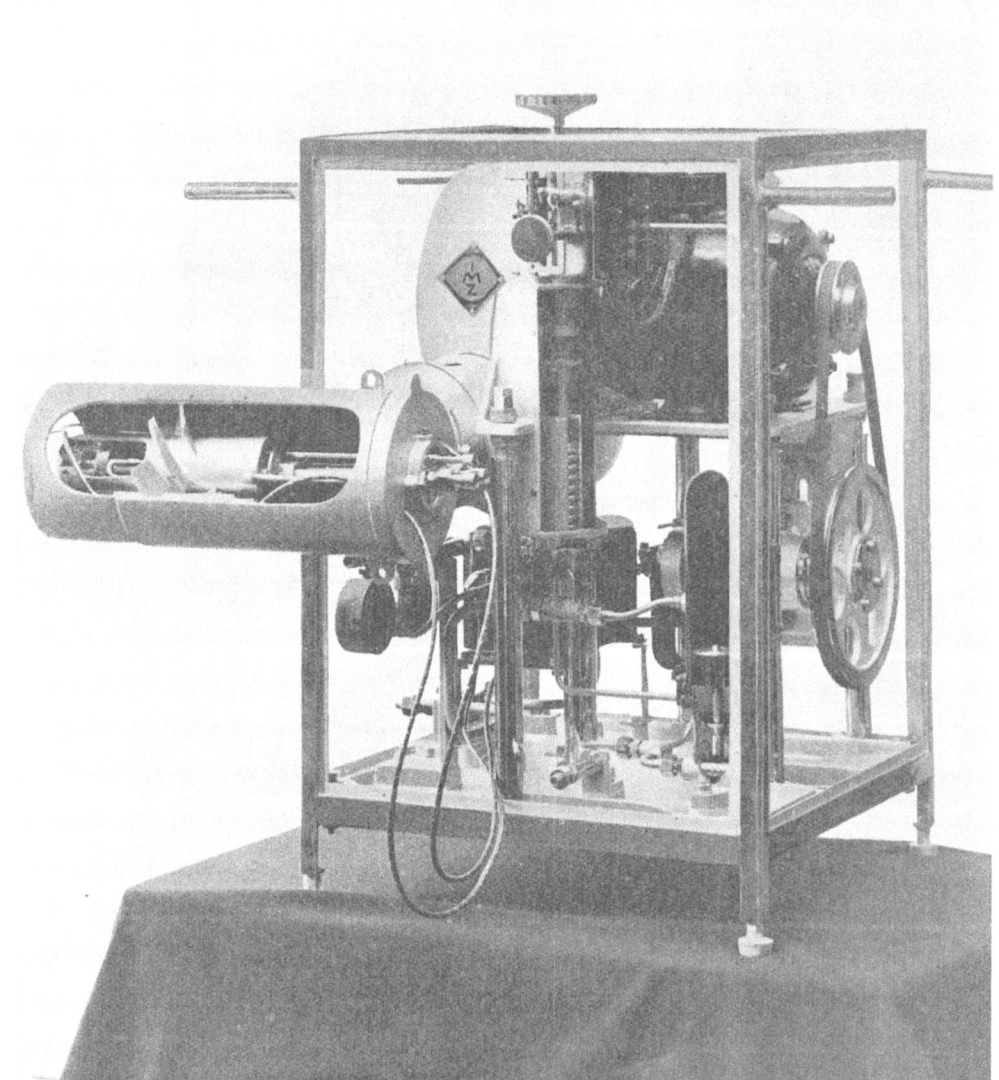

Fig. 244a. Demonstration model of the N. F. automatic domestic oil burner.

expensive solution of the problem, but at the same time it shows how important it is to comfort and economy, as the designers, in order to overcome the drawback to "on-off" regulation, apparently did not shrink from going even as far as this.

With the *N.F. burner* the intensity of the oil flame is automatically adjusted for constant room temperature by means of the thermostat of fig. 240, the principle of

Fig. 244*b*. Demonstration model of the N. F. automatic domestic oil burner.

which has been described before. The regulation is on the "on-off" principle, but the "*on*" *capacity*, instead of being constant, *is automatically readjusted according to the heat desired.*

The burner is started with an "on-off" regulation, but as soon as it is stopped for the first time the oil burner is automatically readjusted by an auxiliary electric motor in such a way that when it is put into operation again it will burn with a flame of somewhat smaller capacity than it did during the first burning period. During the second stop period the burner is readjusted again at a slightly smaller capacity and consequently the periods of burning will gradually last longer whilst the stop periods will gradually decrease, until at last the burner is exactly adjusted to the heat required. If in that case the burner burns *above* its *lowest* flame capacity limit, the flame will burn *continuously*, and if the heat required is *less* than that of the *lowest* flame capacity, it will burn *intermittently with oil flames of the lowest capacity limit*.

Figs. 244*a* and *b* represent general views of this interesting burner equipment, in which the atomization of the oil is obtained by means of the recirculation principle which was discussed in Section III. Automatic adjustment of the air damper by means of the return oil pressure ensures a nearly constant air-oil ratio for all flame capacities.

Fig. 245 gives the Swiss CUÉNOD oil burner equipment, which has *graduated regulation* by means of air operated servomotors. The air pressure required is about 50 cm. (20 inches W.G.) water gauge, which is rather high for domestic burners. A description may be found in D.R.P. 398,651 and in "Le Génie Civil" 7th Febr. 1931 (GREBEL).

It may be clear from the foregoing that, although superficially it may seem to the outsider that the present fully automatic domestic oil burning equipments are practically as perfect as they can be, the experts will certainly not agree with this view, for as a matter of fact, there are several problems still awaiting *simple* solutions. Thus, the avoidance of "cold 70" belongs partly to the task of the heating engineers (e.g. heating by radiation of panels instead of by convection of radiators) and partly to that of the oil burner designers.

An important object to be aimed at in the future development of fully automatic oil burning is to *reduce the number of service calls* per burner and per season. While it may perhaps be rather difficult to make a complicated unit which answers every demand with respect to comfort, it will be far more difficult to make such devices *also commercially profitable*, the main point being a serious risk that the initial profits of the manufacturers are gradually eaten up by the cost of subsequent service, owing to the complicated designs requiring highly paid, skilled attendance.


F. *What temperature must be taken as a basis for automatic regulation?*

The principal methods of automatic regulation are the following: —

*a) Keeping the temperature of the water leaving the boiler constant.* This method is an heritage of the coal-fired boiler. The principle is entirely wrong, as that *water*

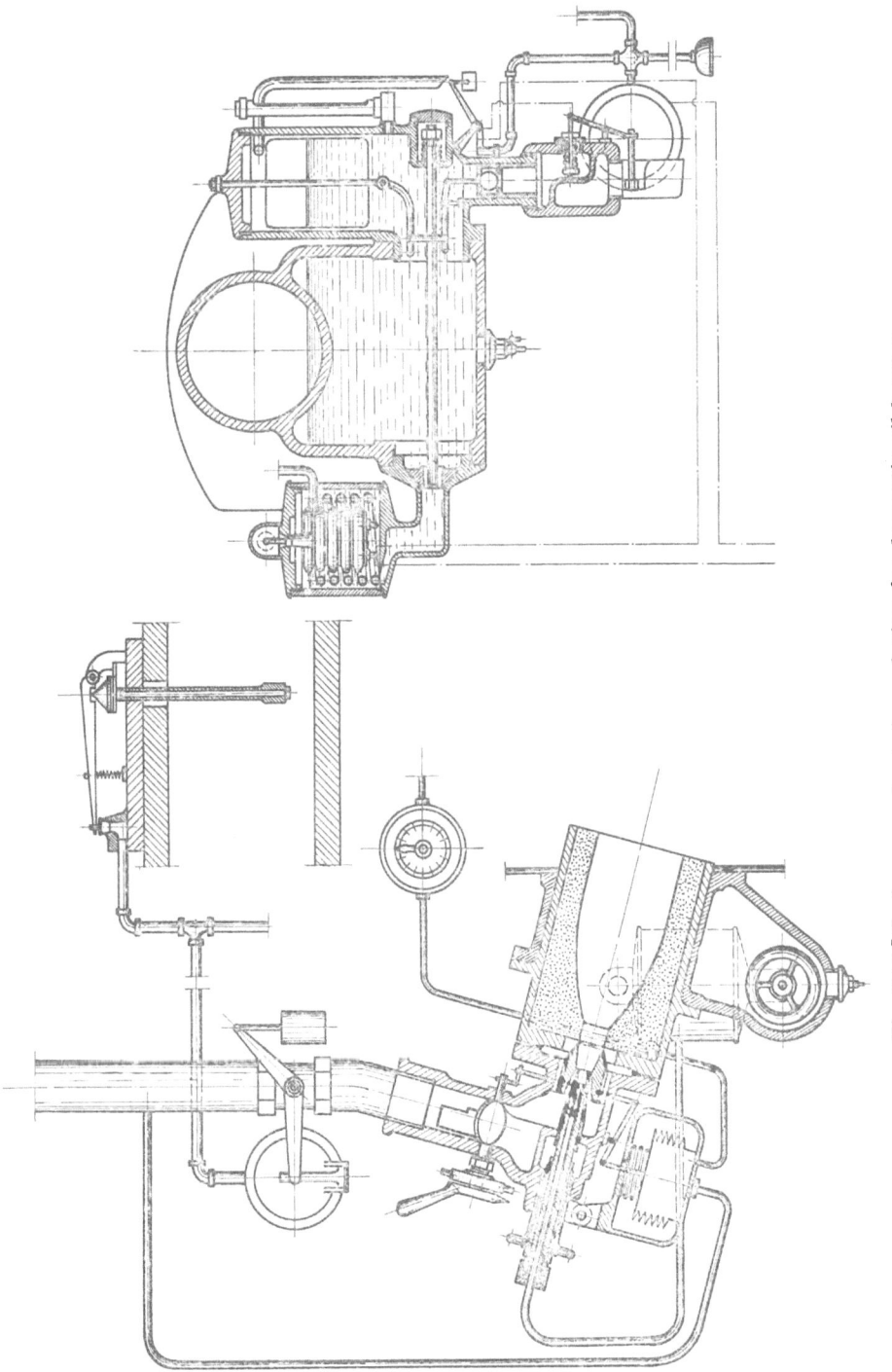

Fig. 245. CUÉNOD graduated regulation for domestic oil burner.

*temperature, instead of being constant, must move up and down in reversed relation to the out-door temperature.* Consequently the burner must be readjusted by hand as soon as the out-door temperature changes and for this reason this method is often called: "semi-automatic", although a better name would be: "quasi-automatic".

*b) Keeping the temperature of the water which is returned to the boiler constant* is but slightly better than (a) because, if the out-door temperature drops, that water temperature will decrease too, because the house will be cooled. As, however, the maximum temperature fluctuations allowable for the house are very limited, viz., only a few degrees F., the temperature variations of the return water will certainly be too small for efficient regulation. To improve this point, the return water line is sometimes purposely exposed to the out-door temperature, which makes conditions a little better, but involves a considerable waste of heat.

*c) Regulating by means of a combination of the water temperature before and after the boiler* is an improvement on (b), necessary because, with that method (b), the temperature of the water leaving the boiler is not limited and, in order to prevent boiling, the above combination is effected, usually by means of adjustable levers.

*d) Keeping the air temperature in the main room constant* is based on a sound principle; however, it has its drawbacks too, namely, if the windows in that main room are opened in the early morning by the servants, the whole heating system may be set up and the house overheated. As a second point may be mentioned the difficulty of finding a correct place for the temperature observer, where it is not exposed to fluctuating currents of alternately cold and warm air.

*e) Keeping the air temperature in every room constant* is theoretically an improvement on (d), which makes the heating system independent of the conditions in one particular room. It, however, includes several practical difficulties, because it will be necessary to install thermostats in every room to *regulate the radiator valves.* Although such devices are available on the market (see fig. 246 FULTON SYLPHON regulator, TAPP, "Handbook of Oil burning" 1931, page 190, STOHN "Temperaturregler" pag. 26 (1933)), they

Fig. 246. FULTON SYLPHON radiator valve.

have found no general application up to the present, mainly because they are too expensive if sufficienty sensitive. Moreover, the place of a radiator valve is a very unsuitable position for a room thermostat feeler.

*f) Regulating the temperature of the water leaving the boiler by the out-door temperature by some predetermined reversed relation* represents a sound principle, although it does not keep the room temperature constant if, for instance, the windows are opened. The radiator temperatures are based on average conditions of draught,

ventilation etc. of the rooms and are constant as long as the out-door temperature does
not change.

A well-known example of this method of regulation is shown by fig. 247, representing the DUO-STAT developed by the RAYMOND Company. According to this
principle the burner is put on or off by a mercury switch operated by a Bourdon
spring, the inside of which is connected to *two* temperature bulbs at a time, one being
exposed to the out-door temperature while the other is kept at the temperature of the

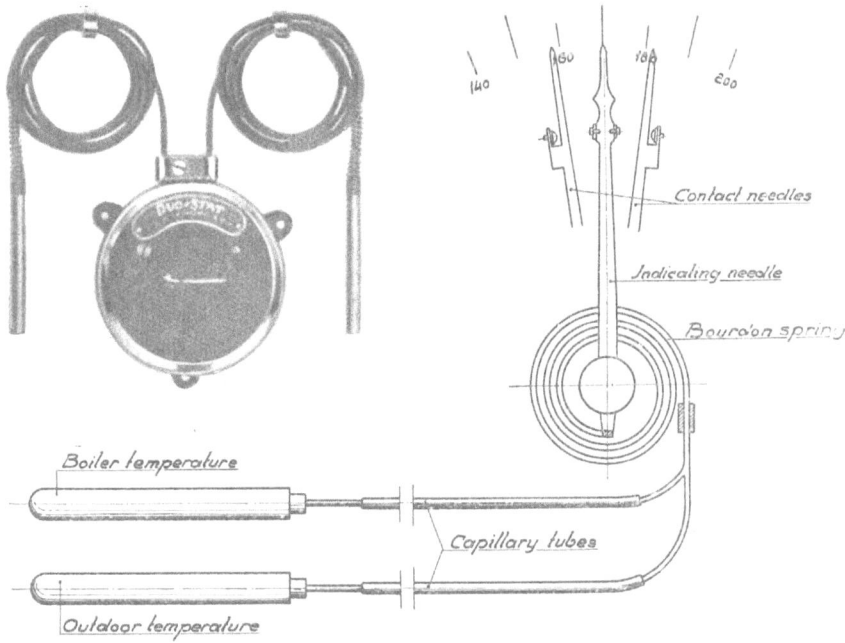

Fig. 247. RAYMOND DUO-STAT principle.

water leaving the boiler or entering an important radiator. If the out-door temperature falls, the pressure in the Bourdon spring or bellows will be reduced, this causing
the burner to operate, but the resulting rise of the water temperature will gradually
increase the pressure again until a point is reached at which the burner is stopped.
Thus a suitable equilibrium may be established between the out-door temperature and
the hot water temperature in some predetermined relation. An important advantage
of the DUO-STAT regulation is that the heating system is independent of the temperatures in the rooms and consequently *it cannot be upset by opening windows somewhere,*
as may happen with room thermostatic control.

It is very easy to speak of *the* out-door temperature, but extremely difficult to
find a satisfactory place to record it, this being one of the most difficult points when
installing a DUO-STAT. Moreover, the loss of heat from a house is not exclusively
dependent on the out-door temperature but also on wind, rain and sunshine, thus a

house when wet will lose more heat than when it is dry, even if the out-door temperature is the same in both cases. Some sunshine falling through the windows from time to time may reduce the heat loss of a house considerably without causing any perceptible change of the ambient temperature. In this respect room thermostats may be better than regulating systems based on recordings of the out-door temperatures, provided the former are sufficiently sensitive.

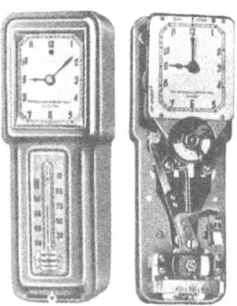

Fig. 248. MINNEAPOLIS-HONEYWELL clock thermostat.

g) *Clock Thermostats.*

An ordinary room thermostat must be readjusted in the morning and evening if it is desired to have the house cooler during the night than during the day. This may be avoided by using a *clock thermostat* (see fig. 248 MINNEAPOLIS-HONEYWELL) which comprises adjustable cams moved by clockwork and which can be set to regulate the thermostat for different temperatures at given times. If such a clock thermostat is set, for instance, to maintain a room temperature of 70° F. from six a.m. to eleven p.m., and 60° F. for the remaining hours, this will not only contribute considerably to the comfort of oil burning — automatically ensuring warm rooms for dressing in the morning — but at the same time will provide for economical oil consumption.

CHAPTER IV

DEVELOPMENT OF AUTOMATIC DOMESTIC OIL BURNERS FOR HEAVIER FUELS

It was mentioned in the first Chapter that an important future development will probably be the adaptation of domestic burners to the use of heavier oils, say, up to 6.5° ENGLER or 200 seconds REDWOOD I, while the majority of the present burners do not go further than, say, 2.5° ENGLER or about 75 seconds REDWOOD I. The reason why this problem may become urgent lies in the fact that the gas oils (viscosity about 1.5° ENGLER or 42 seconds REDWOOD I), which are now the main constituents of most of the domestic fuel oils at present distributed, will be increasingly used for running high-speed Diesel engines for traction.

At present there are already two countries, viz., ENGLAND and FRANCE, in which there has been an appreciable development of domestic burners for heavier fuel oils, resulting in several automatic equipments capable of burning "200 sec. fuels", sometimes even up to 400 sec. The following makes may be mentioned: CLYDE, AUTOMESTIC, QUIET MAY (the English firm), BRITISH OIL BURNER, KALOROIL, LAIDLAW-DREW, URQUHART-VICTORY and the French C.A.T. and S.I.A.M.

Of course such heavier fuels must be preheated, usually to about 65° C. or 150° F., depending on the fluidity, before they can be atomized efficiently. Whereas preheating offers no serious problems with continuous industrial oil burning, especially if waste steam is available, things are far more complicated with domestic installations which are run automatically and discontinuously.

One of the main points to be considered with preheating devices for automatic installations is that the *atomizer must receive fully preheated oil at the very moment it is put into operation* and consequently there must be a certain stock of heated oil present as near to the burner tip as possible, even after a long quiescent period. In British Patent No. 339,314 LUBBOCK and JOYCE describe a thermo-syphonic system for fuel heating for this purpose, whilst, in British Patent No. 403,522 (CLYDE) an expansion chamber is used.

With domestic installations preheating is usually effected by means of electricity, which usually is a rather expensive form of energy and the cost of which must be recovered by the lower prices of the heavier fuels. In order to reduce the cost of electricity, preheating must be adjusted to a temperature which is just sufficient for the purpose, generally the temperature at which the fuel oil has a viscosity of about 2 or 2.5° E. Sometimes a part of the heat required is obtained from the hot water

Fig. 249a. AUTOMESTIC unit for heavy fuels.

system by leading the oil through a coil, the balance being supplied by electricity, a method which proves to be very economical. Yet another system uses the heat radiated from the boiler front for primary preheating.

Oils over 200 seconds REDWOOD viscosity, moreover, require warming coils to be fitted to the storage and service tanks. This, however, is scarcely practicable for domestic installations. As a rule, gravity supply should be avoided when using heavier oils and it is preferable to employ a circulating pump which draws directly from the main tank and circulates the oil through a ring main, which is provided with heaters and strainers and from which the oil to the burners is to be taken.

Preheaters should be made of *corrosion-resisting metals*, thus excluding any brass or copper parts which might come into contact with the oil, this because the corrosive action, especially of the heavier oils which usually contain more sulphur than the lighter oils, increases rapidly with the temperature. For strainers copper gauze must be avoided, *monel metal gauze* being best suited to the purpose. *Aluminium* may be used in most cases, although it may corrode with certain oils. Its high heat conductivity and its low weight make its use very attractive, especially for the thermostat "feeler" pocket. Thus the "laziness" of an aluminium device may be considerably less than one made of stainless steel. For the same reason the mass of the preheater should be kept as small as possible in order that its own heat content should be small.

The heating surfaces of electric preheaters should be designed so that the heat output is not in excess of about *10 Watts per sq. inch*, in order to avoid carbonization of the oil and pitting of the heating surfaces. The heaters should be fitted with *thermostatic control* and with a safety switch (fusible element switch), which cuts out the electric current as soon as the temperature rises above a certain limit, such as might occur if the oil supply were stopped for some reason.

Fig. 249b. LAIDLAW-DREW burner for heavy fuels.

Some further points to be considered with preheaters are:

1. The oil should be *filtered after preheating* in order to remove water and foreign matter which remained in suspension in the cold oil but which settle out as soon as the viscosity is low enough. Filters should be provided with sufficiently large chambers for collecting water and dirt and, if possible, in duplicate in order that they may be cleaned without interruption. The gauzes of pressure strainers should be strongly supported to prevent collapsing when choked, and fitted with a spring-loaded by-pass valve. The multiple disc type of strainers, such as is shown in fig. 91, is to be recommended.

2. Provision should be made for the *expansion of the oil* in the preheater when being heated, so that no drip occurs at the nozzle or leakage anywhere else. It should be remembered that this expansion may give rise to extremely high pressures and that in a reservoir completely filled with oil and tightly shut several hundreds of atmospheres may be developed merely by heating it moderately, say some 20 degrees.

3. *No soft soldered pipe or other joints* should be used on the heater or between the heater and the burner tip.

4. *Nozzle and regulating valves should be kept at a suitable temperature* in order to

ensure their proper operation. If the oil under a valve or a membrane is allowed to cool, it may interfere with their movements. With the preheater of the KALOROIL

Fig. 249c. CLYDE burner unit for heavy fuels.

BURNERS Ltd., which is claimed to burn fuel up to 400 sec. REDWOOD I, the nozzle, filter and shut-off valve are placed in an extension which is made at the top of the heater; consequently these parts are always kept warm by the oil itself. As the burner controls are so arranged that the burner cannot start until the oil is at the right temperature, the jet is always at the correct temperature, and valve and filter are protected from cold, gummy oil which could affect their operation.

5. An *efficient shut-off valve* between the nozzle and the outlet from the heater should be provided in order that no oil shall leak from the burner tip and collect in the combustion chamber. This valve is usually electrically operated.

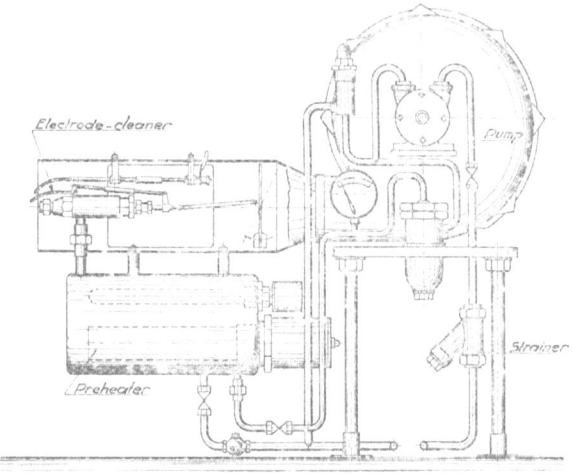

Fig. 249d. C.A.T. fully automatic domestic burner.

As regards the preheating temperature required for a certain oil, no exact figures can be given, as this depends greatly on the type of burner and furnace. Usually a viscosity of 7° Engler may be considered a high limit for *industrial* oil burning but, with favourable conditions, such as in a well-designed combustion chamber of high temperature, perfect combustion may be obtained both with pressure-atomizing burners and with centrifugal burners for industrial use. A viscosity of 2.5 to 4° Engler is common for *industrial* oil burning, whereas for *domestic* oil burning 2 to 2.5° E. is generally preferred.

Finally a remarkable fact may be mentioned, viz., that with pressure-atomizing oil burners the capacity first increases with rising preheating temperature but attains a *maximum*, usually at about 7 to 8° Engler, after which the capacity decreases with increasing temperature and decreasing viscosity. An example of this well-known phenomenon is mentioned by GUILLERMIC in his book "Le Chauffage par les Combustibles Liquides", Chapter X, page 182, giving the following figures:

| Preheating temp. | Viscosity | Capacity |
|---|---|---|
| 40° C. | 12.0° E. | 168 kgs./hr. |
| 50 | 9.0 | 192 |
| 60 | 6.7 | 201 |
| 70 | 5.0 | 192 |
| 80 | 3.9 | 182 |
| 90 | 3.0 | 173 |
| 100 | 2.3 | 166 |

Another example is mentioned in the article "Higher Viscosity raises Capacity" by GLENDENNING, published in the Fuel Oil Journal of April 1935. This shows that it is uneconomical to preheat the oil higher than to the lowest temperature which is just sufficient to obtain good combustion, not only because of the heat thus uselessly applied but also in view of the burner capacity.

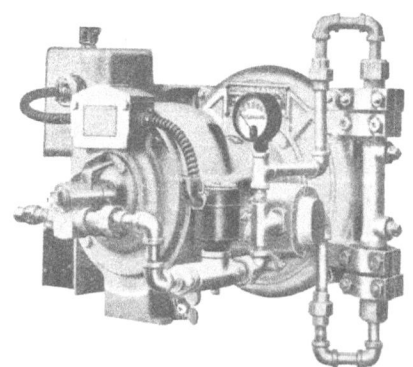

Fig. 249e. The ENTERPRISE horizontal rotary unit with electric preheating for heavy fuels.

On the other hand, with domestic oil burning a larger margin of safety for good combustion will be required than for industrial oil fires which are continuously watched; this is the reason why, for domestic oil burning, the oil is usually preheated to temperatures corresponding to 2 to 2.5° Engler viscosity.

It need hardly be said that the oil must never be preheated to temperatures at which the *first vapour bubbles* would be evolved under atmospheric pressure, as this would spoil the operation of the whirl chamber and tend to cause irregular atomization and incomplete combustion. With respect to this point, however, attention may be drawn to the fact that the same trouble may occur if the ordinary heavy fuel is casually mixed with some light gas oil or kerosene used for flushing choked lines or cleaning tanks or the like.

In the following pictures of oil burners for heavy fuels the preheaters may be easily recognized: CLYDE (fig. 249a), AUTOMESTIC (fig. 249b), LAIDLAW-DREW (fig. 249c) which are all English makes, the French C.A.T. (fig. 249d), and the American EUTERPRISE (fig. 249e).

---

# CHAPTER V

## FUTURE DEVELOPMENT OF THE FIRE POT TYPE

In Chapter VII of the First section, where the development of Domestic Oil burning in America was dealt with, it was explained that the bad results obtained there about 1923/24 with natural draught burners of the "mushroom" type resulted in the development of a new type using so-called "mechanically assisted evaporation". This *fire pot type* (or vaporizing pot type) using "forced draught" was an improvement on the mushroom type, because its operation was no longer dependent on the irregularities of natural draught and many disappointments could thus be prevented. On the other hand, the oil fed by gravity was still prepared for combustion by vaporization.

Figures 57 (LACO), 59 (NORTHERN), 58 (BOCK), 60 (HAAG) and 61 (PRIOR) are sufficiently clear to show the principle (also see figs. 230 and 231). The oil is admitted by gravity into a fire pot, usually of refractory material, in which it evaporates by the heat of the flame itself. A strong current of air is blown on the oil in order to obtain thorough mixing with the oil vapours. The oil is partly evaporated and partly cracked, so it may leave a carbon deposit behind if unsuitable oils are used. The fire is manually adjusted by pinching the oil valve and air damper, the low fire limit being fixed by the temperature of the fire pot. To prevent flooding if the flame should happen to be extinguished, there is a drain pipe fitted to the bottom of the fire pot leading to a *trip bucket* (see fig. 62) comprising a valve which shuts off the oil supply as soon as a certain weight of oil has collected in a bucket. A neatly designed unit, which contains 1) a regulating float for constant level, 2) a safety trip bucket, 3) regulating valves and 4) a strainer, all in one, is the DETROIT device shown in fig. 250.

It is certainly not correct to consider the domestic oil burner of the fire pot type exclusively as a competitor of the fully automatic atomizing type, as in fact there are two different classes of people offering separate prospects for each type. The *fire pot type* might be considered as the *cheap, popular edition of the fully automatic burner*, its main advantages being the elimination of the coal shovel, dust and dirt, whilst it is brought within the reach of a large class of people living in moderate circumstances by low cost, which is secured by sacrificing to some extent the refined comfort of automatic regulation.

Fire pot burners are often claimed to be "fully automatic" if they are able to keep the temperature of the hot water leaving the boiler constant, but this system should be called *semi-* or even better *quasi-automatic*, because, in fact, *this temperature should not be kept constant but should move up and down with the heat loss of the house.* Consequently such systems require *personal* attendance to adjust them according to the heat loss. Such semi-automatic regulation, although theoretically wrong, may

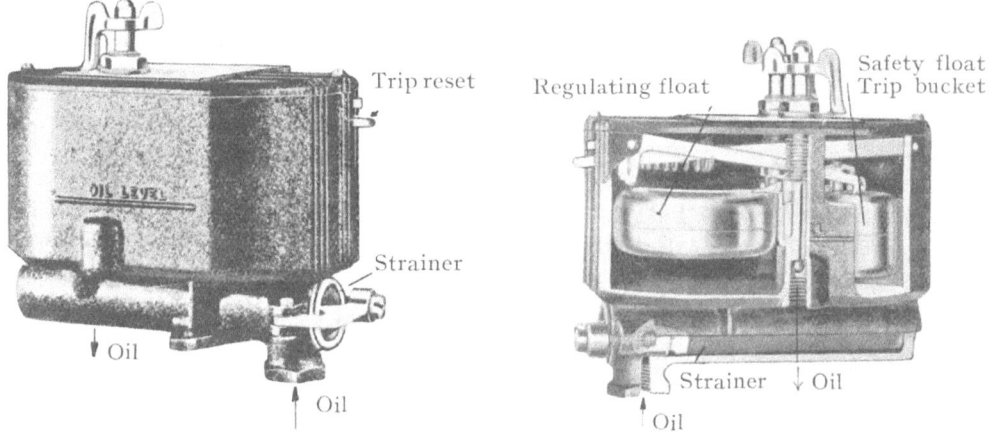

Fig. 250. Detroit safety float for constant level burners.

have some practical use, inasmuch as it avoids extremes of over- or underheating. *Fully automatic regulation* comprises systems which regulate the oil fire *according to the momentary heat loss of the house without any personal attention.*

There are many cases, for instance when hot water boilers for central heating are placed in kitchens, where some personal attention by the servant forms no objection to speak of and semi-automatic fire pot burners are successfully used, in which cases the comfort of oil burning is nevertheless greatly appreciated, also in that it avoids coal dust and the removal of cinders, particularly so because the kitchen is adjacent to the living rooms. Moreover, *this burner type has considerable prospects for purposes lying between industrial and domestic use*, often called *"commercial purposes"*, such as bread-baking, fish-frying, hot-water heating for small laundries, heating of butcher's cauldrons, etc., for which the fully automatic type offers no advantages.

A recent application of such "commercial" oil burning, for which purpose the manually controlled vaporizing pot type has special advantages, is given in fig. 251, showing a *portable equipment developed by the* Royal Dutch Air Lines (K.L.M.) *for preheating aeroplane engines* in winter time. It consists of a PRIOR oil burner and air heater, delivering air of approx. 80° C., which is distributed to the engines by means of a number of canvas hoses. The air is produced by a blower driven by the engine of the truck by means of a pulley on the gear shaft. Part of this air is used for operating the oil burner, while the rest is forced through an air-heater comprising elements with LIESEN needle surfaces (see fig. 251), which allow very large heating surfaces to be obtained in small spaces. This pre-heating method has many advantages over the usual way of idling the engines, as in the former case the complete engine and its full

Liesen needle surface for air preheater.

Fig. 251. Recent application of oil burning for preheating aeroplane engines in winter time (K.L.M. Royal Dutch Air Lines).

oil content is heated up and the engine is started with lubricating oil of the correct viscosity, whilst in the latter case the lubrication during the first few minutes is very often so bad that excessive piston wear may occur.

As regards its *constructive forms*, it may be remarked that the lower prices of the fire pot type compared with the fully automatic burners are secured in the first place because the oil, not as a rule being atomized, requires no pump, pressure regulator or atomizing tip, an inexpensive constant-level device with float being sufficient for the purpose, whilst in the second place no automatic electric ignition is used, thus avoiding coils, electrodes, switches and expensive safety devices, such as protectostats or stackstats, which are necessary safeguards for automatic ignition but are not needed for manual lighting and control, for which an inexpensive safety trip-bucket (see fig. 62 or 250) is usually considered quite sufficient.

For this fire pot type, however, the upper limit for burning heavier oils will be reached sooner than for the atomizing class of burners. It was mentioned before that in the near future the increased application of oil burning for central heating, on the one hand, and the rapid development of Diesel traction, on the other, will probably move the equilibrium of production and consumption towards the used of heavier oils for the former purpose. Although this does not mean that heavier oils will be distributed in the very near future, it is worth while making some preliminary investigation into the possibilities of using heavier, residual fuels purely from a scientific standpoint, without as yet taking any further commercial considerations into account.

The quality of a domestic fuel oil must be determined in such a way that even the most "fastidious" oil burner on the market does not give rise to complaints. With respect to the Dutch market it seems that for its present domestic fuel oil quality the limit has not only been reached but probably even surpassed, anyhow for the *purely vaporizing* burners of the range burner type, which cannot be satisfactorily operated on the present domestic fuel oil owing to carbonization troubles. Consequently, the problem involved in purely vaporizing burners is, not the burning of *heavier* oils but the prevention of troubles with the *present* oil, or, in other words, making this burner type "less fastidious".

In this respect burners of the *fire pot type* are in better case than the purely vaporizing burners of the range burner type, because with the former any oil coke deposit formed will be burnt to dry, brittle carbon, which is partly blown away and anyhow does not seriously interfere with the operation of the burner, the more so because the fire pot copes with the accumulation of comparatively large quantities. As a matter of fact the present domestic fuel oils in Holland do not give rise to complaints when used in *fire pot burners*, but if the residue content should be raised and the oils be made less volatile with higher CONRADSON figures, some way or other of getting rid of the carbon formation will have to be found.

------

# CHAPTER VI

### INFLUENCE OF WATER VAPOUR ON CARBON TROUBLES

Regarding carbon troubles attention may be drawn to the *influence of water vapour*, briefly mentioned before in section II. Steam brought into contact with carbon at sufficiently high temperature (above 900° C.) will be reduced to hydrogen, at the same time oxidizing the carbon to carbon monoxide ("water gas reaction"). According to data in MELLOR's Modern Inorganic Chemistry (1925), the compositions of resulting gases, if steam is brought into contact with carbon, are at different temperatures roughly as follows:

| Temperature | Percentages of: | | | |
| --- | --- | --- | --- | --- |
| | % Steam decomposed | % Hydrogen | % Carbon monoxide | % Carbon dioxide |
| 674° C. | 8.8 | 65.2 | 4.9 | 29.8 |
| 1010° C. | 94.0 | 48.8 | 49.7 | 1.5 |
| 1125° C. | 99.4 | 50.9 | 48.5 | 0.6 |

The water vapour in oil flames may originate not only from steam jets used for pulverization or from the humidity of the air, but also *from the combustion itself* and in this respect it is interesting to calculate the *water vapour concentration* which would theoretically occur in flames of various hydrocarbons with complete combustion. Taking two extreme hydrocarbons, viz., *cetane* ($C_{16}H_{34}$, a paraffin) and *tetralin* ($C_{10}H_{12}$) of the experiment mentioned in Chapter VII (fig. 85), the following figures may be obtained, assuming the following conditions:

  *a*)  Complete combustion.
  *b*)  Volumes calculated with reference to a temperature basis of 1000° C. and atmospheric pressure.
  *c*)  Dissociations disregarded.
  *d*)  25% excess of air.

| Constitutents | Composition of final gas mixture from: | |
| --- | --- | --- |
| | 1000 gms. cetane | 1000 gms. tetralin |
| Water vapour . . . . . . . | 1355 gms. or   7820 lts. | 818 gms. or   4730 lts. |
| Carbon dioxide . . . . . . | 3105   ,,   ,,   7370 ,, | 3330   ,,   ,,   7900 ,, |
| Nitrogen . . . . . . . . . | 14250   ,,   ,, 53200 ,, | 12930   ,,   ,, 48300 ,, |
| Oxygen . . . . . . . . . | 865   ,,   ,,   2800 ,, | 790   ,,   ,,   2540 ,, |
| Gaseous products . . . . . | 71190 lts. | 63470 lts. |
| Water vapour   . . . . . . | 0.0189 gm/litre final gas | 0.0129 gm/litre final gas |

Now in this form these figures do not say much and it is therefore better to express them as: *The quantity of water X which must be added to 1 kg. tetralin to obtain a water vapour content of the final gas equal to that resulting from the combustion of 1 kg cetane without any addition of water*, or in short: to obtain "*cetane equivalence*".

This value X answers to the equation:

$$\frac{818 + X}{63510 + \dfrac{22.41 \times 1273}{18 \times 273} X} = 0.0189 \text{ or solved: } X = 435 \text{ gms./kg.}$$

Thus we have the astonishing fact that, in order to give *tetralin* the same "humidity" of its final combustion products as *cetane* has without any addition of water, it is necessary to burn *a mixture of 1000 gms. tetralin and 435 gms. water*, supposing this were possible. In other words: *Tetralin* requires for its *"cetane equivalency"* an addition of 43.5% by weight of water.

Obviously such considerable differences in humidity of the *final* combustion products of the hydrocarbons considered will also have their influence *during* the combustion process, more particularly as regards the equilibrium of the *"water gas reaction"* during the process and the *tendency to smoke of the flames*.

The above-mentioned assumption that the volumes are related to an arbitrarily chosen temperature basis of 1000° C. cannot have any influence on this conclusion as, disregarding dissociations for a moment, the temperature does not change the composition. As to the influence of dissociation, it is certain that this causes the content of water vapour as such at 1000° C. to be less than those above calculated, but it should not be forgotten that it is not the *water vapour as such* which acts upon the carbon but its dissociation product (oxygen in status nascendi) and consequently with the water gas equilibrium this dissociation has already been taken into account. When comparing for a rough estimate two cases of combustion at equal temperatures, it may be taken that if twice the quantity of water vapour is brought into contact with carbon at equal temperatures it will convert twice as much carbon too, whatever the dissociation may be.

Still greater differences may be observed when comparing the *gaseous hydrocarbons*, such as *methane* and *acetylene*. When burning 1 kg. methane (25% H) and 1 kg. acetylene (7.7% H), the final combustion products will show the following compositions:

| Constitutents | Methane | Acetylene |
|---|---|---|
| Water vapour . . . . . . . | 2250 gms. or 13060 lts. | 693 gms. or 4010 lts. |
| Carbon dioxide . . . . . . | 2750 ,, ,, 6530 ,, | 3390 ,, ,, 8050 ,, |
| Nitrogen . . . . . . . . . | 16450 ,, ,, 61390 ,, | 12700 ,, ,, 47400 ,, |
| Oxygen . . . . . . . . . | 1000 ,, ,, 3270 ,, | 773 ,, ,, 2520 ,, |
| Gaseous prod. . . . . . . . | 84250 lts. | 61980 lts. |
| Water vapour . . . . . . . | 0.0267 gm/litre final gas | 0.0112 gm/litre final gas |

This means that *1 kg acetylene requires an addition of 1138 grams of water or 114% by weight to bring it to "methane equivalency"*, which tallies with the observation that, whereas a methane flame can scarcely be made to smoke, it is almost impossible to make an acetylene flame smokeless without special precautions, such as the application of compressed air or oxygen. If, however, the air around an acetylene flame is moistened with condensing steam, its tendency to smoke may be considerably reduced, obviously due to changing the water gas equilibrium in the flame.

Humidity of combustion products of various hydrocarbon fuels.

| Number: | 1 | 2 | 3 | 4 | 5 | 6 | 7 | 8 | 9 | 10 | 11 | 12 | 13 | 14 |
|---|---|---|---|---|---|---|---|---|---|---|---|---|---|---|
| Hydrocarbon: | Methane | Ethane | Propane | Butane | Pentane | Gasoline | Kerosene | Gas oil | Light fuel oil | Medium fuel oil | Heavy fuel oil | Heavy Tar oil | Tetralin | Acetylene |
| Hydrogen content % by weight . | 25% | 20% | 18.2% | 17.2% | 16.7% | 16% | 15% | 14% | 13% | 12% | 11% | 10% | 9.1% | 7.7% |
| Water formed in gms./kg hydrocarbon ...... | 2250 | 1800 | 1638 | 1548 | 1503 | 1440 | 1350 | 1260 | 1170 | 1080 | 990 | 900 | 819 | 693 |
| Combustion gases at 1000°C. litres (25% excess air) | 84248 | 77830 | 75330 | 74130 | 73210 | 72520 | 71195 | 69730 | 68460 | 67160 | 65990 | 64550 | 63470 | 61980 |
| Water concentration in gm/litres | 0.0267 | 0.0231 | 0.0217 | 0.0208 | 0.0205 | 0.0198 | 0.0189 | 0.0181 | 0.0171 | 0.0161 | 0.0150 | 0.0139 | 0.0129 | 0.0112 |

Grams of water to be added to 1 kg hydrocarbon to make it equivalent to:

| | 1 | 2 | 3 | 4 | 5 | 6 | 7 | 8 | 9 | 10 | 11 | 12 | 13 | 14 |
|---|---|---|---|---|---|---|---|---|---|---|---|---|---|---|
| methane | — | 329 | 443 | 507 | 533 | 587 | 651 | 713 | 776 | 843 | 913 | 974 | 1034 | 1138 |
| ethane | — | — | 118 | 189 | 217 | 218 | 340 | 407 | 476 | 543 | 618 | 682 | 748 | 854 |
| propane | — | — | — | 68 | 97 | 153 | 223 | 289 | 361 | 432 | 506 | 572 | 637 | 746 |
| butane | — | — | — | — | 22 | 77 | 150 | 216 | 290 | 362 | 436 | 502 | 569 | 678 |
| pentane | — | — | — | — | — | 52 | 124 | 192 | 266 | 339 | 411 | 480 | 547 | 655 |
| gasoline | — | — | — | — | — | — | 68 | 137 | 210 | 283 | 357 | 425 | 494 | 603 |
| kerosene | — | — | — | — | — | — | — | 70 | 142 | 218 | 293 | 362 | 432 | 542 |
| gas oil | — | — | — | — | — | — | — | — | 75 | 151 | 228 | 301 | 369 | 483 |

Fig. 252. Humidity of Combustion Products of various Hydrocarbons.

In the table of fig 252 a number of hydrocarbon fuels are arranged according to their C/H ratio, including six more or less hypothetical fuels indicated as "kerosene", "gas oil" etc. up to "heavy tar oil". The values calculated for different "equivalencies" give a good impression of the large quantities of water to be added.

The influence of the *humidity of the air* on the combustion process may in some cases be considerable too, as may be illustrated by the following example: Suppose a boiler is fired with a heavy fuel oil containing 12% H and 88% C one time in *Europe* at an ambient temperature of 10° C. (50° F.) and 70% humidity, and another time in a *tropical country* with an ambient temperature of 30° C. (86° F.) and 90% humidity. In both cases the *quantity of dry air* required for complete combustion of 1 kg oil is equal to *17.73 kgs.*, supposing there is 25% of excess air. A *temperature of 10° C. and 70% humidity* corresponds to a water vapour content of 5.3 gms. per kg dry air (see graph of fig. 253) resulting in a total quantity of water vapour present of *94 gms. per kg fuel burnt.* For a temperature of

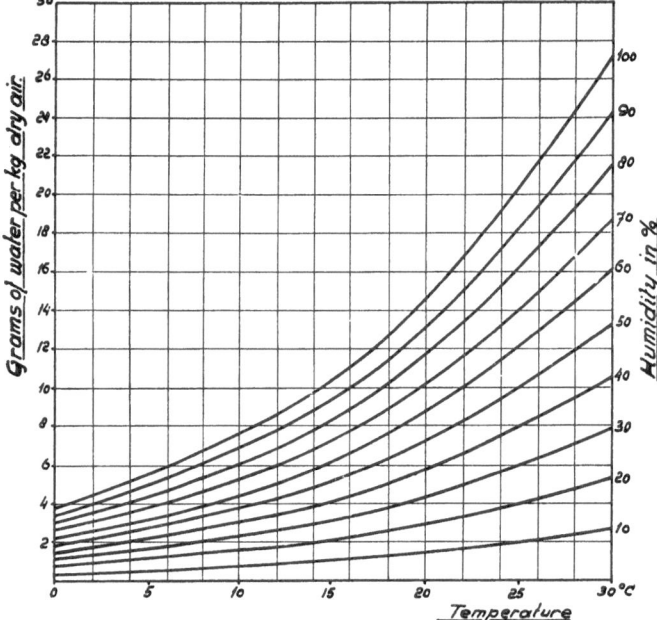

Fig. 253. Humidity calculated on a dry air basis.

30° C. and 90% humidity the graph gives 24.3 gms. per kg dry air or a total quantity of *395 gms. per kg fuel.*

Thus we see that the *natural humidity* of the air gives a difference in water vapour content of *301 gms. per kg fuel or 30% by weight*, which certainly will make conditions in the combustion chambers entirely different too. From this it follows that it must be easier to obtain soot-free and coke-free combustion with the same fuel in tropical, than it is in colder climates and the same different behaviour may be expected in internal combustion engines.

It seems certain that in the near future considerable interest will be concentrated upon what may be called the *humidity control ("air conditioning") of the combustion chambers*, as in fact the early success of the first Russian oil burners of the steam atomizing type (see section I) must be mainly attributed to the *chemical action of steam as a soot-preventor*. If it be considered that the quantities of steam used for pulverization in the industrial application of oil burning range from 15 to 75% of the weight of oil burnt, depending on the oil, steam pressure and temperature and the oil burner design, it is clear that these quantities may certainly have considerable influence.

It may be easily proved by the following experiment that it is the *chemical* action and not the physical pulverizing effect of the steam. Suppose the fire under a steam-boiler is adjusted in such a way that with the pressure atomizing burners using preheated heavy fuel 10% $CO_2$ content is obtained with a slight tinge of smoke at the chimney. When looking into the combustion chamber it will be seen that the flames contain dark smoky streaks and the space between flames and tubes is not clear. As soon as some steam is drawn into the air registers or into the entrances for the secondary air, *taking precautions to prevent any jet effects*, it will be seen that the flames get perfectly clear and the $CO_2$ content may be raised without the slightest trace of smoke. For most industrial fuels *an addition of steam amounting to not more than 5 to 10% of the weight of oil has a remarkable effect*, and the cost of this extra steam consumption is more than balanced by higher efficiencies, due to higher $CO_2$ figures, better heat transfer by radiation (clear combustion chamber) and better heat transfer by convection (less carbon deposit on the tubes and walls). Moreover the flames will be shorter and will no longer reach the tubes, the result being longer life and reduced cost of upkeep.

Evidently water vapour formed in the *last* part of the combustion process will not have so much effect as any water vapour formed at the *beginning* of the flame. In this connection it will be clear why an *aldehydeous* combustion start (which type of combustion produces formaldehyde as an intermediate, which in its turn gives water vapour as a final product before the combustion of the hydrocarbon molecule is finished), will be extremely propitious to soot-free flames. This explains why the addition of water vapour as such to the air will have a greater effect than adding it indirectly by using a fuel which is rich in hydrogen.

As to domestic oil burning, the author is strongly convinced that the principle of adding water vapour to oil flames of fire pot burners holds promising prospects, not only because of the resulting ease with which *soot-free combustion* is obtained with heavier oils, but even more so because of the effective removal by steam of any *coke deposit* in the fire pot. In other words: *Adding water or steam to the combustion process may make burners of the fire pot type less "fastidious" as to the quality of the oil used.*

## CHAPTER VII

### NATURAL DRAUGHT BURNERS

In Chapter VII of the First section it was shown how, in about 1923, a vivid development of natural draught burners, mainly of the "mushroom type" (see fig. 54), took place in the U.S.A. for central heating. It may be remembered that this type of burner (fig. 53), comprising some inexpensive castings, and using no electricity nor having moving parts, did not fulfil the high expectations entertained, with the result that there was an abrupt fall in the sales in about 1925 (see fig. 54). The main reasons for these disappointments were:

1. *Natural draught proved to be unreliable as a motive force*, being too dependent on irregular weather conditions.

2. *Too large an excess of air was needed* to obtain clean combustion, resulting in excessively *high oil consumptions*.

3. *The range of regulation was too small* for continuous operation in central heating plants, requiring at least 1 : 6 or better 1 : 10.

4. *Automatic control by room thermostats was not possible* with these non-electric burners, even "quasi-automatic" control by the boiler temperature involving difficulties.

5. *The burners could not be made sufficiently fool-proof* and often gave rise to difficulties with fire-insurance companies.

When thinking about the further development of natural draught burners it is well to remember these five points, each of which must be solved before any success can be hoped for.

Consequently discussing the various points, the following remarks may be apposite:

Re 1. *Natural draught.*

In Chapter X of the Second section some causes of irregular draught and means of avoiding them were discussed; for the sake of brevity the reader is referred to that paragraph. Obviously, a *constant draught* may be obtained from a variable draught *by cutting off all irregularities occurring above the minimum value* of the variable draught. This may be done by self-adjusting air-leak devices, such as shown in fig. 108, which spoil the excess draught by admitting a variable quantity of air into the chimney. To some extent the irregularities may be levelled by suitable chimney caps, a number of which were shown in figs. 109, 110 and 112. Special attention is drawn to the HOOGENDAM cap, the principle of which is the destruction of the energy of the wind squalls by perforated shells.

Needless to say, this "cutting-off" method of draught regulation entails *weak draughts, say about 3 mms. water gauge* ($\frac{1}{8}$ inch W.G.), and consequently future natural

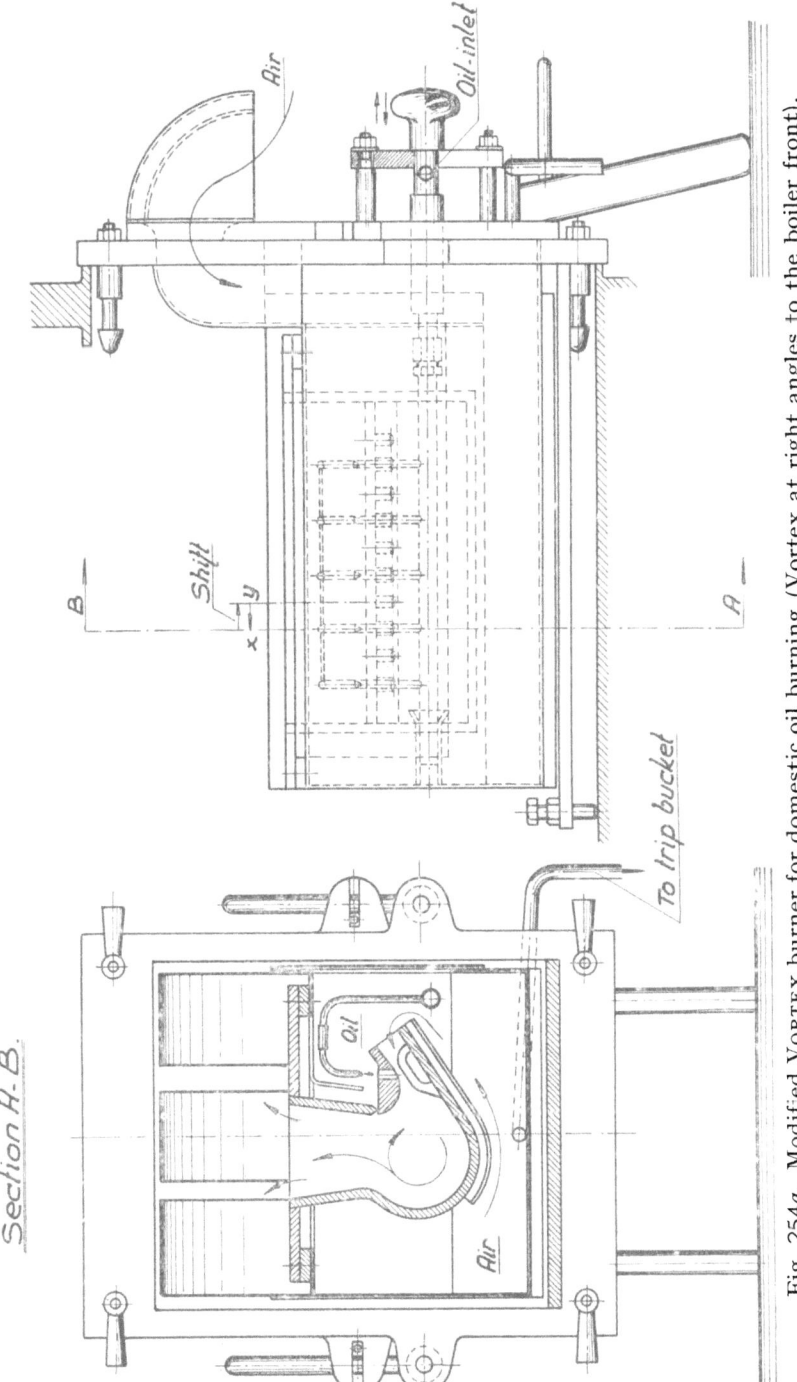

Fig. 254a. Modified VORTEX burner for domestic oil burning (Vortex at right angles to the boiler front).

draught burners must be based on extremely keen "draught economy", viz., by reducing resistance of air passages to the minimum (stream-lined channels and diffusor cones).

Another point to be considered is the difficulty of obtaining *turbulent motion* of the flame gases, which is desirable for good combustion, by such weak draughts.

*Re 2. The excess of air* required for good, clean combustion is directly related to point 1, as cutting down the air volume to a reasonable excess of 25% will usually result in a reduced turbulency and a smoking fire. This is especially true of low loads. In order to avoid this difficulty the best solution was found by adopting "high-low" regulation, which limits the trouble of excess air adjustment to two loads only.

*Re 3.* The *range of regulation* required for domestic heating was fully discussed in the third Chapter of this section. The most simple means of solving this problem is "high-low" regulation, a "graduated" regulation causing all kinds of trouble.

*Re 4. Room-thermostatic control* of the oil fire without using electricity, compressed air or other acting media still awaits solution. *Semi-automatic* regulation by the boiler temperature, however, is possible, especially if "high-low" regulation is used.

*Re 5. Fool-proofness* of natural draught burners is dependent on effective trip-bucket devices (see fig. 62 and fig. 250) which shut off the oil supply as soon as oil happens to leak from the burner into the bucket. This system should be applied at all other joints etc. which involve a risk of leakage.

Moreover, it should be borne in mind that sometimes the leaking oil was unable to reach the bucket because the passages provided for this purpose were clogged up with carbon and soot.

It may be advisable to install as an additional safety device some *flame-guarding instrument* of the type shown in figures 235 and 236. The electric current necessary for shutting the electro-magnetically operated oil valve may be supplied by an accumulator if no other source is available.

An attempt to solve the problems of the natural draught burner by engineers of the Testing Station "DELFT" (Holland) of the Bataafsche Petroleum Maatschappij was mentioned in section III, Chapter II. The first design of fig. 139 (Dutch Patent No. 34,379) proved to have several defects which were gradually emended, the result being a *modified Vortex burner* as shown in figs. 254a and b, incorporating the following improvements:

1) *The flame was made to burn inside the boiler*, the "vortex" being reduced to the limit as shown by fig. 254. This caused the flame to "face" more of the heating surface, resulting in a better heat transfer by radiation and in lower oil consumptions.

NATURAL DRAUGHT BURNERS

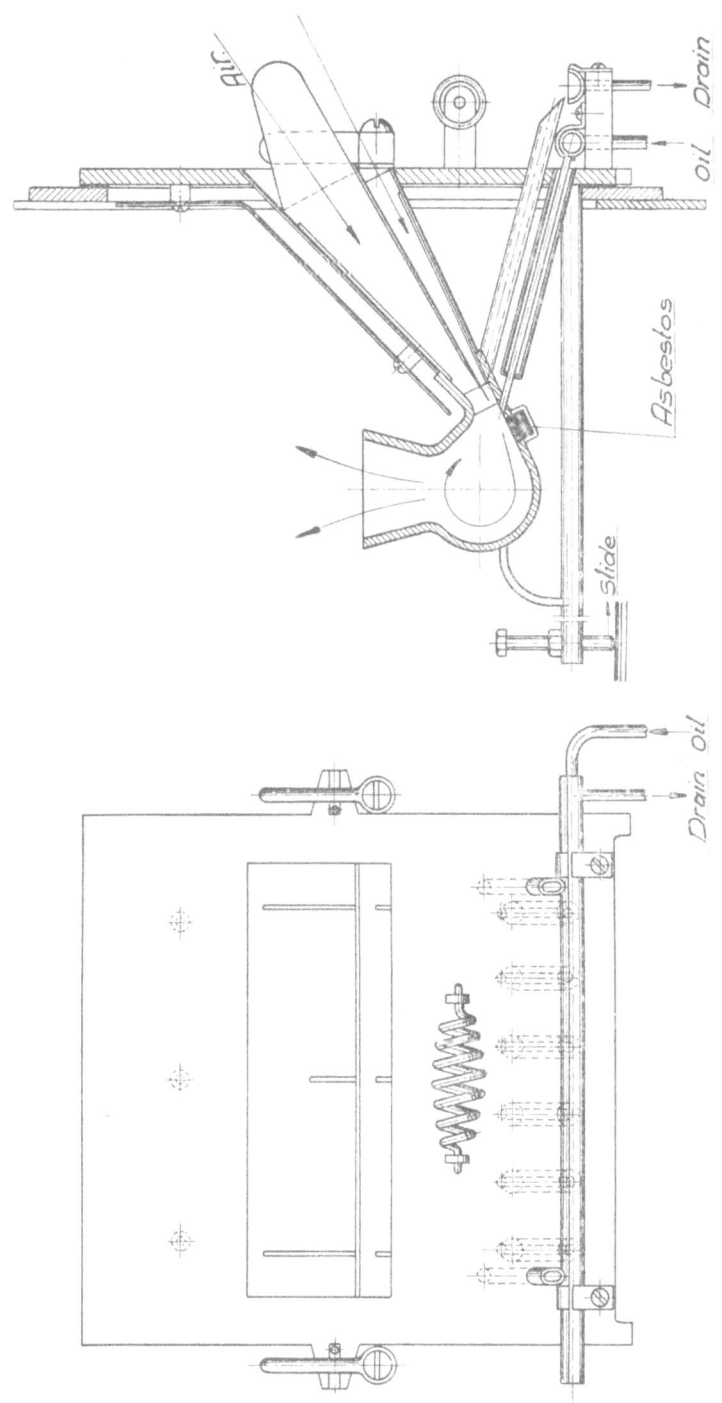

Fig. 254b. Modified VORTEX burner for domestic oil burning (Vortex parallel to boiler front, small capacity).

Fig. 255a. Three VORTEX burners installed at the "DELFT" Testing Station

2) *The combustion chamber was made smaller*, thus permitting of better mixing by turbulency and possibly avoiding laminary flow, which *reduced the amount of excess air* necessary to obtain clean, smokeless combustion. This also led to lower oil consumptions with higher flame temperatures and less chimney loss.

3) *The range of regulation was improved* by adopting the "high-low" system activated by the temperature of the water leaving the boiler. Thus air and oil could be automatically adjusted for efficient combustion for two loads and a *range of regulation of 1 to 7½* could be obtained. In order to stabilize the flame at low load it was found to be useful to line *the hot plate with a layer of asbestos*, as indicated by fig. 254b.

Although its performance was very much improved with respect to oil con-

Fig. 255b. The High-Low regulation device of VORTEX burner.

sumption, combustion, pick-up, range of regulation and semi-automatic control, there still remained one serious objection which made its application for central heating rather doubtful, viz., *a roaring noise at full load*. It was found to be easy enough to stop this noise, e.g., by suitable design of combustion chamber and application of silencers such as used on carburettors of automobile engines, but in that case the combustion could not be kept free from smoke unless the amount of excess air was considerably increased, which again caused high oil consumption, etc. Obviously *this roaring noise*, originating from explosion waves causing thorough turbulent mixing of the flame gases in the combustion chamber, *was inseparable from good combustion.*

The problem of keeping the noise inside the boiler without being audible in the house was found to be extremely difficult; in fact, it was found necessary to place the boiler in a perfectly tight sheet-iron box, the air being drawn in through a tube from the outside. Even small inspection holes, if left open, proved to communicate the noise to the room. Fig. 255a represents the boilers equipped with Vortex burners which are installed to heat the buildings of the above-mentioned Testing Station, and fig. 255b the "high-low" regulation device activated by the boiler temperature.

---

# CHAPTER VIII

### RANGE BURNERS

### A. *Introduction*

The name "range burner" has gradually come to be given to a type of natural draught burner comprising as essential parts: a *vaporizer* to which the oil is admitted in grooves and a number of *perforated shells* which are set concentrically on the grooves as shown in figs. 71, 84, 134 and 135. These devices are described in several patents from about 1916 onwards, out of several the following being quoted:

SOME BRITISH PATENTS ON RANGE BURNERS

Br. Pat. 141,410 (CHADWICK, Cleveland Metal Prod. Co, 1919)

,,   ,,   147,078 (Central Oil & Gas Stove Co, 1916)

,,   ,,   150,083 (HOBDEN, 1919)

,,   ,,   159,467 (Central Oil & Gas Stove Co, 1916)

,,   ,,   188,871 (SEBBELOV, 1921)

,,   ,,   247,833 (SEIFERT, 1925)

,,   ,,   269,698 (HARRISON & SMITH, 1926)

,,   ,,   274.721 (SEIFERT, 1926)

Br. Pat. 276,667 (Lampen und Metallw. Fabr., 1926)

,,   ,,   315,658 (FETTER, 1928)

,,   ,,   328,226 (FETTER, 1928)

,,   ,,   329,554 (JOHNSON & LEACH, 1929)

,,   ,,   338,757 (TAKIMIZU & IDEMITSU, 1929)

,,   ,,   341,052 (ZANROLI, 1929)

,,   ,,   345,578 (LYNN Products Co, 1930)

,,   ,,   354,654 (GARRI & BASSO IN ROSSO, 1930)

Br. Pat. 375,647 (Perfection Stove Co, 1930)

,,   ,,   378,710 (LYNN Products Co, 1931)

,,   ,,   380,050 (Perfection Stove Co, 1931)

Br. Pat. 388,657 (LUDOLPHI & WIEDE-MANN Nachf., 1932)

,,   ,,   389,503 (STANLEY, 1931)

,,   ,,   393,083 (Lampen und Metallw. Fabr. 1931)

,,   ,,   393,284 (THIBERT, 1931)

They are based on two fundamental principles, viz.,

1) The burner is operated by the *"natural draught of the red hot shells"*, which draught is very constant, while the irregularities of the *"natural chimney draught"* have very little influence on the combustion process.

2) The combustion secured is of the *aldehydeous type* by using the *"reverse flame" principle*, i.e., by burning air in an atmosphere of hydrocarbon vapours. This feature ensures soot-free blue-flame combustion (see page 73).

From about 1931 on, initially in the Eastern States of the U.S.A. and later on in Europe too, these burners found widespread application for domestic use in ranges, etc., which explains the name. The fuel mainly used was kerosene and gradually the market was developed to such an extent that special brands known as "range oils" were distributed. Attempts to use gas oils for this kind of burners mostly failed because of carbonization of the oil in the vaporizer grooves, and a typical property of the "range oils", usually consisting of *less* or *un-refined* kerosenes (the T.T.S. values for this purpose being of no importance), was a *low CONRADSON figure*, which is determined by means of the apparatus of fig. 77.

When dealing with the future development of range burners special attention must be given to the *causes of carbonization troubles* and to the means of avoiding them, particularly when burning *gas oils*.

It has been explained in sections II and III that a *gas oil*, usually being a distillate containing no residue, *may be completely evaporated if it is heated to or above its average dew point*, which follows from the fact that the oil was originally obtained at the refinery by cooling oil vapours to *below* the average dew point in atmospheric condensers, this being a physical process which may be reversed and repeated ad libitum (see Lecture II, Ch. II, page 44 and section III, Ch. III, page 177).

It is curious, however, that in oil burning practice *complete* evaporation of gas oil is so rarely obtained and oil burner designers are so accustomed to finding carbon deposits that the "carbon residue content" of an oil is usually considered to be as inevitable as an ash content. *This "carbon content", however, is entirely caused by heating the gas oils too far above their dew points*, to temperatures at which carbon deposits are formed as a result of chemical destruction, but, from the fact no carbon is ever found in the gas oil condensers of crude oil distilling plants, it follows that there must be a margin between the dew point and the temperatures at which the carbon formation occurs. The heavier the gas oil, the smaller its margin will be, and theoretically speaking, the heaviest gas oil obtainable by distillation at *atmospheric* pressure will be characterized by a disappearing margin.

It has been found in practice that for the ordinary gas oils this margin is about 25 to 50° C., which means to say that, if the dew point is 325° C., the formation of carbon will start between about 350° C. and 375° C.; it is therefore clear how highly important *a narrow control of the temperature of the vaporizing device* is in order to prevent carbonization. For kerosene this margin is considerably wider, which explains why carbon troubles scarcely ever occur with this kind of oil.

Three methods for controlling the vaporizer temperature will be discussed consecutively, viz.,

1. *By keeping the oil vapours present in the vaporizer completely saturated* (ordinary range burner).
2. *By using a thermostatic electric device* (RÖDL burner).
3. *By using a thermostatic boiling bath* ("Au bain Marie" burner).

This Chapter will be devoted to the first method only.


### B. *Using Saturated Oil Vapours*

As long as the oil is in contact with its saturated vapour its temperature cannot be raised above its dew point and consequently carbonization will not take place.

This method is very old indeed; in fact, the ordinary kerosene wick lamp is based on the same principle, viz., an increased supply of heat to the wick surface, instead of raising the temperature, causes an increased evaporation, thus keeping the wick surface temperature automatically constant and equal to the average dew point of the oil. This thermostatic regulation only holds good a long as the *supply of oil is ample enough to meet every demand of evaporation*. If the oil is not able to reach a certain heated part of the wick, as may occur if the viscosity of the oil is too high or if the pores of the wick are filled up with heavier residual oils, this thermostatic regulation must fail and result in a local superheating and carbonization of oil vapours and wick.

It is a well-known fact that a cotton wick adjusts itself if it is too long, as the parts which cannot be reached by the oil are burnt away, such in contrast with asbestos wicks, which must be made the correct length beforehand.

As the transportation power of a wick is rather limited, especially for more viscous oils, this burner part has been replaced by some other devices in cases where oils heavier than kerosene were to be used. Thus range burners, for instance, have *open V-grooves with a free oil level* for this purpose. For ignition by means of a match an asbestos wick is placed inside, which is only intended to serve as such during the heating-up period. If the burner base and shells are preheated by other means, e.g., by a gas flame, the burner will be able to burn just as well without wicks, as evaporation then takes place directly from the free oil surface.

A serious objection to this use of *wicks for priming* is that they may locally spoil the thermostatic control of the oil temperature and thus cause carbonization if heavier oils are used. On an asbestos wick taken from a range burner which has been burning on gas oil, three different layers may be clearly observed, viz. (see fig. 256),

1) The *lower layer below the oil level*, soaked with oil without any carbon deposit.

2) The *middle one from the oil level to about 0.5 cm. above it*, heavily loaded with carbon soaked with oil (oil coke), and

3) The *upper one*, which is perfectly dry, burnt white and without any carbon deposit.

As may be clearly seen in the pictures of figs. 256 and 258, the carbon formation is obviously started from the vaporizer wall and wick *at or slightly above the oil level*, and gradually grows therefrom until the groove is completely covered. No carbon would be formed if the oil were able to reach every part of the wick quickly enough to compensate for any excess heat by a reinforced evaporation, and consequently *the wick should not reach out of the oil level higher than about 1 or 2 mm.*, which, however, in some cases may cause lighting troubles.

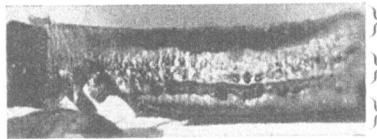
} dry carbon-free zone
} carbonization zone
} wet carbon-free zone

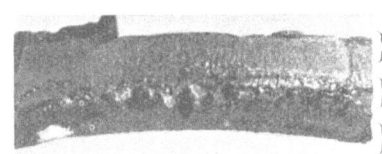
} dry carbon-free zone
} carbonization zone
} wet carbon-free zone

Fig. 256. Carbon formation on asbestos wicks of range burners.

Another solution of the problem is to light the burner with a very low level, the wick reaching about 5 mm. above it, and, after the shells are heated up, shift over to the higher oil level of normal operation, the wick being completely immersed and thus protected against carbonization. Of course the same effect may be obtained by means of some device which keeps the wick turned down during normal operation.

A bad thing with the *formation of oil coke* is that it is *a process which grows of itself*, because the layer of oil coke formed acts like a wick with rather bad transportation qualities, the pores being gradually filled up with high boiling hydrocarbons which tend to be superheated and are easily cracked.

A wick is an assembly of capillary tubes which raise the oil above the oil level. The same capillary effect may also be observed with ordinary walls causing the oil to creep along above the oil level. It is not astonishing that such a slightly wetted wall contains the same conditions which are so favourable to carbon formation as are contained in wicks, and consequently *oil coke formation may be started also from heated walls which are covered with a film of oil.* As soon as the first trace of carbon is formed the process rapidly increases. Even an oxidized surface or blisters of metal sulphides may promote the first carbon formation and, consequently, should be avoided.

Regarding the design of range burner vaporizers the following rule may be formulated: *Oil films* caused by oil creeping along the walls by capillary action at places where the oil level touches the surrounding walls *should not be exposed to a strong radiation of heat,* as this may lead to local superheating and carbon formation. Consequently the oil level where it touches the walls should not "face" any part of the red-hot glowing perforated shells. Moreover, the heat necessary for evaporating the oil should preferably be supplied through walls which are *below* the oil level and not above, in order to prevent superheating and cracking of the oil vapours. A similar rule is contained by most Steam Acts prescribing that no fire should be allowed to touch boiler walls *above* the water level.

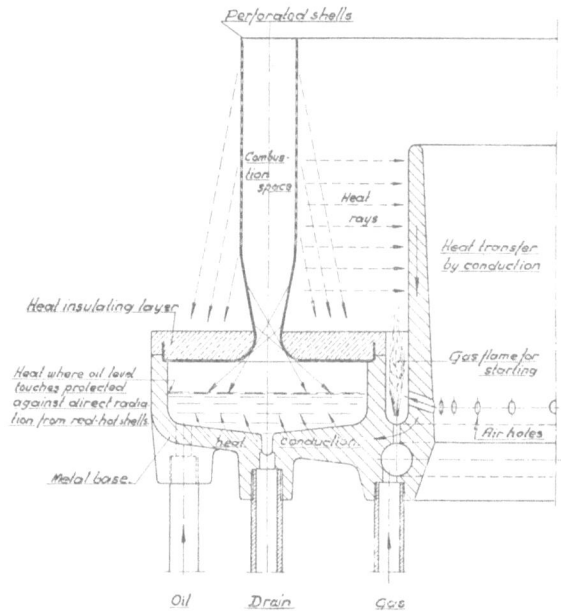

Fig. 257. Scheme of a saturated vapour burner.

An example of a burner base, incorporating the above-mentioned points, is shown in fig. 257, which must be regarded as a suggestion. With an experimental burner of this kind good results were obtained.

For a better grasp of the matter it is necessary to enter a little more deeply into the subject of carbon formation. Carbon may be formed by the exposure of hydrocarbons to high temperatures but it should be borne in mind that this may happen in a *vaporous just as well as in a liquid state.* The carbon of such *"vapour-phase cracking"* is as a rule softer and more evenly distributed than that of *"liquid-phase cracking"* and the former does not, therefore, interfere as soon with the regular operation of the burner as does "oil coke". The majority of the carbon troubles encountered with range burners originate from a *choked oil supply*, which is caused by liquid-phase carbon (= oil coke).

Oil coke is probably formed in three successive stages, viz.,

1. A vapour bubble evolved on a heated spot of the wall pushes the liquid aside and sticks to the wall for a moment. The oil vapours, being in contact with the wall, are superheated because the heat transfer from wall to vapour is considerably less

than that from wall to liquid. The cracked oil vapours deposit a thin velvet-like layer of soft vapour-phase carbon, consisting of particles without much coherence.

2. The vapour bubble leaves the wall and the liquid soaks the layer of carbon particles.

3. Another vapour bubble is evolved at that spot and the wet layer of carbon is cracked dry again, a process intermediate between vapour and liquid cracking.

Fig. 258. Carbon formation on range burner bases.

Heavy asphaltic consistuents may be formed which act as a cement between loose carbon particles.

This cycle may be repeated ad libitum resulting in a steadily growing layer of hard coke, which, by its insulating effect, reduces the heat transfer and causes permanent superheating of the corresponding spot on the wall. This mechanism explains why oil coke is preferably formed at the place of separation between liquid and vapour where vapour bubbles are collected by capillary action and why this coke is often so hard that it can scarcely be removed even with a chisel.

In order to reduce the tendency to carbonization, the evaporators should be made of materials of *high conductivity for heat*, which makes local superheating of the wall difficult. IRINYI proposed in his French Patent No. 515,192 (1920) and British Patent No. 144,298 (1919) the use of *needle surfaces* for evaporators, as the metal needles, when piercing through the vapour bubbles, on the one hand stimulate the formation of bubbles at lower temperatures and, on the other, prevent superheating of vapour and wall by forming some kind of a "bridge" for the heat transfer. Others recommend the use of quartz sand in vaporizers because its sharp edges are favourable for evaporation (GIBBS, fig. 132).

Finally an important point to be considered is the *distribution of the carbon deposit*. Figure 258 illustrates an instructive experiment in this respect. A well-known make of range burner was run on a *gas oil* heavier than the American range oil prescribed as fuel for this kind of burners, resulting in the formation of coke on the burner bases, which stopped the burner *after 175 hours* of operation due to a choked oil supply. The burner base is shown by the picture in fig. 258A.

As the burner could be operated on *kerosene* for an *unlimited* length of time, it was suggested that the burner should be run on a *mixture of 50% kerosene and 50% gas oil* but, curiously enough, several experiments showed that with this *lighter fuel* the burner stopped *after 65 hours* of operation due to a choked oil supply. The reason for this unexpected difference in behaviour may be seen when comparing the photograph A with B. The *lighter 50/50 fuel*, being more volatile, obviously caused a kind of fountain of oil and vapour at the central inlet of the oil, the result being a more *concentrated* formation of oil coke which choked the oil supply sooner than with the heavier fuel, the carbon formed with the latter being *more evenly spread out* over the surface of the vaporizer.

---

# CHAPTER IX

### SUGGESTIONS FOR BURNERS BASED ON THERMOSTATIC CONTROL OF THE VAPORIZER TEMPERATURE

A more decisive measure to prevent the formation of carbon with vaporizing burners is *controlling the vaporizer temperature at or slightly above the average dew point of the oil by means of some thermostatic device*. Of course this method can be applied to distillate oils only, as with residue-containing fuels, i.e., oils which consist not entirely of oil vapours condensed at atmospheric pressure, the dew point lies above the cracking limit. For most gas oils this critical limit will be between about 325 and 350° C., depending on the composition, and for most cases a temperature of 300° or 320° C. will be sufficient to evaporate the oils completely without any deposit being left behind (see fig. 73).

### A. Rödl's burner

The first example to be discussed is a burner developed by an Austrian engineer called Rödl. His burner, which is illustrated in figure 259 consists of a cylindrical vaporizing chamber into which the oil is admitted by means of an open level control (MARIOTTE flask or reversed bottle, see fig. 92). This chamber is heated by two sources of heat, 1) *by an electric heat element around it* and 2) *by a small pilot range burner in the centre*. To start the burner the electric circuit is switched on and at first heat is only applied by electricity. As soon as the temperature is high enough to vaporize the lighter parts of the oil, the small pilot range burner is ignited by means of an electric glow plug heated by the same electric current. Thereafter the heat supply is increased rapidly, causing the oil to vaporize completely and its vapours to issue from the holes, where they are ignited by a second glow plug. As

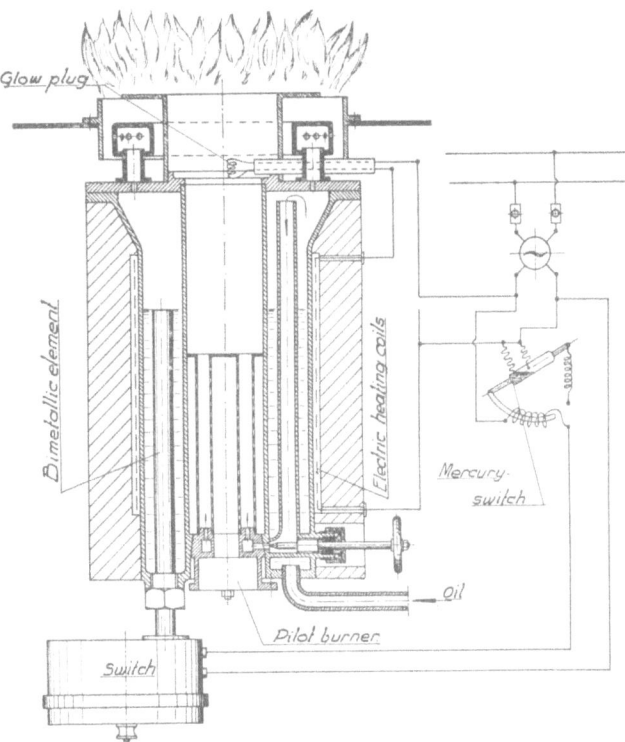

Fig. 259. Rödl's burner.

soon as the desired temperature (slightly above the average dew point of the oil) is reached, an electric relay activated by a bimetallic element cuts off the electric heating. The pilot range burner is adjusted in such a way that its heat is almost sufficient to keep the vaporizer at that temperature, whilst a small quantity of electric heat is automatically supplied to keep it constant within narrow limits.

A special advantage of this arrangement is that the consumption of electricity is only small, *the greater part of the heat required for vaporization being supplied by the small pilot range burner inside*. Once this has been correctly adjusted to full load it needs no further adjustment for other loads, because if the oil supply is decreased, which may be done by lowering the oil level, the pressure in the vaporizer is slightly reduced too, which also causes the pilot burner to burn somewhat lower, so that its heat follows the oil consumption of the large burner.

Another advantage is that this burner may be put into operation merely by turning a switch, and consequently it may be operated automatically by a room thermostat. Slight traces of residue in the fuel, however, are left behind in the vaporizer, and this gradually raises the average dew point, causing a reduction of the burner capacity if the vaporizer temperature is not re-adjusted. This drawback may be avoided by *draining off the traces of residue* from the vaporizing chamber after the burner has been stopped, which is done by means of an electro-magnetically operated three-way cock in the oil feed line. Moreover, this manipulation has the advantage of preventing the burner from fuming after it is extinguished.

As far as the author knows, this burner has not yet been developed on a commercial scale, but it is not impossible that a use has been found for it, viz., in ordinary hearths built for coal.

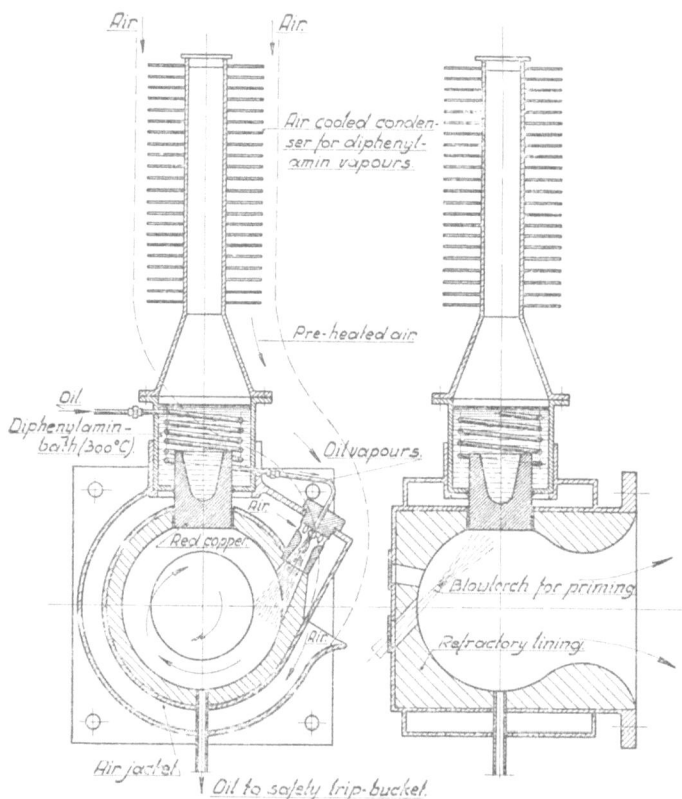

Fig. 260. Diphenylamin controlled vapour jet burner

### B, *Boiling Bath Burner*

A second example of thermostatically controlled vaporizer temperature is a suggestion of the author, who used a *bath of boiling liquid* to keep the vaporizer temperature constant at about 300° C.

A diagrammatic sketch of this type of burner is shown in fig. 260. The flame of this burner, which is first preheated and started by a flame of gas or methylated spirit, is forced to travel vortex-wise through a spherical combustion chamber of refractory material, thus heating a bath before leaving the combustion chamber in an axial direction. The oil gravitates into a coil of copper tube immersed in a bath of *diphenylamin* (boiling at about 300° C.), whereupon it is vaporized and its vapours, issuing from the nozzle of a Bunsen or Meker burner, are mixed with air and burnt.

Any excess of heat delivered by the flame to the diphenylamin and carried away by its vapours is given off to the incoming combustion air by means of a small air-cooled condenser, and in this way it is not only possible to prevent all dangers of superheating, cracking or carbonization of the oil, but also all the heat transferred to the diphenylamin bath is utilized for the combustion process, either as latent heat of vaporization of the oil, of for preheating the combustion air.

Diphenylamin is a relatively cheap, inactive, thermally stable substance, which does not readily burn and may be boiled for hundreds of days without any appreciable chemical changes. This substance and other similar ones (diphenyl oxide) have been proposed for use in some power plants (Dow Chemical Co., Midland Mich.) as a substitute for water or mercury. If this substance is used, any oil having an average dew point below or equal to 300° C. will be completely evaporated. However, oils with an average dew point higher than 300° C. will leave a certain quantity of liquid behind

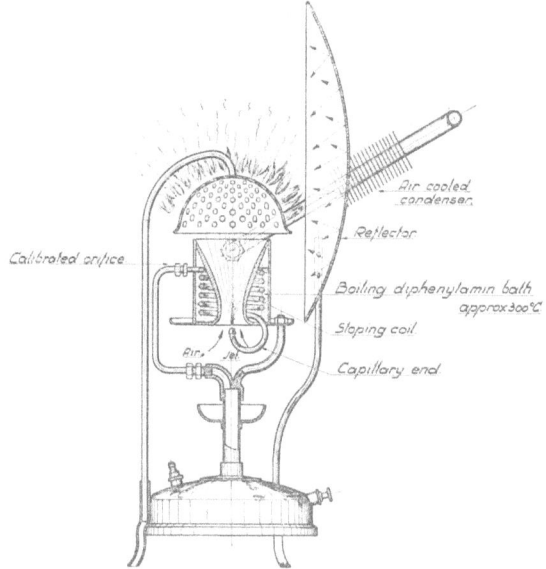

Fig. 261a. Modified vapour jet burner (Diphenylamin bath).

and, in order to prevent its accumulation, the coil is made draining down to the nozzle. Any residual liquid is thus continuously removed and pulverized by the vapour jet.

Fig. 261 shows the same principle applied to the well-known *vapour jet burner*, which speaks for itself. If the oil has previously been thoroughly filtered, the small vapour hole can scarcely ever get choked, because no solid particles can be formed from the oil. It is advisable to make the vaporizer coil of a suitable metal which is not attacked by possible sulphur of the oil, as otherwise choking might be caused by blisters of metal sulphides.

With the burner of fig. 261 it has been found possible to burn without any trouble, not only kerosene, and several kinds of domestic fuel oils, but also a *mixture of 80% of domestic fuel oil of the Dutch market and 20% of a heavy paraffin residue*. This residue-containing fuel was burned with *perfectly blue flames* for several hundreds of hours without any choking of the vapour hole, notwithstanding the fact that it showed a CONRADSON test of 0.8% (carbon residue) and even 0.04% of ash.

Considering that the quantity of oil burnt during this test amounted to approx. 200 litres or 175 kilos, according to the CONRADSON figure there was about *1400 grams of carbon*, and as the volume of the vaporizer coil was only equal to 20 cub.cm., it

is obvious that in this particular case the "CONRADSON carbon" *was not formed*, evidently because cracking conditions in the vaporizer are less favourable than they are in the crucible during the CONRADSON carbon residue test (see fig. 77). This example shows clearly how very important it is to control the vaporizer temperature of burners of this class.

Even the *ash content, corresponding to 70 grams of solid material* present in the above-mentioned total quantity of oil burnt, did not choke either the coil or the orifice, apparently because it is continuously removed in suspension by the non-vaporized part of the oil. It is therefore to be regarded as a fortunate circumstance that ash-containing fuels as a rule also contain enough heavy hydrocarbons not evaporating at 300° C., which automatically remove the ash.

As, moreover, the residual liquid preheated to 300° C. was as thin as water (viscosity < 1° E.) and as volatile as gasoline, its pulverization, even by such simple means as a plain jet, did not give rise to any trouble whatever and resulted in a perfectly blue "gas flame".

Fig. 261*b*. Vapour-jet burner with thermostatic control of the vaporizer temperature by boiling diphenylamin.

---

# CHAPTER X

### ON THE COST OF OIL BURNING

#### A. *General Remarks*

The last Chapter of this book may be devoted to some remarks on the cost of oil burning, which is an important factor in future development. The discussion, however, will be restricted to the *technical side* of the problem. These remarks are more particularly made because it is the author's experience that this point is imperfectly understood, especially by commercial people who are not sufficiently conversant with the technical peculiarities and factors governing the cost of oil burning.

It is often said that "oil burning is expensive" but, although the author will not deny that there are cases, e.g., in coal mine districts, where oil is not able to compete with coal, this is usually not the case, *provided all the advantages obtainable with oil as a fuel are utilized to the full* and comparisons are made under fair and approximately "normal" circumstances with respect to taxes, import duties, etc.

The steady increase in the sales of fuel oil for industrial purposes, especially for marine and naval use, speaks for itself and the following discussion, therefore, will be confined to the *domestic* application of oil burning.

Whereas with industrial oil burning the cost may be one of the main factors when considering conversion from solid fuel to oil, this is certainly not the case with domestic oil burning and in this connection it is interesting to quote the results of an inquiry made by the OIL HEATING INSTITUTE (mentioned by TAPP in his "Handbook of Oil burning") which contained the question: *"What do you consider the most attractive feature of oil heating?"*

The answers from oil burner users were:

1. Labour saving . . . . . . . . . . . . 22.1%
2. Cleanliness . . . . . . . . . . . . . 20.1%
3. Uniform temperature . . . . . . . . . 18.1%
4. Less responsibility . . . . . . . . . . 16.2%
5. Healthfulness . . . . . . . . . . . . . 8.1%
6. More space in stokehold or cellar . . . 7.8%
7. *Economy* . . . . . . . . . . . . . . . *7.3%*

In literature it is often stated as a rough rule that for comparison *one ton of oil is equal to about 2 tons of high grade coal or anthracite and to 2½ tons of coke*, but this is a very rough estimate indeed, as such comparisons depend on many factors, which may be entirely different for individual cases. Consequently it is absolutely necessary to consider every case on its own merits.

The cost of oil burning is composed of the following items:

1. Cost of fuel oil.
2. Cost of auxiliary supplies, such as gas, electricity, steam, lubricating oil, etc.
3. Cost of labour for attendance.
4. Cost of upkeep of the equipment.
5. Interest on the capital investment, deductions for depreciation, taxes for insurance, etc.

Each of these cost factors will be briefly discussed.

## B. *The cost of fuel oil*

The cost of fuel oil comprises the product of the *quantity* of oil consumed and its *price*.

The *quantity* of oil consumed may be very easily measured and that received verified; this is in direct contrast to solid fuel, which causes serious difficulties in this respect. Whereas with oil practically the only mistakes likely to be made may be due to the temperature in the storage tank differing somewhat from that at which the specific gravity was determined, which mistakes will seldom exceed 1%, enormous

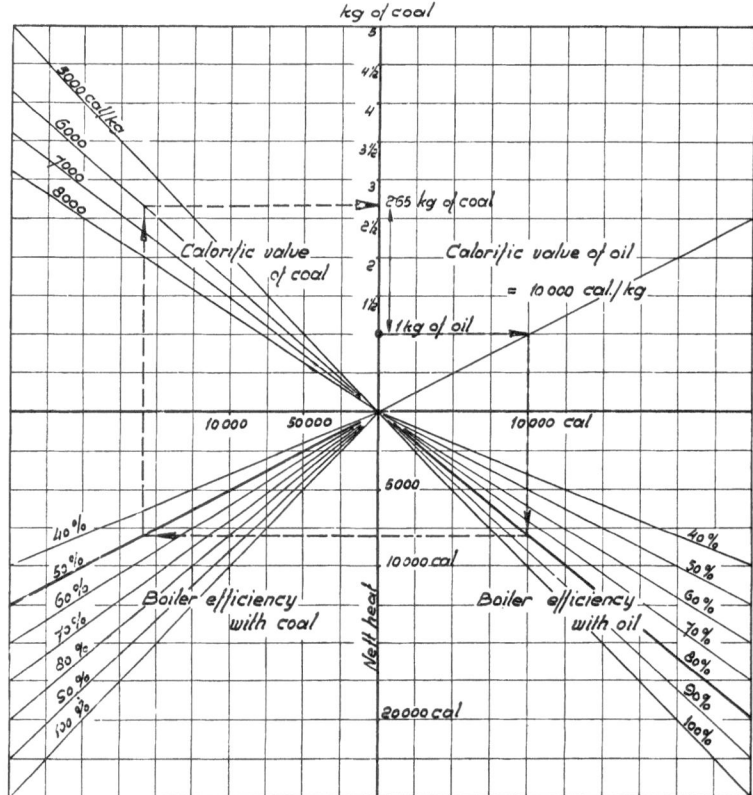

Fig. 262. Comparison of coal and oil.

discrepancies may accur with the measurement of coal deliveries and consumptions, especially if two or more grades of coal are used mixed together. The weight per unit of volume, which must be known for the calculation of the gross heat intake of the installation, depends not only on the specific gravity of the coal itself but also on its granulation. Moreover, the humidity of coal and coke may make a very great difference indeed to the calorific value per unit of weight.

All these questions are quite simple matters with fuel oils, which only contain negligible quantities of water and ash, and which are so homogeneous that it is almost impossible to draw a sample which is not a good average. Particular attention is drawn to these points because it is usually not realized that the determination of the thermal efficiencies with fuel oil is carried out with much more accuracy than the average determination with coal and consequently comparisons should be made cautiously.

Then again it is common practice to compare the fuel bills of two heating seasons, one with coal and the other with oil, but this too is scarcely allowable because two years, though superficially considered as "fair averages" and "almost equal", will usually show greater differences between the total quantities of heat required per season according to the recorded out-door temperatures, wind and rain than those sought between oil and coal.

As to the *price* of the fuel oil, it is important to note that for the comparison of two fuels it is necessary to calculate the price on the basis of *one heat unit usefully delivered for the purpose in question*, which calculation not only involves the *heat content* of the fuel but also the *efficiency of the heat transfer* obtained with it under prevailing conditions. In English speaking countries it is common to take as a basis *one therm*, which equals 100,000 B.T.U., while in the metric system figures are usually based on 10,000 kg cals. (1 therm = 25,200 kg cals.).

By the graph of fig. 262 it is easy to see that if *1 kg of oil*, which has a calorific value of 10,000 cals. (for heavy oils about 300 cals. less and for light oils about 500 cals. more), is burnt with a thermal efficiency of 80% (a figure which may be obtained by good oil burning in most domestic cases), this same net heat delivery of 8000 cals. would require *2.66 kg of coal* having a calorific value of 6000 cals./kg (which figure applies to coke or low-grade coal) and burnt with a thermal efficiency of 50%, which figure is a common average for small domestic installations fired with low-grade coal.

The higher thermal efficiency of oil burning is due to lower excess air, more complete combustion, lower chimney losses by better heat transfer of cleaner heating surfaces, no loss of combustibles through cinders, better regulation of the intensity according to the requirements.


C. *The cost of Auxiliary Supplies, such as Gas, Electricity, etc.*


A point which did not receive enough attention during the first period of development of the fully automatic electric oil burner was the *consumption of electricity* and thus it is recorded that early installations sometimes showed at the end of a heating season a bill for electricity which was almost equal to that of the fuel oil!

One of the main reasons for this disagreable feature was the use of excessive air pressures; for, although superficially it may seem to make little difference, a reduction of the air pressure from say 4" W.G. to 1" W.G. (from 100 mm. to 25 mm. water gauge) effects considerable reduction in the power required if we take the total quantity of air electrically compressed during a heating season.

Moreover, this reduction in the air pressure agreeably assisted in making the domestic installations quieter, noise being one of the arguments of opponents to oil burning. It was found that reducing the resistance of air ducts by avoiding sharp edges, etc., at the same time removed causes of noise and thus "air pressure economy" proved to be a double advantage. Of course, lowering the air pressure put higher demands on the atomization and on the design of the combustion chambers so that good turbulent mixing should nevertheless be obtained and, as this often met with difficulties where an oil burner was to be installed in a boiler originally designed for coal, the manufacturers gradually resorted to building complete burner-boiler units (figs. 69, 70 and 241).

Besides excessive consumption of electricity by continuous sparks, another reason often advanced was that the oil burning equipments used were often not of a *suitable size* for the job, resulting in unduly lengthy operation during the heating season (too high a figure for the "on-off" ratio). This adverse circumstance has gradually been overcome by the burner manufacturers, who now make series of oil burners of various sizes or provide means for ample adjustments.

The electricity consumption of modern, small, fully automatic oil burner equipments for domestic use in a central heating plant of, say, 35,000 cals./hr. (140,000 B.T.U./hr.) usually does not exceed that of a 100 c.p. bulb. How much this means in money compared with the cost of oil depends of course on the prices of oil and electricity and cannot be answered here.

A very interesting comparison, though not for domestic but for industrial oil burning on a basis of 100 lbs. of oil per hour, is made by GOLLIN in his excellent publication: "Fuel oil for Central Heating" (The Heating and Ventilating Engineer, 1931) (English paper) from which the following data are taken.

GOLLIN makes a comparison of the *additional cost per ton of oil* burnt for various principles of oil burning on a basis of 100 lbs. of oil per hour or a steam production of 1,500 lbs. per hour, assuming an oil price of 70/– per ton and a charge for electricity of 1 d. per k.w.hr., and calculates the following figures:

ADDITIONAL COST FOR AUXILIARY POWER FOR OIL BURNING

1. *Pressure atomizing system* . . . . . . . . . . . . . . . . . . . . . . . . . *5.3 d./ton*
   *Assuming:* Oil pump delivers 150 lbs. of oil at 120 lbs./sq. inch requiring 0.023 H.P. or steam-driven 1.4 lbs./hr. = 0.1% of steam generated. From pump delivery 50 lbs. are returned and 100 lbs. heated from 50° F. to 200° F. preheating temperature corresponding to 60 sec. viscosity. Preheating requires 7500 B.T.U. per lb., say 8.0 lbs. steam per hour = 0.53% of steam generated. Air drawn in by natural draught.

2. *Low pressure air atomizing (all the air through burner)* . . . . . . . . . *30.0 d./ton*
   *Assuming:* Total quantity of air needed = 23,000 cub.ft. per hour at
   say 12″ W.G. blown through burner, requiring fan power = 1.3 H.P.
   or 1.56% of steam generated. Oil heated from 50° F. to 150° F. = 5000
   B.T.U. per lb., say 5.5 lbs. of steam or 0.365% of steam generated. Oil
   pumped from storage tank, say, 150 lbs. at 5 lbs./sq. inch = 0.005 H.P.

3. *Low pressure air atomizing (25% of air through burner)* . . . . . . . *19.8 d./ton*
   *Assuming:* Fan delivers 6,000 cub.ft. of air/hr. at say 16″ W.G. requiring 0.8 H.P. or 0.96% steam generated. The oil is atomized by air
   or air turbine, 150 lbs. pumped from storage tank at 5 lbs./sq.inch and
   preheating as under 2.

4. *Mechanically spun rotary atomizing* . . . . . . . . . . . . . . *11.0 d./ton*
   *Assuming:* 15% of air around cup at say 6″ W.G. = 3450 cub.ft. per
   hour delivered by fan, which takes 0.4 H.P. or 0.48% of steam generated.
   The oil 150 lbs. pumped from storage tank at 5 lbs./sq.inch and 100
   lbs. given velocity of 28.3 ft./sec. = 0.006 H.P.; preheated from 50° F.
   to 130° F. = 4000 B.T.U./lb., say 4.5 lbs. steam = 0.3% of steam
   generated.

5. *Steam atomizing* . . . . . . . . . . . . . . . . . . . . . . . *35.0 d./ton*
   *Assuming:* 0.6 lb. of steam per lb. of oil = 4% of total steam generated.
   150 lbs. pumped from storage tank at 5 lbs./sq.inch. Oil heated from
   50° F. to 130° F. = 4,000 B.T.U. per lb. = 0.3% of steam generated.
   Natural draught.

HASLAM and RUSSELL in their book "Fuels and their Combustion" (1926,
Chapter XVI) give a similar comparison for industrial plants and come to the
conclusion that the cost of mechanical atomizing and of low pressure air atomizing oil
burning are almost equal, the cost of medium pressure air (2 lbs/sq.inch) and steam
atomizing being about twice as high, whereas high pressure air (80 lbs.sq.inch) costs
about four times as much per ton of oil, interest, depreciation and repairs included.

Of course it must be borne in mind that for smaller plants, such as domestic installations, the additional cost per ton of oil is considerably higher than the figures for
the above-mentioned industrial installations, which tendency is also evident from the
graph given by HASLAM & RUSSELL on the subject.

D. *The cost of labour for attendance*

The cost of labour for attending coal-fired central heating plants forms a rather important item on the budget, which may be completely avoided by adopting fully automatic oil burning. Besides the wages of the man who attends to the fire and removes the cinders, additional taxes are charged by the governements of most countries for having this man in service, usually including an insurance on accidents and disease.

If for small installations people themselves take care of the fire, this saving of human labour does not show in terms of money but in additional comfort, which is usually even more appreciated than the money.

### E. *The Cost of Upkeep of, and Repairs to the Equipment*

This item was often under-estimated during the early stages of development of the fully automatic oil burners and it was soon felt that the slogan "Buy and install an oil burner and forget it" did not hold true for the dealers, who were often reminded of its existence by numerous service calls during the ensuing years. It has therefore been gradually recognized as necessary to draw up a contract for subsequent service, at the same time making the installations as fool-proof and simple as possible and, by using interchangeable parts, reducing the time of repairs to the utmost.

The same happened in the automobile industry some years ago and a powerful factor in the stimulation of sales was found to be the development of *an extensive organisation for the rendering of good, rapid, but inexpensive service* to buyers. In this respect the oil burner industry still has to make up some arrears.

### F. *Interest on Capital Investment, Deduction for Depreciation, etc.*

While for industrial oil burning this item, of course, plays an important rôle when considering the installation of new equipment, it is far less important where most domestic installations are concerned, as the majority of people buy them like automobiles and furniture, viz., writing off the total sum at once.

---

Finally, the author closes this series of essays with the remark that it is his conviction that the future development of oil burning depends largely on promoting the knowledge of the principles of the subject and rousing the interest of those concerned with the sale, production and consumption of fuel oil in the many promising prospects of oil burning, and he will be satisfied if he has succeeded in his intention to some extent at least.

---

# BIBLIOGRAPHY

BOOKS:

AMERICAN SOCIETY of HEATING and VENTILATING ENGINEERS GUIDE, (yearly edition N. Y.).

BELL, „Petroleum Oil Fuel" (1902, London, historical).

BLOOMER, „Petroleum Fuel" (1888, St. Petersburg, historical).

BOOTH, „Liquid Fuel and its Combustion" (1902, London, historical).

BOYD, „Petroleum, its Development and Uses" (1895, London, historical).

BRANNT, „Petroleum its History and Uses" (1895, London, historical).

BRITISH PATENT OFFICE, „Abridgements of Specifications etc." (Class book of British patents from 1909 on, class 75).

BUTLER, „Oil Fuel" (1914, Griffin's Scientific Textbooks, London, historical).

COLOMER & DORDIER, „Les Combustibles Industriels" (1921, Dunod-Paris, historical).

DUNN, „Industrial Uses of Fuel Oil" (1926, San Francisco).

ESSICH, „Die Ölfeuerungstechnik" (1927, Springer-Berlin).

FUEL OIL JOURNAL, „Oil burner Survey 1932" (American market).

GEHLHOFF, „Theoretische Physik" (1924, Barth-Leipzig, temperature and radiation measurements).

GRUME GRIMAILO, „The Flow of Gases in Furnaces" (Chapman-Hall, London, theoretical flame temperatures and furnace design).

GUILLERMIC, „Le chauffage par les combustibles liquides" (1935, Béranger, Paris).

GULISCHAMBAROV, „Petroleum Heating of Steam ships and Lokomotives" or „Die Naptha-heizung der Dämpfer und Lokomotiven" (1894, historical).

HASLAM & RUSSELL, „Fuels and their Combustion" (1926, McGraw Hill, N. Y., historical development of combustion theories).

HERMANSEN, „Industrial Furnace Technique" (1929, Benn-London, furnace design.).

HÖHN, „Die Verfeuerung flüssiger Brennstoffe" (1921, Schweiz. Verein von Dampf-kesselbesitzern, especially air atomization).

KLEIN, „Het Olie-stoken" (1921, Vereeniging van Scheepswerktuigkundigen, practice of marine installations).

LEW, „Die Feuerungen mit flüssigen Brennstoffen" (1925, Kröner–Leipzig, historical, especially Russian development).

LEWES, „Oil Fuel" (1913, Collins–London, historical).

LÖFFLER, „Ölfeuerung" (1934, Winkler–Leipzig, present domestic development on the European Continent).

NATIONAL BOARD OF FIRE UNDERWRITERS, „List of Inspected Gas, Oil and Miscellaneous Appliances" (N. Y. half-yearly publ.).

NATIONAL ELECTRIC LIGHT ASSOCIATION, (N. Y.), „Domestic Oil-burners" (Publication No. 289–8 Dec. 1928, description design).

NELSON, „Petroleum refinery engineering" (1936, McGraw–Hill, N. Y.).

NOBEL, „On the extended Use of Oil Residues in Furnaces without Spraying" (Moscow, 1882, historical).

NORTH, „Oil Fuel" (1905, London, historical).

OSTWALD, „Beitrage zur graphische Feuerungstechnik" (1920, Leipzig, principles of the OSTWALD diagram).

POUNDER, „Oil burner Installations" (1924, Emmot–London, marine inst.).

ROSZMÄSZLER, „Die flüssigen Heizmaterialiën" (1910, Hartleben–Wien, historical).

SABATIER, „Die Katalyse" (1927, Akadem. Verlag, Leipzig, catalysts).

SCHNITZER, „Die automatische Ölfeuerung" (1934, present development domestic use in Austria).

SCHULZ, „Die Ölfeuerung" (1925, Knapp–Saale, general survey).

SIEBERT, „Petroleum-heizung und Rohöl-feuerungsanlagen" (1921, Wolff–Dresden, historical).

SPIERS, „Technical Data on Fuel Oil" (1932, British National Committee World Power Conference, London).

STOHN, „Temperaturregler" (1933, Marhold–Saale, automatic regulation).

STEPANOFF, „Die Grundlagen der Lampen theorie" (1906, Leipzig).

SYNDICAT D'APPLICATIONS INDUSTRIELLES, „Les Combustibles Liquides et leurs Applications" (1921, historical).

TAPP, „Handbook of Oil burning" (1932, Am. Oil burner Association, N. Y., exhaustive study on domestic oil burning, America).

TELECZINSKI, „Petroleum and its Use as a Domestic and Commercial Fuel" (Lemberg, 1870, historical, Russian development).

THWAITE, „Liquid Fuel" (1887, London historical development).

TRINKS, „Industie-öfen" (1931, V. d. I. Verlag, Berlin, German development, industrial burners and furnaces).

TURIN, „Les Foyers de Chaudières" (1925, Dunod–Paris, historical).

VERSENEV, „The Management of Petroleum-fired Boilers" (1891, Moscow, historical, Russian).

WILLIAMS, „Fuel, its Combustion and Economy" (1879, London, historical).

WINKLER–BRUNCK, „Lehrbuch der technischen Gasanalysen" (1927, Felix–Leipzig, flue gas analyses).

WIRTH, „Brennstoff Chemie" (1922, Stilke–Berlin, combustion theories).

OTHER PUBLICATIONS:

ADAMS, „Liquid Fuel", (Journal Royal Soc. Arts, 1868, XVI page 432).

ALTUKHOV, „Heating Furnaces with Oil and its Residues" (Tekhnik, 1883, No. 18, page 8, Russian history).

AMMERS, „Warmte-overdracht door Straling" (Warmtetechniek, Oct. 1930, page 50, Heat-transfer by radiation).

BARGEBOER, „Het Verband tusschen Rookgastemperatuur en Luchtovermaat" (Warmtetechniek, Dec. 1931, page 144, Relation between flue gas temperature and excess air).

BICKFORD, „Petroleum Vapour Burners" (Horseless Age IX, page 632, hist.).

BUERK, „Petroleum as a Fuel at Sea" (Iron Age, 1878, page 24, hist.).

CALLENDAR, „Dopes and Detonation" (Engineering, 4 Febr. 1927 etc. page 147, Development hydroxylative combustion theory).

CROSS & LYMAN, „A Study of Oil-fired Heating Boilers" (Heating & Ventilating, Oct. 1931, Influence draught variation).

DE BREY, „Considérations pratiques concernant le résidu combustible de pétrole et ses applications" Chimie et Industrie, mai 1923.

DUZAN, „Le chauffage aux huiles lourdes", (Science et Industrie, No. 222, 1932, Oil burning in France).

DUZAN, „Le chauffage industriel aux huiles lourdes" (Technique moderne, 15 juin, 1930, page 422).

ECKERT & PEW, „Economic Development of Furnace Oils and their Effect on Burner Design" (Am. Petr. Inst., Nov. 1932, Am. market).

(Editorial), „The Kerosene burner comes to life again in New England" (National Petr. News, 21 Sept. 1932, page 37, range burners).

GLENDENNING, „Higher Viscosity raises Capacity" (Fuel Oil Journal, 1935).

GOLLIN, „Fuel oil for Central Heating" (Heating and Ventilating Engineer).

GREBEL, „L'Emploi des huiles résiduaires de pétrole au chauffage central" (Génie Civil, 7 Febr. 1931, page 129).

HASE, „Het bepalen van de Warmteoverdracht in Haarden" (Archiv. f. Wärmewirtschaft, Dec. 1932, Warmtetechniek, Juli 1933, Heat transfer in furnaces).

HODGETTS, „Liquid Fuel" (Trans. Civ. Mech. Soc., 1888, page 63, hist.).

HOTTINGER, „Ölfeuerungen bei Dampfkesseln und Zentralheizungen" (Schw. Bauzeitung, Bnd. 83, Nr. 25, page 292, Survey Switzerland).

KÜHN, „Über die Zerstäubung flüssiger Brennstoffe", (Diss. Danzig, '27, Measurement magnitude of oil drops).

LEDUC, „Le chauffage industriel aux huiles lourdes" (Technique moderne, 15 juillet, 1930, page 498).

LIER, „Über Öl und Koksfeuerungen in Zentralheizungen" (Monatsbulletin Schw. Verein Gas und Wasserfachmännern, No. 5, May 1924).

LUBBOCK, „Gas flow and Radiation" (Journal of Heating and Ventilating Engineers, Aug. 1933, heat transfer).

LUBBOCK, „Oil Fuel in the Brick Industry" (Transactions of the Ceramic Society, Febr. 1931 and November 1932, Clowes & Sons, London).

LUBBOCK, „Industrial Uses of Fuel Oil" (Journal of the Institute of Fuel, Dec. 1930).

LUBBOCK, „The Firing of Metallurgical Furnaces" (Lecture Sheffield Branch of the Institute of Metals, 11th March 1927).

LUHN, „Koksbildung bei Ölfeuerungen und ihre Beseitigung" (Brennstoff Chemie, 15 Jan. 1933, page 681).

KEISER and SCHMIDT, „Nieuwe Pyrometer" (Ref. Warmtetechniek, Oct. 1931).

KRAPF, „Ölfeuerungen für Zentralheizungen" (Schweiz. Techn. Zeitschrift, 10 Nov. 1932, pag. 681).

MENGERINGHAUSEN, „Amerikanische Ölfeuerungen für Wohnhäuser" (Zeitschr. des Vereins d. Ing., 10 May, 1930).

MEYER, „Der Mechanismus der Primär-reaktion zwischen Sauerstoff und Graphit" (Zeitschr. f. phys. Chemie, Abt. B, Bnd. 17, Heft 6, April 1932).

MINNE, „Les Appareils de Combustion de Mazout" (Revue des Combustibles liquides, May 1932, page 157).

MONDAIN–MONVAL and QUANQUIN, „Recherches sur l'oxydation directe des Hydro-carbures par l'air" (Annales de Chimie, Tome XV, April 1931, Survey of hydroxylation theories, reversible oxydation).

MORRISON, „Efficiency of Yellow Oil flames" (Power, 13 Oct. 1931).

ORDE, „Liquid Fuel for Steamships", (Proc. Inst. Mech. Eng. 1902, page 417, historical).

(Editorial, POWER), „Burning Boiler Oil" (Power, 4 Sept. 1923 and following issues, historical development).

ROOSA, „Warmte-overdracht door Straling" (Warmtetechniek, Nov. 1930).

SCHMIDT, „Strahlung von Gasräumen" (Zeitschr. d. Vereins d. Ing. 28 Oct. 1933).

SCHOU, „Om Oliefyring" (Ingeniøren, (Danish), 6 Nov. 1920, page 681).

SEELY and TAVANLAR, „Study of Performance Characteristics of Oil burners"(Heating Piping and Airconditioning, May 1931).

SENNER, „Burners tested for Domestic Heating" Oil and Gas Journal, 14 January 1926, page 80).

TER LINDEN, „Vuurhaarden en Vuurhaardwanden" (Warmtetechniek, Sept. 1933, page 93, on heat transfer).

THUSTON, „The technical Features of the New G. E. Oil furnace" (General Electric Review, Dec. 1932, burner-boiler-unit).

VOGT, „Ölfeuerungen zur Raumheizung" (Archiv für Wärmewirtschaft, March, 1932)

VROOM, „Burners for firing waste Fuels" Refiner, April 1931).

WILLISTON, „Liquid Fuel for Power Purpose" (Engin. Mag. XXV, 1903, hist.).

WILKE, „Die Ölfeuerung in Zentralheizungen" (Bericht über III. Kongres für Heizung und Lüftung, Dortmund 1930).

PERIODICALS:

ANNALES DE L'OFFICE NATIONAL DES COMBUSTIBLES LIQUIDES (Paris).

ARCHIV FÜR WÄRMEWIRTSCHAFT UND DAMPFKESSELWESEN (V. d. I.–Berlin).

BRENNSTOFF UND WÄRMEWIRTSCHAFT (Knapp–Saale).

FEUERUNGSTECHNIK (Spamer–Leipzig).

FUEL OIL JOURNAL (Becker, N. Y.).

GESUNDHEITSINGENIEUR (Oldenbourg–München).

HEATING AND VENTILATING (Industrial Press, N.Y.).

HEATING AND VENTILATING ENGINEER (London).

HEATING, PIPING and AIRCONDITIONING (Keeney–Chicago).

LIST OF INSPECTED GAS, OIL AND MISCELLANEOUS APPLIANCES (National Board of Fire Underwriters).

OIL HEAT (Heating Journals, N. Y.).

POWER (McGraw–Hill, N. Y.).

LA REVUE DES COMBUSTIBLES LIQUIDES (Paris).

L'USINE (Journal de l'Industrie et de la Métallurgie, Paris).

WARMTETECHNIEK (Moorman, The Hague).

# INDEX

Additional material from *Oil burning,*
978-94-017-5721-8, is available at http://extras.springer.com

# ERRATA

Page 73 fig. 83: reserve, *read* reverse

„   94 2nd line: (see page 81)

„   211 12th line from below: ROTAN, *read* ROTAN pump

ROMP, OIL BURNING